Lecture Notes in Civil Engineering 461

Lecture Notes in Civil Engineering (LNCE) publishes the latest developments in Civil Engineering—quickly, informally and in top quality. Though original research reported in proceedings and post-proceedings represents the core of LNCE, edited volumes of exceptionally high quality and interest may also be considered for publication. Volumes published in LNCE embrace all aspects and subfields of, as well as new challenges in, Civil Engineering. Topics in the series include:

- Construction and Structural Mechanics
- Building Materials
- Concrete, Steel and Timber Structures
- Geotechnical Engineering
- Earthquake Engineering
- Coastal Engineering
- Ocean and Offshore Engineering; Ships and Floating Structures
- Hydraulics, Hydrology and Water Resources Engineering
- Environmental Engineering and Sustainability
- Structural Health and Monitoring
- Surveying and Geographical Information Systems
- Indoor Environments
- Transportation and Traffic
- Risk Analysis
- Safety and Security

To submit a proposal or request further information, please contact the appropriate Springer Editor:

– Pierpaolo Riva at pierpaolo.riva@springer.com (Europe and Americas);
– Swati Meherishi at swati.meherishi@springer.com (Asia—except China, Australia, and New Zealand);
– Wayne Hu at wayne.hu@springer.com (China).

All books in the series now indexed by Scopus and EI Compendex database!

Paolo Schito · Alberto Zasso
Editors

Proceedings of the XVII Conference of the Italian Association for Wind Engineering

IN-VENTO 2022

Springer

Editors
Paolo Schito
Dipartimento di Meccanica
Politecnico di Milano
Milan, Italy

Alberto Zasso
Dipartimento di Meccanica
Politecnico di Milano
Milan, Italy

ISSN 2366-2557 ISSN 2366-2565 (electronic)
Lecture Notes in Civil Engineering
ISBN 978-3-031-53058-6 ISBN 978-3-031-53059-3 (eBook)
https://doi.org/10.1007/978-3-031-53059-3

This Springer imprint is published by the registered company Springer Nature Switzerland AG
The registered company address is: Gewerbestrasse 11, 6330 Cham, Switzerland

Paper in this product is recyclable.

Preface

This volume contains the full papers of most of the contributions presented at IN-VENTO 2022, the XVII Conference of the Italian Association for Wind Engineering (ANIV), held in Milano, Italy, on September 4–7, 2022.

The XVII edition of IN-VENTO conference comes after the online XVI edition held in 2020 during the most difficult part of the COVID-19 pandemic situation. It has been a nice occasion to meet again colleagues and friends after the long period with limited trips and many online meetings. The number of participants was unexpected, so it has been necessary to split the conference into parallel sessions.

The first day of the conference was dedicated to ANIV-G, the ANIV Young Researchers, that were able to meet on Sunday September 4th. In the afternoon of September 4th a lecture to the memory of Prof. Giovanni Solari was organized, and a following conference ice-breaking.

Invited lectures to the conference were Prof. Francesco Ricciardelli of University of Campania "Luigi Vanvitelli", Prof. Horia Hangan of Western University, and Prof. Gang Hu of Harbin Institute of Technology.

The morning of Tuesday September 6th was dedicated to the ANIV Award 2022, with the presentation of the three shortlisted papers, and the price was awarded ex-aequo to Niccolò Barni and Federico Canepa.

Starting September 5th, the conference had 14 sessions, aiming to provide a major forum for the exchange of ideas and discussion on the latest advances in the field of wind engineering, bluff-body aerodynamics, aeroelasticity, structural dynamics and reliability, wind modeling and forecast, wind tunnel and full-scale testing, computational fluid dynamics, and wind energy.

The previous sixteen editions of the IN-VENTO conference series were held biennially since 1990 in Firenze, Capri, Roma, Trieste, Perugia, Genova, Milano, Reggio Calabria, Pescara, Cefalù, Spoleto, Venezia, Genova, Terni, Napoli, and Milano in online; in its last six editions, the conference has become international, receiving attention from a large number of researchers worldwide.

IN-VENTO 2022 had an attendance of about 110 researchers, 45% of which coming from outside Italy. Out of the 77 papers presented in the conference, 28 were selected by the editors for publication in the proceedings book after a double-blind peer-review process.

The editors thank all participants for their contributions to the conference program and to these proceedings.

We also acknowledge the support received from the Italian Association for Wind Engineering, from the Conference Scientific Committee, and from the Local Organizing Committee.

We thank our partner Fondazione Politecnico di Milano and our sponsor Pininfarina.

Organization

Scientific Committee

Gianni Bartoli	Università degli Studi di Firenze
Luca Bruno	Politecnico di Torino
Massimiliano Burlando	Università degli Studi di Genova
Luca Caracoglia	Northeastern University, USA
Giampiero Manara	Permasteelisa
Claudio Mannini	Università degli Studi di Firenze
Sara Muggiasca	Politecnico di Milano
Francesco Petrini	Università degli Studi di Roma "La Sapienza"
Maria Pia Repetto	Università degli Studi di Genova
Francesco Ricciardelli	Università degli Studi della Campania "Luigi Vanvitelli"
Daniele Rocchi	Politecnico di Milano
Ilaria Venanzi	Università degli Studi di Perugia
Alberto Zasso	Politecnico di Milano

IN-VENTO 2022 Organizing Committee

Alberto Zasso (Conference Chairman)
Tommaso Argentini
Marco Belloli
Filippo Calamelli
Giorgio Frontini
Sara Muggiasca
Giulia Pomaranzi
Daniele Rocchi
Paolo Schito
Gisella Tomasini

Contents

Wind Load Assessment of Unitized Double Skin Facades

Francesco Ricciardelli[1(✉)], Giampiero Manara[2], and Guido Lori[2]

[1] University of Campania "Luigi Vanvitelli", Aversa, Italy
fricciа@unicampania.it
[2] Permasteelisa SpA, Vittorio Veneto, Italy

Abstract. Load sharing in multiple skin facades depends on several parameters, main of which are the geometry of the units and the permeabilities and stiffnesses of the skins. Experimental and numerical tools exist, which are commonly adopted by the industry in the design and testing of specific products. Code specifications are also available, facing the difficulty of covering a quite broad range of solutions, whose physical behaviour may differ substantially. In this paper, the evolution of Eurocode specification concerning the wind loading of multiple skin facades and roofs is first traced, from ENV 1991–2-4:1995 to the current proposed update prEN 1991–1-4:2020. Then, the mechanism of static and dynamic pressure propagation into the cavity is illustrated, for the case of double skin facades and roofs; a possible approach for future Code development is shown. Finally, the results of numerical simulations and experimental tests are compared for the particular case of unitized double skin facades.

Keywords: Wind loads · pressure sharing · double skin facades · full scale experiments · design codes

1 Introduction

Global wind loads on buildings are not affected by the details of the envelope, and in particular by whether this is made of a single or by multiple skins. However, when designing multiple skin envelopes the additional question arises of how wind loads are shared among the skins, which influences the design of the skins themselves and of the fixing elements. To this purpose, double skin facades can be classified based on the absolute and relative permeability of the skins, based on the absolute and relative stiffness of the skins, and based on the depth of the cavity and air circulation mechanism. In particular, cavity depths in the wide range of 100 to 2000 mm are used, and the term large cavity is used for values of the depth in the range of 600 to 2000 mm, whereas the term small cavity is used for smaller values of the depth. More recently, the concept of unitized compact double skin facades has been developed, with cavity depths in the range of 100 to 300 mm. Ventilation can be either natural of artificial, and in both cases it is governed by the size and location of openings, and by unintentional leakage. Closed cavity facades, on the other hand, have no openings and a very limited

P. Schito and A. Zasso (Eds.): IN VENTO 2022, LNCE 461, pp. 1–12, 2024.
https://doi.org/10.1007/978-3-031-53059-3_1

unintentional leakage; in that case, a micro-ventilation is established, in response to the artificial injection of a small flow of dry air, which is mostly provided to make sure no condensation occurs.

Experimental and numerical tools exist, which are commonly adopted by the industry in the design and testing of specific products [1]. Few Code specifications are only available, facing the difficulty of covering a quite broad range of solutions, whose physical behaviour may differ substantially; most of these refer to the Eurocodes [2–4], e.g. [5]. In this paper an attempt is made to identify the possible tools needed to fill the gap between the current practice adopted by the industry and Code approaches.

2 Eurocode 1 Specifications

2.1 ENV 1991–2-4:1995

The first, preliminary European Code on wind actions was released in 1995 [2]. Provisions for the load sharing between skins in multiple skin facades are given in Sect. 10.2.10 "*Pressure on external walls or roofs with more than one skin*". In spite of the fact that in the title reference is made to general case in which the number of skins is larger than one, then the section in fact gives provision only for the more limited, though widespread case of double skins.

The proposed approach is based on the quantification of a pressure coefficient for calculation of the cavity pressure; then, once the cavity pressure is known, net pressure on the two skins can be calculated as the difference between the external and cavity pressures for the external skin, and between the cavity and internal pressures for the internal skin. Criteria for calculation of the cavity pressure coefficient are given in Table 10.2.6, and are based on porosity except for the case in which both skins can be considered as impermeable, where stiffness of the skins and cavity depth govern the load sharing. Indeed, a limit permeability below which a skin is to be considered as impermeable is not given. No limits are given for the cavity thickness for the procedure to be applicable.

One first case is that in which one skin is clearly more permeable than the other, i.e. its porosity is at least three times larger than that of the other skin. If the most permeable skin is on the inside face, then the external skin shall be loaded with the entire pressure difference between the outside and the inside of the building, whereas the internal skin shall be loaded with one third of the internal pressure. If the most permeable skin is on the outside face, then the cavity pressure is set to only one third the external pressure, and the inside skin is loaded with the difference between that pressure and the internal pressure, which seems an unjustified reduction of the load on the impermeable skin. As to the external, more permeable skin, this shall be loaded with a pressure ranging between the external pressure and one third that, depending on its porosity and on whether the external pressure is positive or negative. As an effect of the reduction above, the sum of the pressure on the two skins can turn out to be less than the entire pressure difference between the outside and the inside of the building.

A second case is that in which both skins can be considered as permeable (though a limit value for permeability is not given), and the permeability difference is not large, i.e. the permeability of the external skin is in the range of one half to three times that of the internal skin. In this case, the cavity pressure is calculated from flow equalization, and it is the weighted sum of internal and external pressures, using as weights the squares of the porosity of each skin.

Finally, there is the case in which both skins can be considered as impermeable, where sharing is mainly based on stiffness. If one skin is clearly stiffer than the other, then this shall be loaded with the entire pressure difference between the outside and the inside of the building. The less stiff skin is loaded with a pressure corresponding to the difference between the external and the minimum external pressure on the face, if it is located outside, and between the internal and the minimum internal pressure on the face, if it is located inside. In fact, in most cases the internal pressure is uniform, and this brings that the internal skin, if more flexible, receives no load. For very small cavities, with a depth not exceeding 5 mm, a "*suction effect*" is expected to take place, but no provision is given to account for it.

A special case is that in which there are entries of air at the extremities of the cavity. In that case the cavity pressure is equalized with the pressure on the side of the wall or roof with which the cavity communicates. This means, e.g. that if the cavity is put into communication with the outside, then the pressure inside the cavity is equal to the external pressure, therefore the internal skin shall be loaded with the entire pressure difference between the outside and the inside of the building, and the external skin receives no load, and vice versa.

It must be noted that when referring to external pressure coefficients, the Code does not specify whether these shall be $c_{pe,1}$ or $c_{pe,10}$. In the absence of specific indications, it would be reasonable to assume that the rule in 10.2.1 applies. On the other hand, it is the authors' belief that being equalization potentially deriving from localized leakage, it would be more appropriate and conservative to use $c_{pe,1}$ values.

2.2 EN 1991-1-4:2005

In the current version of EN 1991–1-4 [3], provisions for the load sharing between skins are given in Sect. 7.2.10, also in this case limited to double skins.

The section starts with a misleading Clause (1), stating that "*The wind force is to be calculated separately on each skin*", which is clearly true when designing the single skin and its fixing, but does not apply to the overall design of the building hosting the multiple skin wall or roof, nor its individual members, like the sentence seems to indicate. No cavity pressure coefficient is defined, and the net pressure on each skin is given directly from the external and internal pressures.

Differently from ENV 1991–2-4:1995, in Clause (2), a limit permeability is established of 0.1%, separating permeable from impermeable skins.

For the case in which one of the two skins is impermeable and the other is permeable, Clause (3) prescribes that the load on the impermeable skin is "*determined from the difference between the internal and external wind pressure*". It is noted that in the following text such difference is expressed as that between the external and the internal pressure; therefore, even though the statement is clear from the physical point of view, an issue of inconsistency can be raised. On the other hand, no provision is given for the loading of the permeable skin, at least for the general case.

Again in Clause (3), it is then stated that if more than one skin is permeable, then besides the permeability of the skins, the load sharing between the skins depends also on their relative rigidity, on the gap width, on possible openings at the extremities of the layer between the skins, and on "*the external and internal pressures*", the latter dependency leaving the door open to a possible non-linear behaviour of the pneumatic system. Flow equalization of ENV 1991–2-4:1995 has disappeared, and geometric and mechanical parameters are claimed to govern the sharing, like ENV 1991–2-4 [2] did for the opposite case in which both skins are impermeable. Choice of the load sharing between the skins is left to the National Annex, though it is recommended that to the most rigid skin the full load in the absence of the less rigid skin is applied, regardless of the ratio between the rigidities; in any case, no provision is given for the least rigid skin. No provision is also given for the case in which both skins are impermeable.

Limited to cavity depths of less than 100 mm, and in the absence of leakage at extremities of the layer between the skins, further provisions are given, as follows, leaving in any case the possibility to the National Annex to give rules. In particular, it is reiterated that when one skin is permeable and one not, the impermeable skin takes the entire pressure difference between the external and the internal faces of the wall or roof. In addition to that, for the common case in which the permeable skin occupies the external face, then this is to be loaded with one third or two thirds of the net pressure difference across the wall or roof, depending on whether this is negative or positive; this largest loading of the most permeable skin was found also in ENV 1991–2-4:1995. When, on the other hand, the internal skin is the permeable one, this is to be loaded with one third of the internal pressure, regardless of its sign. When both skins are impermeable, sharing depends on rigidity. Regardless of the rigidity ratio between the skins, the most rigid is called to carry the entire pressure difference between the external and the internal faces, whereas only for the case in which the less rigid skin is the external one, then this is called to carry the external pressure, but no provision is given for the case in which the less rigid skin is the internal one. No specific provision is given for the case in which both skins are permeable.

Same observation concerning the choice between $c_{pe,1}$ and $c_{pe,10}$ coefficients as in EN 1991–2-4:2005 applies, whose values have not changed with the revision of the Code. As to internal pressure coefficients, on the other hand, different criteria are given in EN 1991–1-4:2005 with respect to those of ENV 1991–2-4:1995, the discussion of which is beyond the scope of this paper.

2.3 PrEN 1991–1-4:2020

With Mandate M/515 of the European Commission to CEN, in 2012 a revision aimed at the release of the Second Generation of the Eurocodes was initiated. In April 2020 the new draft prEN 1991–1-4:2020 [4] was submitted to TC250/SC1 by the Project Team in charge of its writing. The document deals with the issue of multiple skin facades in Section C.2.9. Modification to the 2020 version have been, and are still being suggested by the Task Group on Wind Actions under TC250/SC1. The modified version of the document will be submitted for approval by TC250, so to start the process of the Public Enquiry and of the Formal Vote, expected to take place in the falls of 2023 and 2024, respectively.

To overcome possible misunderstandings of the current version, in Clause (1) it is specified that wind actions on principal structural members, including those having tributary areas in excess of 10 m^2, and on foundations shall be evaluated regardless of the fact that the wall or roof has one or more skins.

Clause (2), on the other hand, "*gives criteria for the evaluation of the wind action on each skin of multiple skin walls or roofs, for the purpose of the design of the skin itself and of its bearing members*". The limit permeability of 0.1% for a skin to be considered permeable or not is confirmed, but then it is not used, and simpler specifications then in the current version are given, as follows. The least permeable skin shall again be loaded with the entire pressure difference between the outside and the inside of the building, in this case regardless of whether the skins are permeable or impermeable, as per definition given in Clause (2), and also regardless of the ratio of the two permeabilities, as in ENV 1991–2-4:1995. On the other hand, for the pressure on the more permeable skin with approximately uniformly distributed openings the same rule as in EN 1991–1-4:2005 is given; it is two thirds and one third of the external pressure in the case of overpressure and underpressure, respectively. All limitations concerning the gap width and the arrangement of the skins have been removed.

2.4 Comparison

It must be noted that in contrast with a document that has grown in size (122 pages for ENV 1991–2-4:1995 including wind maps, 146 pages for EN 1991–1-4:2005, 300 pages for prEN 1991–1-4:2020), and in the amount of information provided, the problem of the load sharing in double skin walls and roofs has been dramatically simplified over the years. This is the symptom of the fact that the provisions given in ENV 1991–2-4:1995 have not been perceived as appropriate by the Engineering community, and on the other hand that no updated data or methods have arrived from research in that last three decades or so.

In the current unitized facade market, interest is growing more and more for double skin designs with very low values of external porosity, the so called Closed Cavity Facade (CCF) concept. Then a wide range of possible situations is becoming available, from the fully equalized rainscreen, featuring a relatively large total porosity ($\mu_e + \mu_i$) and a much larger porosity of the external skin with respect to that of the internal one ($\mu_e >> \mu_i$), aiming at reducing as much as possible the pressure difference through the external barrier, and the CCF, which aims at a reasonably closed cavity, hence with much smaller values of $\mu_e + \mu_i$, and different possible sharing between the skins, on the basis of the specific design solution. Therefore, it is desirable that Code specifications could reflect the market needs and provide more tailored solutions for the industry and for the consultant.

3 Static and Dynamic Behaviour of Double Skin Facades

When a pressure difference is applied statically between the exterior and the interior of a double skin facade, the ratio of the cavity pressure p_c to the external pressure p_e can be calculated from the continuity equation as:

$$\frac{p_c}{p_e} = 1 - \frac{(\mu_i/\mu_e)^2}{1 + (\mu_i/\mu_e)^2}\left(1 - \frac{p_i}{p_e}\right) \tag{1}$$

When the internal pressure p_i equals the atmospheric pressure, then Eq. (1) depends only on the ratio of the internal to the external porosities. Equation (1) can be used to evaluate the net pressures on the external and internal skins, which turn out to be:

$$\Delta p_e = p_e - p_c = \Delta p \frac{(\mu_i/\mu_e)^2}{1 + (\mu_i/\mu_e)^2} \tag{2a}$$

$$\Delta p_i = p_c - p_i = \Delta p \frac{1}{1 + (\mu_i/\mu_e)^2} \tag{2b}$$

where $\Delta p = p_e - p_i$ is the total pressure difference across the facade. Equations (2) are shown in Fig. 1, and can be used also in the case in which the pressure difference slowly varies in time. On the other hand, when the time variation of the pressure difference is fast, then the dynamic and thermodynamic behaviour of the system needs to be accounted for. This is mainly affected by (a) the ratio of the internal to the external porosity, (b) the total porosity, (c) the volume and the depth of the cavity, (d) the relative stiffness of the skins. Equations governing the behaviour are (a) the continuity equation, (b) the flow equation, (c) the plate bending equation, (d) the gas law, and (e) the thermal balance equation. Finally, a non-linear behaviour of the pressure transfer can be observed, deriving from nonlinearities in the mechanical properties of the components; examples can be the loss of friction between components or the leakage of gaskets with increasing pressure.

Generally speaking, the cavity pressure can be obtained as the product of the external pressure and of a Frequency Response Function (FRF) [6]:

$$P_c(f) = H(f)\,P_e(f) \tag{3}$$

where $P_c(f)$ and $P_e(f)$ are the Fourier spectra of the cavity and external pressures, respectively, and $H(f)$ is the FRF across the external skin. Then, the Power Spectral Density Function (PSDF) of the net pressure fluctuations across the external skin is:

$$S_{\Delta p_e}(f) = S_{p_e}(f)\left[1 + |H(f)|^2 - 2\,|H(f)|\,cos(\varphi(H(f)))\right] \tag{4}$$

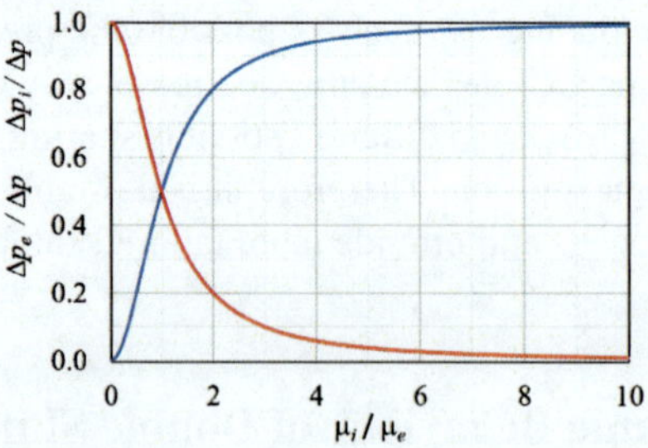

Fig. 1. Static load sharing between the skins as a function of the relative permeability; blue curve is the nondimensional pressure difference on the external skin (Eq. 2a); red curve is the nondimensional pressure difference on the internal skin (Eq. 2b).

where $S_{p_e}(f)$ is the PSDF of the external pressure, and $|H(f)|$ and $\varphi(H(f))$ are the modulus and phase of $H(f)$. Notice that for very low frequencies static pressure sharing applies, therefore for $f \rightarrow 0$ the modulus of $H(f)$ tends to the value given by Eq. (2a) and shown in Fig. 1, and the phase tends to zero. The PSDF of the net pressure fluctuations across the internal skin is:

$$S_{\Delta p_i}(f) = S_{p_e}(f)\,|H(f)|^2 \tag{5}$$

Once the PSDF of the external pressure fluctuation and the FRF of the external skin are known, then eqs. (4) and (5) allow evaluating the fluctuating load on the two skins separately, the static one being given by eqs. (2).

The theory above applies when no Helmholtz resonance occurs in the cavity, therefore when the Helmholtz resonant frequency is much larger than the frequency range of the atmospheric turbulence. This is always the case for compact double skin facades.

Figure 2 shows a sketch of limiting cases of the modulus and phase of $H(f)$, evaluated by means of a Permasteelisa proprietary assessment tool developed by the Department of Building Physics of the Ghent University. The blue curve represents the case in which the porosity of the external skin is much larger than that of the internal skin and, vice versa, the red curve represents the case in which the porosity of the internal skin is much larger than that of the external skin. Two frequency ranges can be identified; at the low frequencies, porosity dominates, and the sharing depends on the ratio of the internal to external porosities as well as on the total porosity; at higher frequencies, the effect of the relative porosity becomes negligible and the ratio of the stiffnesses of the internal and external skins governs the pressure transfer. The frequency around which the transition between the porosity-driven behaviour and the stiffness-driven behaviour takes place depends on the total porosity.

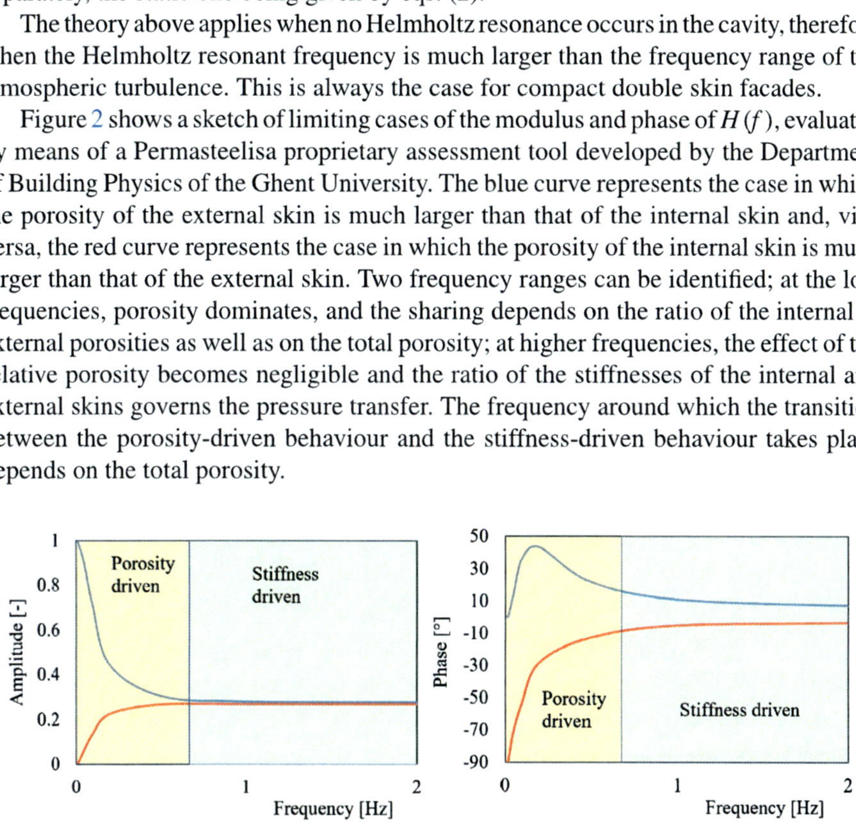

Fig. 2. Sketch of the Frequency Response Function of the pressure transfer across the external skin; blue curve is the case in which the porosity of the external skin is much larger than that of the internal skin ($\mu_e = 5 \cdot 10^{-6}$, $\mu_i = 1 \cdot 10^{-7}$); red curve is the case in which the porosity of the internal skin is much larger than that of the external skin (($\mu_e = 5 \cdot 10^{-7}$, $\mu_i = 1 \cdot 10^{-7}$)).

4 Experimental Tests and Numerical Simulations

4.1 Experimental Campaign

A unitized double skin facade specimen was tested at the Permasteelisa SpA Test&Lab in Vittorio Veneto, Italy [7, 8]. The unit has a nominal width of 1,500 mm and it is divided in a vision area and a spandrel area; the vision area inner skin is provided with an operable inward glazing (Fig. 3). Main characteristics of the unit are given in Table 1. The spandrel cavity was ventilated from the external environment through two, 1,452 mm^2 slots located in the mullions. The vision cavity ventilation was varied during the tests; three cases were considered, the first in which the vision cavity was nominally sealed, the second and the third in which the vision cavity was ventilated from the external through either one or two, 1,452 mm^2 slots, respectively, located in the transom.

First, air leakage tests were carried out by pressurising the cavity. In the case in which the vision cavity is nominally sealed, an unintended porosity of about 100 mm^2 was found, which starts being activated when the cavity pressure exceeds 100 Pa, and becomes fully activated when the cavity pressure reaches 300 Pa. From the tests, it is impossible to know what percentage of the unintended leakage is in the external skin and what in the internal skin. In addition, it was found that when the cavity pressure approaches 500 Pa the operable glazing starts leaking towards the inside, and the cavity pressure doesn't increase any further. Clearly, no similar behaviour was found when the cavity is subjected to suctions.

In a second stage, harmonic tests were performed in which a cyclic pressure build up was applied to the wooden box, and pressure in the cavity of the facade was measured. The procedure was repeated at pressure steps of 50 Pa up to the maximum pressure of 1,000 Pa, controlling the maximum pressure in the cycle. The test was performed both in pressure and suction.

Table 1. Details of the facade unit.

		Outer	Cavity [mm]	Inner
Vision	Dimensions	1,420 x 2,700 mm	160	1,300 x 2,600 mm
	Glazing panel (out)	2 x 10 mm glass		10 mm glass
	Cavity			16 mm
	Glazing panel (in)			12 mm glass
Spandrel	Dimensions	1,420 x 1,200 mm	137	1,420 x 2,700 mm
	Glazing panel (out)	10 mm glass		2 mm steel
	Cavity	137 mm		
	Glazing panel (in)	2.5 mm aluminium		

4.2 Numerical Simulations and Comparison

Numerical simulations of the same facade unit were carried out using a numerical model originally developed by Ghent University [9]. The model may be regarded as an extension of the well-known numerical model of the insulating glazing units adopted in [10]. The key principle for the solution of the system consists of the pressure equilibrium between the volume expansion (or contraction) undergone by the gas into a cavity and

Fig. 3. Experimental setup.

the variation of the volume of the skins under the net pressure between inside and outside the cavity. The model allows also to account for the two major features of a closed cavity double skin, i.e. (a) the dry air flow possibly pumped into the cavity to control the relative humidity (active ventilation), and (b) the skin internal and external permeability (passive ventilation).

First, the case of the nominally sealed unit was considered. Being it unknown from which skin(s) the unintended leakage derives, three possibilities were investigated in which (a) the unintended porosity is entirely on the external skin, (b) the unintended porosity is entirely on the internal skin, and (c) the unintended porosity is half on the external and half on the internal skin. Figure 4 shows the results of the simulations in comparison with the experimental results, and the latter tend to agree with the simulation in which the unintended leakage is entirely on the external skin, corresponding to an external porosity $\mu_e = 2.6 \cdot 10^{-5}$. The three cases significantly differ in terms of magnitude of the FRF up to about 0.5 Hz, whereas some difference in the phases can be observed also for frequencies in excess of 1.0 Hz.

Then, the two cases in which one slot or two slots are left open in the transom of the vision cavity, corresponding to values of the external porosity of $3.8 \cdot 10^{-4}$ and $7.6 \cdot 10^{-4}$, respectively, which is the case of a ventilated facade. In Fig. 5 the FRFs of the pressure transfer across the external skin. In this case the decay of the FRF magnitude is much slower, and a full transfer of pressure fluctuations inside the cavity, and a small phase shift are observed up to 0.5 Hz when one slot is open, and up to 1.0 Hz when two slots

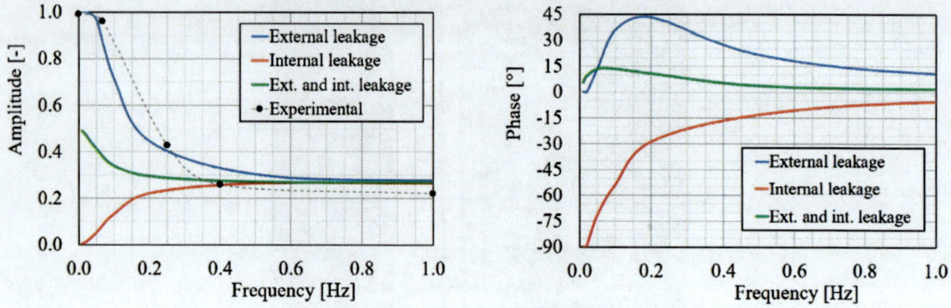

Fig. 4. Frequency Response Function of the pressure transfer across the external skin for the unit with an unintended leakage of 100 mm^2; positive pressure.

are open. The experimental results, where applicable, are in rather good agreement with those coming from the numerical simulations, even though they tend to underestimate the pressure transfer by some 10% on average.

Finally, in Fig. 6 three cases of very low porosity are shown, for which the porosity driven range extends only to 0.1 Hz (case in which the total porosity is $5.5 \cdot 10^{-6}$), or even to less than 0.02 Hz for the two cases in which the total porosity is $6.0 \cdot 10^{-7}$.

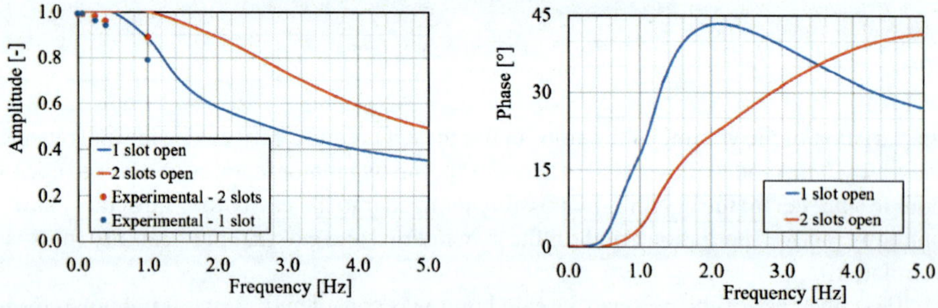

Fig. 5. Frequency Response Function of the pressure transfer across the external skin; one or two 1,452 mm^2 slots are open in the transom of the vision cavity.

4.3 Loading of Skins

Availability of the FRF of the pressure transfer across the external skin allows the use of Eqs. (4) and (5) to evaluate the PSDF of the pressure difference across the two skins, once the spectrum of the external pressure fluctuations is also known. Under the assumption of a linearized quasi-steady behaviour, the latter can in turn be derived from the spectrum of the oncoming turbulence. Such approach applies reasonably well to the case of facade elements directly exposed to the oncoming flow; it might prove inaccurate when applied to roof or facade elements exposed to separated flow, in which case signature turbulence plays a strong role on the overall surface pressure fluctuations; in the latter case, wind tunnel pressure measurements are needed to characterize $S_{p_e}(f)$.

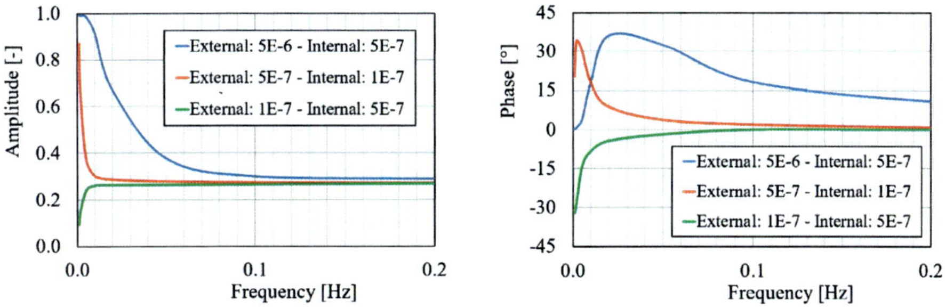

Fig. 6. Frequency Response Function of the pressure transfer across the external skin for three different cases of very low porosity.

Finally, to evaluate the peak load on each skin, an estimation is needed for the peak factor of the pressure difference across the skin. This is also to be derived from the PSDF of the pressure fluctuations.

Both aspects are beyond the scope of this paper, but deserve consideration in future studies.

5 Conclusions

Design of multiple skin facades or roofs requires the evaluation of the total peak pressure difference across the facade, which is used for the design on the main supporting elements, and of the peak pressure difference across each skin, which is used to design the skins themselves and the secondary fixing elements. Total pressure differences are the same as those acting on single skin envelopes, therefore do not need specific attention.

Pressure differences on each skin have been dealt with by Eurocode 1 since its first appearance in 1995; however it is noticed that degree of detail has decreased from ENV 1991–2-4:1995 to EN 1991–1-4:2005, and then to prEN 1991–1-4:2020. This is considered as a symptom of the fact that the approach used has not been perceived as appropriate by the Engineering community, and on the other hand that no updated data or methods have arrived from research in that last three decades or so.

Static load sharing can be derived directly from knowledge of the porosity of the two skins, using continuity equation. On the other hand, both experimental and numerical tools are available to evaluate the Frequency Response Function of pressure transfer across the external skin, and the result appears to be in reasonable agreement; the latter, combined with the Power Spectral Density Function of the pressure fluctuation on the external skin allows evaluation of the Power Spectral Density Function of the pressure difference across each skin, together with an estimate of the associated peak factor.

References

1. Van Den Bossche, N.: Watertightness of building components. Principles, testing and design guidelines. Ph.D. Thesis, University of Gent (2013)

2. European Committee for Standardization (CEN): ENV 1991–2–4, Eurocode 1: Basis of design and actions on structures, Part 2–4: Actions on structures – Wind actions (1995)
3. European Committee for Standardization (CEN): EN 1991–1–4, Eurocode 1: Actions on structures, Part 1–4: General actions - Wind actions (2005)
4. European Committee for Standardization (CEN): prEN 1991–1–4, Eurocode 1: Actions on structures, Part 1–4: General actions - Wind actions (2020)
5. COPREC: Tableaux des Pressions de vent W50 applicables aux façades légères et déterminées selon l'Annexe Nationale NF EN 1991–1–4/NA. - Fiche Technique n. 45. (2010)
6. Straube, J.F.: Pressure moderation and rain penetration control. Proc. OBEC PER Seminar, pp 1–51 (2001)
7. Permasteelisa Group: Multiple skin load sharing, Part 1. Internal Research Report (2018)
8. Permasteelisa Group: Multiple skin load sharing, Part 2. Internal Research Report (2018)
9. Van Den Bossche, N., Janssens, A., Moore, T., Lacasse, M.: Development of a numerical model for pressure equalization in facades during wet conditions. Build. Environ. **178**, 106919 (2020)
10. SJ Mepla - Software for Structural Glass Designhttps. http://www.mepla.net/en/ Accessed 20 July 2022

Global Pressure Coefficients for Building Roofs Revisited

Antonio Malasomma(✉), Vincenzo Picozzi, Alberto M. Avossa, and Francesco Ricciardelli

University of Campania "Luigi Vanvitelli", Aversa, Italy
antonio.malasomma@unicampania.it

Abstract. In the assessment of structural reliability for design against wind actions, Eurocode 1 provides two sets of pressure coefficients, one for fixings and cladding elements, the other for main structural elements. In addition, the Italian CNR Guidelines introduced a third set of coefficients for a quicker and simpler evaluation of the overall effects of wind on main supporting structures and on foundations. These are denoted as *global pressure coefficients*. However, despite some common features, the three sets turn out to be inconsistent from a statistical point of view. In this paper, a revision of the global pressure coefficients is made, in view of a possible reconsideration of the values given by Eurocode 1. To make it consistent from a statistical point of view, Cook and Mayne's statistical definition of the pressure coefficient is adopted. Then, based on wind tunnel measurements available from aerodynamic databases, global pressure coefficients for flat and gable roofs are derived. In particular, the aerodynamic database of the Tokyo Polytechnic University is used, providing measurements on model-scale buildings with roof pitches up to 45°.

Keywords: Pressure Coefficients · Wind Load · Aerodynamic Database · Low-Rise Buildings · Gable Roofs

1 Introduction

Current version of EN1991-1-4 [1] contains two sets of pressure coefficients, for the design of fixings and cladding elements ($c_{p,1}$), and of main structural elements ($c_{p,10}$) of buildings. However, the two sets derive from different sources, therefore introducing an inconsistency in the statistical definition of the design pressure coefficients [2]. For low-rise buildings, $c_{p,1}$ values were derived from the work of Stathopoulos [3], who extracted the largest peak value of the measured pressure coefficient time series. These coefficients were first adopted by the National Building Code of Canada (NBCC) [4]. Instead, $c_{p,10}$ values were taken from British Standards [5]. They derive from the work of Cook [6], that defined the design pressure coefficient as the 78% fractile of the Extreme Values (EV) distribution within the period T of 10min utes or 1hr, corresponding to the averaging period of the mean wind speed [7]. This and other differences were found by Geurts et al. [2], highlighting the need for a consistent basis for pressure coefficients datasets. In

P. Schito and A. Zasso (Eds.): IN VENTO 2022, LNCE 461, pp. 13–22, 2024.
https://doi.org/10.1007/978-3-031-53059-3_2

addition to the two sets of pressure coefficients, a third set was introduced in the Italian CNR Guidelines since its first version [8], to make it simpler the evaluation of the overall effects of wind load on main structures and foundation elements. These are denoted as *global pressure coefficients*, and we will refer to them as c_p. Global pressure coefficients have also been adopted by prEN1991-1-4:2020, draft version of the Eurocode 1 Part 1–4 on wind actions [9], currently going through the approval process at CEN. However, the way global pressure coefficients were derived introduces further inconsistency.

In the last years, Cook and Mayne's definition of the pressure coefficient has been adopted worldwide, recognizing the 78% fractile of the EV distribution as a stable statistic [10]. This should create a background for the revision of the pressure coefficients adopted in EN1991-1-4, so to make them consistent from a statistical point of view. To this aim, in this paper criteria are presented for a revision of global pressure coefficients. First, available coefficients are discussed, together with the way they were derived. Then, a methodology is outlined defining global pressure coefficients for buildings roofs. Finally, coefficients for flat and gable roofs are derived based on available aerodynamic datasets.

2 Methodology

2.1 Current Global Pressure Coefficients

Global pressure coefficients proposed in CNR Guidelines were derived as a simplification of $c_{p,10}$ adopted in EN1991-1-4, that were in turn obtained as an envelope of the pressure coefficients proposed by Cook [6]. Original pressure coefficients $c_{p,10}$ were evaluated as a function of the wind angle of attack, and were meant for application to a more detailed roof zoning. To derive the $c_{p,10}$ currently used by the Eurocode, first an envelope was made with respect to the wind attack angle in order to evaluate the worst values for each zone. Then, the roof zoning was redesigned to make it simpler. As an example, in Fig. 1 a comparison is made between the different roofing zones for flat roofs: Fig. 1a shows the original roof zoning proposed by Cook; in Fig. 1b a first simplification is made by considering the envelope within $\mp 45°$ of pressure coefficients applied to the roof zoning of Fig. 1a; finally, in Fig. 1c the simplified roof zoning of the Eurocode is shown. To fit Cook's coefficients to the simplified roof zoning, an area-average was made. Analysing Fig. 1b and Fig. 1c, it is clear that the weighted average was made between zones a and b to obtain the coefficient for zone F, between zones c and d to obtain the coefficient for zone G, and between zones e and f to obtain coefficient for zone H. Instead, coefficient for zone I corresponds to that of the original zone g.

The same area-averaged operation was done starting from coefficients $c_{p,10}$ of Eurocode for the derivation of global pressure coefficients, whose zoning for flat roofs is shown in Fig. 1d. By comparison with Fig. 1c, it is clear that global pressure coefficients for zone A are derived by area-averaging the coefficients of zones F, G, and H. Instead, global pressure coefficient for zone B corresponds to pressure coefficient for zone I.

2.2 The Aerodynamic Database

Pressure coefficients used in this work are taken from the aerodynamic database of the Tokyo Polytechnic University (TPU) that includes the results of wind tunnel tests on

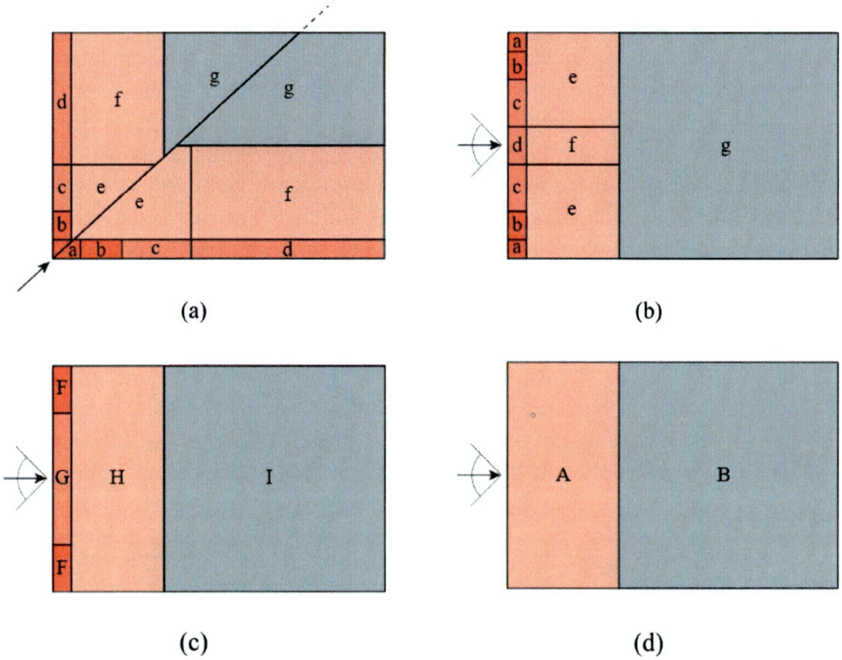

Fig. 1. Flat roofs zoning: (a) after Cook, (b) after the envelope of the Cook's coefficients, (c) after EN1991-1-4, (d) for global coefficients after CNR Guidelines.

low-rise buildings with rectangular plan [11]. The tests were performed in the Boundary Layer Wind Tunnel at TPU (2.20m wide and 1.80m high) on building models with length scale $\lambda_l = 1 : 100$ and velocity scale $\lambda_v = 1 : 3$. An exponential mean wind profile with gradient height of 450m and exponent 0.20 was measured, corresponding to Category II exposure in EN1991-1-4. The wind speed at 10m full scale is 22m/s with a turbulence intensity of about 0.25. Tests involve wind angles from 0° to 90° with steps of 15°. Flat and gable roofs are analysed in the current work. In details, buildings with four different height-to-breadth ratios, three width-to-depth ratios, and eight roof pitches were tested, for a total of 96 configurations. Further information can be found in [12]. The time histories provided by TPU are sampled at frequency of 500Hz and their duration is 18s at model scale, corresponding to $f_s = 15$Hz and a duration of 10minutes at full scale. The data are also moving-averaged every 0.2s at full scale. In order to provide pressure coefficients, the measured pressures over the building surfaces are normalized with respect to the mean velocity pressure $q_{\overline{H}}$ at the mean roof height $\overline{H}$:

$$c_p^*(t) = \frac{w(t)}{q_{\overline{H}}} \tag{1}$$

2.3 Definition of Roof Zoning

In the CNR Guidelines [8, 13], global pressure coefficients for flat and gable roofs are provided, and for the latter the two cases of wind orthogonal and parallel to the ridge are given. In all cases, the coefficients are the result of an envelope in the range ∓45° with respect to the perpendicular to the building face, i.e. the value causing the largest effects. In fact, if we suppose to have a linear relationship between the wind loading and the effects on structural members, then the maximum effects derive from the maximum global pressure coefficient. In Fig. 2, roof zoning is shown. When the wind is perpendicular to the ridge, the roof zoning is governed by the location of the ridge, usually half way along the short side of the building. The roof is thus divided into a windward surface, over which both suction and pressure can occur (depending on pitch angle), and a leeward surface over which only suction is expected. When the wind is parallel to the ridge, the roof zoning is governed by the parameter $e = \min\{h; b/2\}$, with h corresponding to the building height and b to the building breadth, i.e. the dimension of the windward surface. The roof is thus divided into two zones: Zone A, on which the maximum suction is expected, and Zone B with lower suction values and possible positive pressures. Notice that in the following, the case of flat roofs will be considered as a special case of gable roofs having pitch angle $\alpha = 0°$.

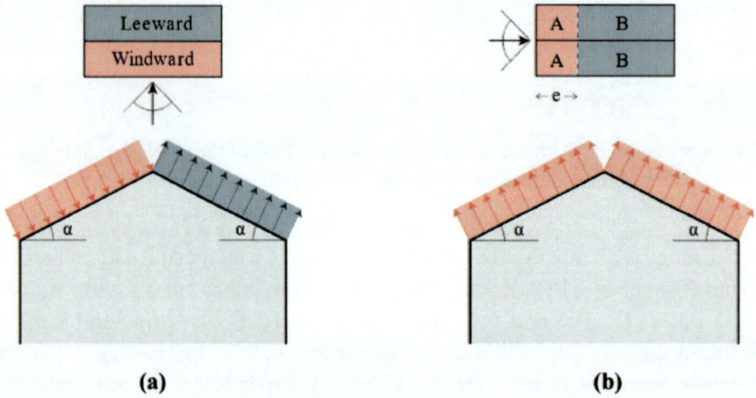

Fig. 2. Roof zoning with example of maximization of the wind loading effects for (a) wind perpendicular to the ridge and (b) wind parallel to the ridge.

The two loading conditions aim at maximizing the effects of the wind loading. When frames compose the main structural system of low-rise buildings, then it is expected that the loading pattern of Fig. 2a maximizes the effects in the elements of middle frames. On the other hand, the loading pattern of Fig. 2b maximizes the effects in the elements of frames near to the windward wall of the building. Notice that in the case of flat roofs there is no ridge line. According to CNR Guidelines, in this case the loading pattern of Fig. 2b is used for wind coming from both directions.

2.4 Extreme Value Analysis

The TPU database provides time histories of pressure coefficients at a number of taps over the building surfaces. In order to derive the global pressure coefficient $c_{p,i}$ for zone

i of the roof, an area-average is first made of the time-histories $c_p^*(M_j; t)$ at points M_j belonging to zone i:

$$c_{p,i}^*(t) = \frac{\sum_j A(M_j) \cdot c_p^*(M_j; t)}{\sum_j A(M_j)} \tag{2}$$

where $A(M_j)$ is the tributary area of the pressure tap corresponding to point M_j. Then, according to the EN1991-1-4 approach, the TPU pressure coefficients must be normalized with respect to the gust velocity pressure $\hat{q}_H = [1 + 7I_u] \cdot q_H$, evaluated at the roof height H [14]. The following transformation is thus applied:

$$c_{p,i}(t) = \frac{1}{[1 + 7I_u]} \cdot \frac{q_{\overline{H}}}{q_H} \cdot c_{p,i}^*(t) \tag{3}$$

According to Cook and Mayne [7], the design value $c_{p,i}$ of the pressure coefficient for the zone i is defined as the 78% fractile of maximum (minimum) values in the reference period $T = 10$minutes. When the Gumbel distribution is assumed to describe the extremes, then it turns out to be:

$$c_{p,i} = U_T + 1.4 \cdot a_T \tag{4}$$

where U_T and a_T are the location and the scale parameters in the reference period T, respectively, and 1.4 corresponds to the reduced variate $y = -\ln(-\ln 0.78)$.

Regardless of the method used to estimate the Gumbel parameters, analysis of maxima is usually applied to the maximum (minimum) values in a reference period T. This means that from N different samples with length T, the corresponding maximum values $\hat{c}_{p,j}$ ($j = 1, \ldots, N$) are retained and the EV method of analysis applied to them. However, in the case of the TPU online database only one sample is provided, and a different strategy is needed. To evaluate the Gumbel parameters, first the sample time history $c_p(t)$ of the pressure coefficient is divided into N sub-samples $c_{p,j}(t)$ of length $\tau = T/N$. For the j-th sub-sample, the maximum (minimum) values $\hat{c}_{p,j}$ are derived. The Method of Moments (MoM) is then applied to estimate the location U_τ and scale a_τ parameters in the reference time τ:

$$\begin{aligned} U_\tau &= \mathrm{E}[\hat{c}_p] - 0.45 \cdot \mathrm{STD}[\hat{c}_p] \\ a_\tau &= \frac{\sqrt{6}}{\pi} \cdot \mathrm{STD}[\hat{c}_p] \end{aligned} \tag{5}$$

where $\mathrm{E}[\hat{c}_p]$ is the expected value (sample mean) of the N maximum values and $\mathrm{STD}[\hat{c}_p]$ is their standard deviation. Finally, the Gumbel parameters in the reference period T are evaluated by using the following equation:

$$U_T = U_\tau + a_T \cdot \ln N; \qquad a_T = a_\tau \tag{6}$$

Equation (6) holds in the case of independence of the maximum (minimum) values. For this to apply, the period τ must be longer than the time lag τ_0 for which the Auto Correlation Function (ACF) approaches zero [15]. For the analysed cases, it was found

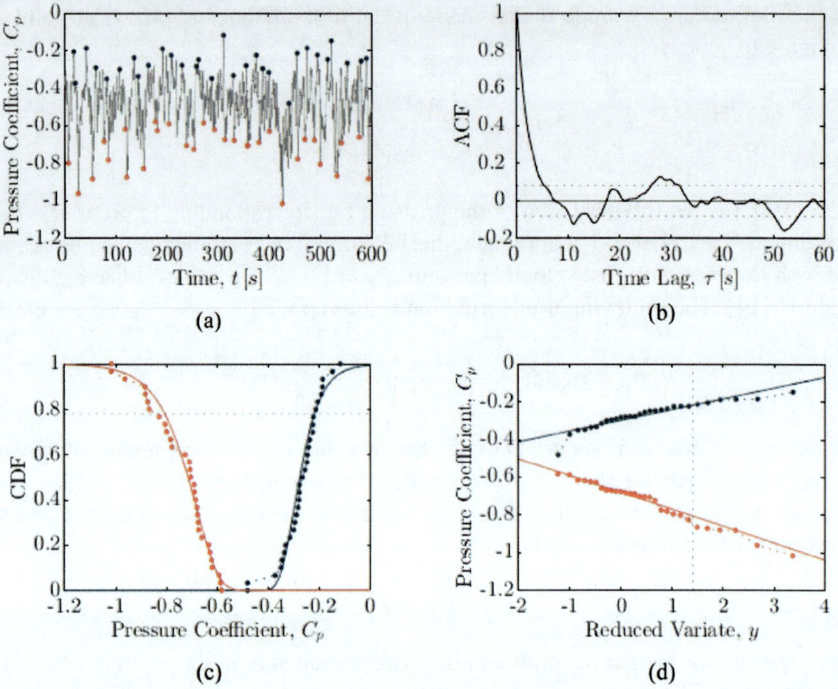

Fig. 3. Steps for the derivation of the design values of pressure coefficients: (a) picking of N maximum (blue dots) and minimum (red dots) values from the time history C_(p,i) (t); (b) Auto Correlation Function of C_(p,i) (t); (c) empirical (dots) and fitted Gumbel (continuous line) CDF of maximum and minimum values of C_(p,i) (t); (d) Gumbel paper with empirical (dots) and Gumbel fitted values (continuous line) of maximum and minimum values of C_(p,i) (t).

that a duration $\tau = 20$s satisfies the above condition. So doing, $N = 30$ sub-samples are thus derived from each time series of the pressure coefficients, and EV analysis is applied to the corresponding maximum (minimum) values.

As an example, in Fig. 3b the ACF for the time history of Fig. 3a is shown for which $\tau_0 = 7$s. In Fig. 3c the empirical CDFs of the maximum values (blue dots) and minimum values (red dots) extracted from the time series of Fig. 3a are shown, together with their Gumbel fitting (continuous lines). Within the approximation in modelling EVs, it is found that the Gumbel distribution well fits both maximum and minimum values (Fig. 3d).

3 Results and Discussion

3.1 Global Pressure Coefficients for Flat and Gable Roofs

The results of the application of the proposed methodology to flat and gable roofs are summarized in Fig. 4, where the global pressure coefficients c_p are plotted against the roof pitch angle α. Notice that according to its definition, the roof pitch angle α is positive for both windward and leeward surfaces in the case of wind flow orthogonal to the ridge. Moreover, flat roofs are considered as a special case of pitched roof where $\alpha = 0°$. The shaded areas in Fig. 4 represent the range of pressure coefficients evaluated for each of the tested buildings, while dots represent their mean value.

When the wind is perpendicular to the ridge (Fig. 4a), the mean value of the minimum pressure coefficient on the windward surface decreases from -1.0 ($\alpha = 0°$) to -0.2 ($\alpha = 45°$) as the pitch angle α increases. For the leeward surface, it increases from -0.5 to -0.82 for pitch angles from 0° to 22°, then decreasing to -0.66 for $\alpha = 45°$. On the other hand, maximum (positive) pressure coefficients are almost zero on the leeward surface and on the windward surface when $\alpha < 15°$. Instead, it increases almost linearly for pitch angles larger than 15°, reaching a value of 0.6 for $\alpha = 45°$.

When the wind flow is parallel to the ridge (Fig. 4b), the mean value of the minimum pressure coefficient for Zone A is at an almost constant value of -1.0 for all analysed pitch angles. In Zone B, it slightly increases from -0.47 ($\alpha = 0°$) to -0.58 ($\alpha = 15°$), then reaching a value of -0.44 for $\alpha = 45°$. Positive pressure coefficients are approximately zero in both zones, reaching its largest value of about 0.15 in Zone A for a pitch angle $\alpha = 45°$.

Notice that for both cases of wind orthogonal and parallel to the roof ridge, two different values of pressure coefficient exist for flat roofs. This derives from the choice of dividing the roof into two different areas; one single value would have been obtained, if the entire flat roof had been considered as one single area.

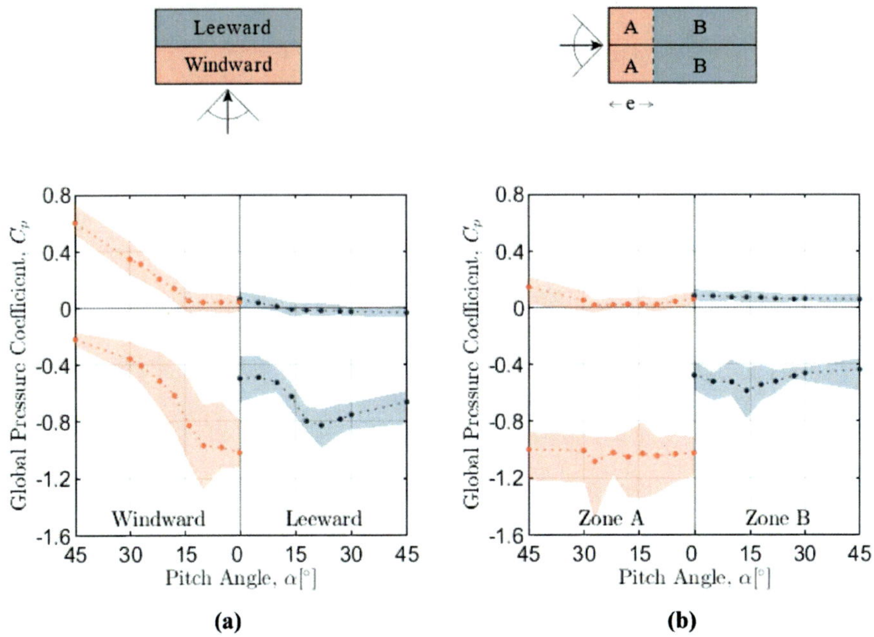

Fig. 4. Global pressure coefficients against roof pitch angle for (a) wind orthogonal to the ridge and (b) wind parallel to the ridge. The shaded areas represent the interval of pressure coefficients evaluated for each of the tested buildings; dots represent their mean value.

The values discussed above refer are mean values over the tested buildings. Indeed, a scatter exists, and this is represented by the shaded areas in Fig. 4. Overall, it is evaluated a coefficient of variation V[c_p] of the minimum (negative) pressure coefficients ranging from 0.05 to 0.25. Very large and meaningless values are evaluated for the maximum (positive) pressure coefficients, due to their means approaching zero.

3.2 Proposal and Comparison with Current Global Pressure Coefficients

Codification of pressure coefficients requires simple and transparent values, which can be used by practitioners minimising the chances of misinterpretation. To this aim, a simplification of the coefficients shown in Fig. 4 is required. In Table 1 the proposed coefficients for flat and gable roofs are reported in the case of wind orthogonal to the ridge. A constant value is suggested for pitch angles between 0 and 10°, equal to −1.0 for the windward surface and to −0.5 for the leeward surface. Instead, a linear variation is proposed for pitch angles between 10° and 20°, and from 20° to 45° for both surfaces. Moreover, for windward surfaces a second loading case is appropriate, with positive values of the pressure coefficient. In this case, a linear variation between 0 and 0.5 is proposed for pitch angles ranging from 10° to 45°. On the other hand, when the wind is parallel to the ridge, constant values of the negative pressure coefficient of −1.0 and −0.5 are proposed for Zone A and Zone B, respectively.

Table 1. Proposed Global Pressure Coefficients for wind orthogonal to the ridge.

Pitch Angle α[°]	Wind orthogonal to the ridge		
		Windward	Leeward
10 – 45	Positive	$0.6 \cdot \frac{\alpha-10}{35}$	
0 – 10	Negative	-1.00	-0.50
10 – 20		$-1.0 + 0.5 \cdot \frac{\alpha-10}{10}$	$-0.5 - 0.4 \cdot \frac{\alpha-10}{10}$
20 – 45		$-0.5 + 0.3 \cdot \frac{\alpha-20}{25}$	$-0.5 + 0.2 \cdot \frac{\alpha-20}{25}$

The global pressure coefficients proposed in this paper for flat and gable roofs are also shown in Fig. 5 (in red) together with values suggested by the CNR Guidelines (in blue). Notice that the coefficients from both the original CNR Guidelines [8] and from the updated version [13] are reported. The global pressure coefficients from the two versions correspond, except for the leeward surface when the wind is orthogonal to the ridge (dashed line in Fig. 5).

Globally, it is found that when the wind is perpendicular to the ridge, the coefficients given in the CNR Guidelines tend to underestimate suctions and overestimate pressures on the windward roof surface. For the leeward surface, they slightly overestimate suctions for pitch angles between 0° and 15°, and underestimate them for larger pitches. For wind parallel to the ridge, the pressure coefficients given in the CNR Guidelines tend to slightly underestimate suctions, as evaluated with the present approach.

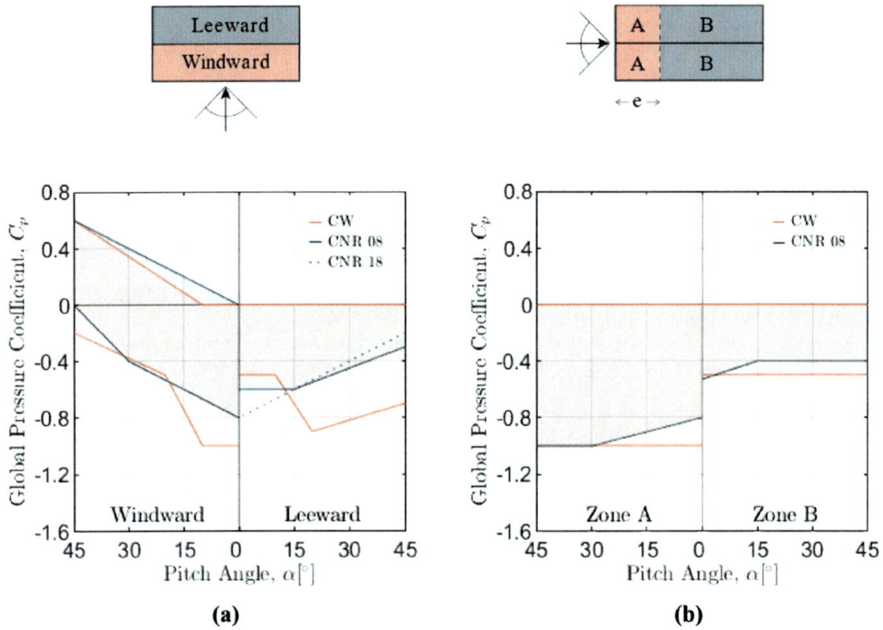

Fig. 5. Comparison between the proposed global pressure coefficients and those currently given by the CNR Guidelines for (a) wind perpendicular to the ridge and (b) wind parallel to the ridge.

4 Conclusion

In this paper, the definition of global pressure coefficient is revised together with the procedure for its assessment to make it consistent with the Cook and Mayne approach. In fact, the coefficients given by EN1991-1-4 are inconsistent both in their definition and from a statistical point of view. An application is made for flat and gable roofs with pitch angles up to 45°, by analysing samples of pressure coefficients provided by the aerodynamic database of the Tokyo Polytechnic University. Extreme Value analysis is applied, and the design values of global pressure coefficients are derived. In doing this, the Gumbel EV distribution is used to fit the empirical distribution of maximum (minimum) values. As a result, it is shown that the global pressure coefficients provided in CNR Guidelines result to be overall smaller than the new values. Only in the case of positive pressure coefficients for windward surface, the new calibrated pressure coefficients turn out to be smaller.

The analysis of the global pressure coefficients of this work is limited to gable roofs and to the analysis of tests from a single database. However, different roof configurations exist including gable roofs with different roof pitches, mono-pitch roofs, hip roofs, etc. Excluding very complex cases, current results should be extended to both different roof types and walls providing consistent global pressure coefficients. On the other hand, including test results from other wind tunnels would improve the accuracy of pressure coefficient.

Acknowledgement. The authors of the paper thank Prof. Gianni Bartoli for sharing his contribution to the drafting of CNR Guidelines and to the derivation of the current global pressure coefficients reported in CNR Guidelines. We also thank Prof. Yukio Tamura for sharing the complete TPU aerodynamic dataset and providing information about it.

References

1. CEN: Eurocode 1: Action on structures – Part 1–4: General actions – Wind actions, EN1991-1-4:2005, Comité Européen de Normalisation (2005)
2. Geurts, C., et al.:Transparency of Pressure and Force Coefficients. Proceedings of Third European & African Conference on Wind Engineering EACWE3, Eindhoven,The Netherlands (2001)
3. Stathopoulos, T.: Turbulent wind action on low-rise buildings. Ph.D. Thesis. The University of Western Ontario, Ontario, Canada (1979)
4. NRCC: National Building Code of Canada, NBCC1980, National Research Council of Canada. Associate Committee on the National Building Code (1980)
5. BSI: Loading for buildings – Code of practice for wind loads, BS 6399–2:1997, British Standards Institution (1997)
6. Cook, N.J.: The designer's guide to wind loading of building structures – Part 2. Butterworths, Building Research Establishment (1990)
7. Cook, N.J., Mayne, J.R.: A novel working approach to the assessment of wind loads for equivalent static design. J. Wind Eng. Ind. Aerodyn. **4**, 149–164 (1979)
8. CNR: Guide for the assessment of wind actions and effects on structures, CNR-DT 207/2008, Consiglio Nazionale delle Ricerche (2008)
9. CEN: Eurocode 1: Action on structures – Part 1–4: General actions – Wind actions, prEN1991-1-4:2020, Comité Européen de Normalisation (2020)
10. Gavanski, E., Gurley, K., Kopp, G.: Uncertainties in the estimation of local peak pressures on low-rise buildings by using the gumbel distribution fitting approach. J Struct. Eng. **142** (11) (2016)
11. Tokyo Polytechnic University (TPU) Online Aerodynamic Database Homepage. http://wind.arch.t-kougei.ac.jp/system/eng/contents/code/tpu, Accessed1 July 2022
12. Quan, Y., Tamura, Y., Matsui, M. Cao, S.Y., Yoshida, Y.: TPU aerodynamic database for low-rise buildings. Proceedings of the 12th International Conference on Wind Engineering (ICWE12), Vol.2, Cairns, Australia, pp. 1615–1622 (2007)
13. CNR: Istruzioni per la valutazione delle azioni e degli effetti del vento sulle costruzioni, CNR-DT 207/2018 (*in Italian*), Consiglio Nazionale delle Ricerche (2018)
14. Picozzi, V., Malasomma, A., Avossa, A.M., Ricciardelli, F.: The relationship between wind pressure and pressure coefficients for the definition of wind loads on buildings. Buildings **12**(2), 225 (2022)
15. Quan, Y., Wang, F., Gu, M.: A method for estimation of extreme values of wind pressure on buildings based on the generalized extreme-value theory. Math. Prob. Eng. **2014**, 1–22 (2014). https://doi.org/10.1155/2014/926253

Gust Factor of the Alongwind Response of Structures Subjected to Thunderstorm Outflows

Luca Roncallo(✉) and Federica Tubino

University of Genoa, Genoa, Italy
luca.roncallo@edu.unige.it

Abstract. This study investigates the possibility of generalizing the well-established Davenport gust factor technique to the evaluation of the maximum alongwind dynamic response of single-degree-of-freedom systems to thunderstorm outflows. In particular, the sensitivity of the gust factor to the analytical model adopted for the modulating function defining the slowly-varying mean wind velocity is first assessed. Then, the intervals of variation of the parameters shaping the wind velocity (the intensity of the background wind and the duration of the intense phase) are obtained from the analysis of 129 full-scale thunderstorm records. Finally, the gust factor is analytically evaluated on varying said parameters and its sensitivity to such parameters is studied. Results show that the dependence of the gust factor on the intensity of the background wind and the duration of the intense phase is not negligible.

Keywords: Alongwind Response · Gust Factor · Non-stationarity · Thunderstorm outflows

1 Introduction

Over the last decades, a significant amount of research in the wind engineering field has been dedicated to the study of thunderstorm outflows and their impact on structures, aiming to develop models and tools to pursue a safer and cost-efficient structural design (e.g. [1–4]). However, the complexity of the phenomenon and the lack of data over which to validate the proposed models have prevented the development of an efficient and handy approach for the estimate of the maximum dynamic response as the one developed by Davenport for synoptic winds. A first generalization of the gust factor technique was introduced by Kwon and Kareem [5] adopting suitable equivalent parameters [6] and under the assumption of long pulse duration. Successively, Kwon and Kareem [7] outlined the computation framework of this procedure and recently they derived a closed-form solution for the equivalent parameters [8]. Although the approach is suitable for rapid engineering calculations, it has the shortcoming of being based on an ideal Evolutionary Power Spectral Density (EPSD) model of the thunderstorm wind velocity. Moreover, the derivation of the gust response factor can be rather conservative due to the assumption of long pulse duration which neglects the effects due to the transient dynamic

P. Schito and A. Zasso (Eds.): IN VENTO 2022, LNCE 461, pp. 23–30, 2024.
https://doi.org/10.1007/978-3-031-53059-3_3

response [9]. Starting from a large number of wind velocity records collected in the High Tyrrenian Sea, Roncallo and Solari [10] developed an EPSD model for thunderstorm wind speed, introducing a novel analytical representation for the slowly-varying mean wind velocity characterized by two parameters which possess a physical interpretation: the duration of the intense phase of the thunderstorm wind velocity and the intensity of the background wind. Successively, the model has been adopted to estimate the maximum alongwind response of Single-Degree-of-Freedom (SDOF) systems generalizing the formulation of the Davenport's gust factor using the technique of the equivalent parameters [6, 8] and accounting for the transient dynamic effects, showing a satisfying agreement with the gust factor obtained from full-scale thunderstorm records [9].

On the basis of these previous studies, the present paper investigates three main aspects: the sensitivity of the generalized gust response factor of SDOF systems to the modulating function of the slowly-varying mean wind velocity; the analysis of the range of variability of the parameters characterizing said modulating function and finally a sensitivity analysis of the gust response factor on varying these parameters, outlining their importance in the estimation of the maximum response.

2 Analytical Formulation

The thunderstorm wind velocity is modelled as a uniformly modulated process [10]:

$$v(t) = \overline{v}_{max}\gamma(t)\left[1 + I_v\tilde{v}'(t)\right] \tag{1}$$

where $\overline{v}_{max}$ and $\gamma(t)$ are, respectively, the maximum value and the deterministic modulating function of the slowly-varying mean wind velocity, $t \in [-T_{max}/2, T_{max}/2]$ the time and $T_{max} = 600$ s, I_v is the mean value of the turbulence intensity (assumed constant [10]) and $\tilde{v}'(t)$ is the so-called reduced turbulent fluctuation, dealt with as a zero-mean stationary Gaussian random process with unitary standard deviation. This latter is statistically characterized by its Power Spectral density (PSD) [11]:

$$S_{\tilde{v}'}(n) = \frac{1}{n}\frac{6.868L_v/\overline{v}_{max}}{\left[1 + 10.302nL_v/\overline{v}_{max}\right]^{5/3}} \tag{2}$$

where n the frequency and L_v is the integral length scale [10, 12]. It is worth to notice that the model in Eq. (2) was originally defined for stationary winds [11] and its extension to thunderstorms was assessed by [10] based on the statistical analysis of a large number of wind speed records.

Let us consider a linear elastic SDOF system characterized by mass m, fundamental frequency n_0, damping ratio ξ and alongwind displacement x, schematized as a point-like surface with area A perpendicular to the wind velocity direction and constant drag coefficient c_D. Invoking the hypothesis of small turbulence and assuming the maxima of the mean and fluctuating part of the response as simultaneous [9], the maximum response can be expressed as [9]:

$$x_{max} = G_x\overline{x}_{max} \tag{3}$$

where $\overline{x}_{max}$ and G_x are, respectively, the maximum value of the time-varying mean part of the response (assumed as static) [9] and the gust response factor in its generalized form for thunderstorm winds [9]:

$$\overline{x}_{max} = \frac{\rho \overline{v}_{max}^2 A c_D}{2m(2\pi n_0)^2} \tag{4}$$

$$G_x = 1 + 2I_v g_x\left(\nu_x \tilde{T}_{eq}\right)\tilde{J}\mathcal{C} \tag{5}$$

where $\mathcal{C}$ and $\tilde{T}_{eq}$ are, respectively, the non-dimensional equivalent standard deviation and equivalent period [6, 8, 9], g_x the Davenport peak factor, ν_x the expected frequency of the stationary part of the response and $\tilde{J}$ is the non-dimensional standard deviation of the dynamic response to the stationary part of the turbulence:

$$\tilde{J}^2 = \int_0^{+\infty} |H(\tilde{n})|^2 S_{\tilde{v}'}(\tilde{n}) d\tilde{n} \tag{6}$$

$$\mathcal{C}^2 = \frac{1}{(2I_v)^2\tilde{J}^2} \frac{\int_{-\tilde{T}_{max}/2}^{+\tilde{T}_{max}/2} c_{00,\tilde{x}'}^5(\tilde{t})d\tilde{t}}{\int_{-\tilde{T}_{max}/2}^{+\tilde{T}_{max}/2} c_{00,\tilde{x}'}^4(\tilde{t})d\tilde{t}} \tag{7}$$

$$\tilde{T}_{eq} = \int_{-\tilde{T}_{max}/2}^{+\tilde{T}_{max}/2} \exp\left[4 - \frac{4(2I_v)^2\tilde{J}^2\mathcal{C}^2}{c_{00,\tilde{x}'}(\tilde{t})}\right] d\tilde{t} \tag{8}$$

with $\tilde{n} = n/n_0$ a dimensionless frequency, $\tilde{t} = t/T_0$ a dimensionless time (being $T_0 = 1/n_0$ the fundamental period), $\tilde{T}_{max} = T_{max}/T_0$, $S_{\tilde{v}'}(\tilde{n}) = n_0 S_{\tilde{n}'}(n)$, $\tilde{x}'(t) = x'(t)/\overline{x}_{max}$ the dimensionless fluctuating part of the response (i.e. $x'(t)$). In Eqs. (6)–(8) $H(\tilde{n})$ is the complex frequency response function and $c_{00,\tilde{x}'}(\tilde{t})$ the Non-Geometrical Spectral Moment (NGSM). They are defined as:

$$H(\tilde{n}) = \frac{1}{1 - \tilde{n} + 2i\tilde{n}\xi} \tag{9}$$

$$c_{00,\tilde{x}'}(\tilde{t}) = (2I_v)^2 \int_0^{+\infty} \left|Z(\tilde{n},\tilde{t})\right|^2 S_{\tilde{v}'}(\tilde{n}) d\tilde{n} \tag{10}$$

where $Z(\tilde{n},\tilde{t})$ is the dimensionless Evolutionary Frequency Response Function (EFRF) defined as:

$$Z(\tilde{n},\tilde{t}) = \int_{-\tilde{T}_{max}/2}^{\tilde{t}} h(\tilde{t}-\tilde{\tau})e^{-i2\pi\tilde{n}(\tilde{t}-\tilde{\tau})}\gamma^2(\tilde{\tau})d\tilde{\tau} \tag{11}$$

with $h(\tilde{t})$ the dimensionless impulse response function:

$$h(\tilde{t}) = \frac{2\pi}{\sqrt{1-\xi^2}} e^{-2\pi\xi\tilde{t}} \sin\left(2\pi\sqrt{1-\xi^2}\tilde{t}\right)\mathcal{H}(\tilde{t}) \tag{12}$$

being $\mathcal{H}(\tilde{t})$ the Heaviside function.

3 Gust Factor Sensitivity to the Modulating Function

Two analytical expressions were proposed for the modulating function $\gamma(t)$ of the slowly-varying mean wind velocity (Eqs. (1), (11)) by Roncallo and Solari [10]:

$$\gamma(t) = \frac{1-\gamma^*}{\left[1+\left(\frac{t}{T}\right)^2\right]^{1/2}} + \gamma^* \quad Model I \tag{13}$$

$$\gamma(t) = (1-\gamma^*)\exp\left\{-\left(\frac{t}{T}\right)^2\right\} + \gamma^* \quad Model II \tag{14}$$

where the parameters γ^* and T are, respectively, a measure of the intensity of the background wind velocity and a measure of the duration of the intense phase of the thunderstorm outflow. In addition, a new model is proposed in this study that allows for the derivation of an analytical solution for the EFRF in Eq. (11) and reads:

$$\gamma(t) = \begin{cases} \frac{1-\gamma^*}{2}\left[\cos\left(\frac{2\pi t}{T}\right)+1\right]+\gamma^*, & |t| < \frac{T}{2} \\ \gamma^* & , \ |t| \geq \frac{T}{2} \end{cases} \quad Model III \tag{15}$$

It is worth noticing that, in principle, the parameters γ^* and T are respectively bounded between 0 and 1 and 0 and T_{max}. However, it is clear that the case $\gamma^* = 0$ and $T = 0$ s describes an unrealistic scenario, where the downburst is occurring in absolute absence of background wind and the duration is so short that it is instantaneous. Instead, the case $\gamma^* = 1$ describes exactly the synoptic event in which there is no modulation in the mean wind velocity (and in the turbulent fluctuations) but it is constant within 10 min, hence returning to the stationary case.

The gust response factor in Eq. (5) constitutes a generalization of the gust factor from Davenport [13], being function of the parameters C and $\tilde{T}_{eq}$ that account for the transient dynamic effects due to the nonstationary characteristics of the thunderstorm loading. The estimate of these parameters (Eqs. (7) and (8)) requires the definition of the modulating function $\gamma(t)$ for which the models in Eqs. (13)–(15) can be adopted. Since the models are mathematically different from one another, it is worth to investigate the effects of choosing one model or another on the gust response factor. Hence the sensitivity of the gust response factor is studied by comparing Eq. (5), evaluated for a set of SDOF systems characterized by fundamental frequency n_0 in the range [0.05–3] Hz and damping ratio ξ in the range [0.2–5]%, employing the three models in Eqs. (13)–(15) and setting the couples of parameters γ^* and T in order to best trace the mean trend of 129 modulating functions extracted from a database of 129 thunderstorm time histories collected in full-scale in four different ports in the High Tyrrenian sea, i.e. Genoa, La Spezia, Savona and Livorno [9, 10, 12, 14]. Figure 1 plots the ensemble, the mean trend and the fitting of the modulating function $\gamma(t)$ (Fig. 1a) along with the comparison between the gust response factor (Fig. 1b) estimated employing the three models in Eqs. (13)–(15).

Figure 1a shows that model III furnishes a very similar representation of the mean trend as the one provided by model II. From the fitting the couples of parameters γ^* and

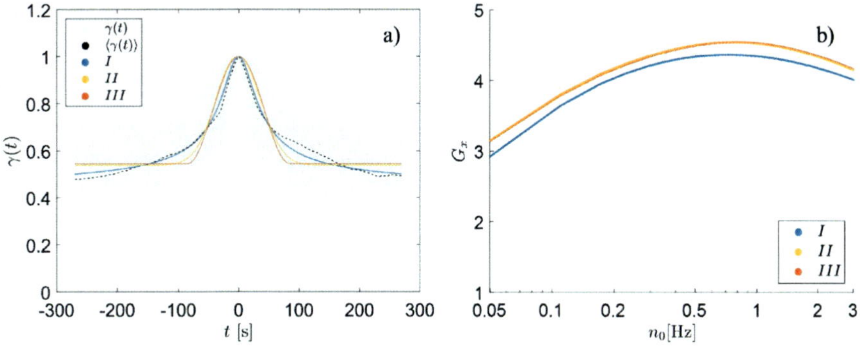

Fig. 1. (a) Samples of the modulating function γ(t), mean trend and fitting; (b) Comparison of the gust response factor estimated employing the three models in Eqs. (13)–(15) (model I blue lines, model II yellow lines, model III orange lines) for $\xi = 0.2\%$.

T are extracted for each model: $\gamma^* = 0.45$ and $T = 26.45$ s for model I; $\gamma^* = 0.54$ and $T = 51.34$ s for model II; $\gamma^* = 0.55$ and $T = 169.81$ s for model III. Figure 1b shows that, although the models for the modulating function of the slowly-varying mean wind velocity are mathematically different, they provide very similar results in terms of gust response factor. In particular, model II and model III furnish identical values of the gust factor, slightly higher than the ones provided by model I which furnishes the best representation of the mean trend of the function $\gamma(t)$ (Fig. 1a). This allows to employ the closed-form solution of the EFRF from model III to carry out the analyses without affecting the quality of the results while drastically reducing the computational burden [15].

4 Gust Factor Sensitivity to the Modulating Function Parameters

The sensitivity of the gust factor to the parameters of the modulating function is investigated in this section. According to results in the previous Section, analyses are carried out employing model III (Eq. (15)) in non-dimensional terms by defining the following dimensionless quantities:

$$\tilde{T} = \frac{T}{T_0} \tag{16}$$

$$\mathcal{T} = \frac{T}{T_{max}} \tag{17}$$

where $\tilde{T}$ is duration of the intense phase of the outflow normalized with the fundamental period of the structure and plays the role of a non-dimensional fundamental frequency whereas $\mathcal{T}$ is the duration of the intense phase with respect to the whole duration of the phenomenon T_{max}. It is important to notice that, while $\mathcal{T}$ is strictly a property of the thunderstorm, $\tilde{T}$ relates one property of this latter, i.e. the duration of the intense phase, with a mechanical property of the dynamical system, i.e. the fundamental period of the structure.

Starting from the thunderstorm records available, the parameters γ^* and T are estimated from each thunderstorm realization in order to get information on their variability. Due to the differences between one thunderstorm and another, the trend of the related modulating function may be, in general, different and so the values assumed by the two parameters. These latter are estimated by fitting each of the 129 records of the modulating function employing model III in Eq. (15). Figure 2 plots the couples of parameters γ^* and T obtained from the fitting.

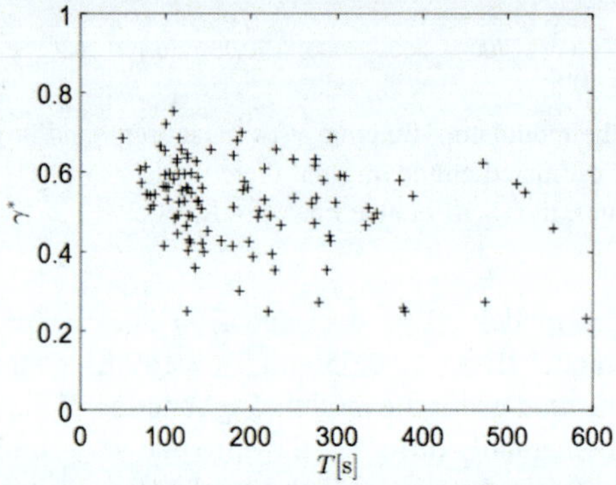

Fig. 2. Values of the parameters γ^* and T extracted from the time-histories.

Results show that the range of variability of the parameter γ^* is between 0.2 and 0.8 while the range of variability of T is between 50 s and 600 s, although most of the values lie in a closer range between 100 s and 300 s. The pdfs of the parameters are reported in Fig. 3.

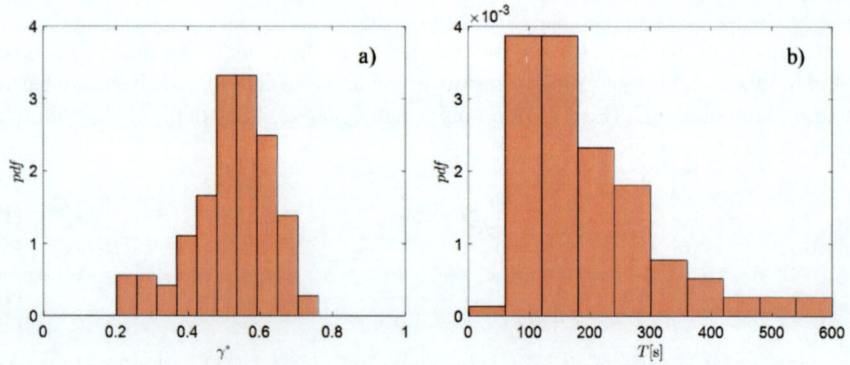

Fig. 3. Pdfs of the parameters γ^* (a) and T (b).

According to the values of the parameters previously reported in Figs. 2 and 3, the values $\gamma^* = [0.3, 0.5, 0.7]$ and $\mathcal{T} = [1/6, 1/4, 1/2]$ are considered in this analysis. As an example, Fig. 4 plots the tendencies of the gust factor evaluated for a damping ratio of 0.2% varying the parameters γ^* and $\mathcal{T}$.

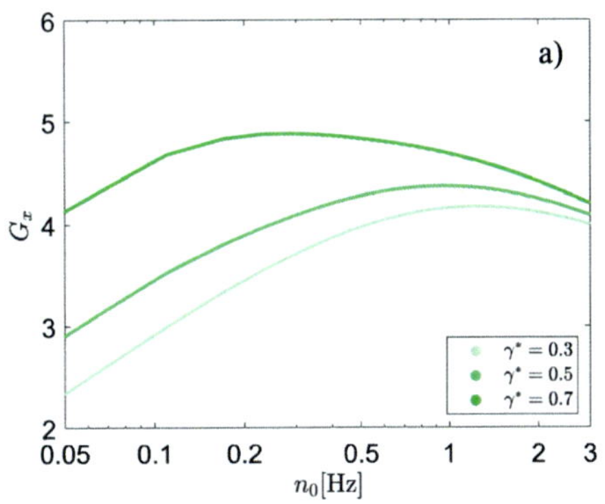

Fig. 4. Gust factor variation on varying the non-dimensional background wind speed intensity γ^* ($\mathcal{T} = 1/4$) (a) and the non-dimensional duration of the intense phase $\mathcal{T}$ ($\gamma^* = 0.5$) (b), for systems with damping ratio $\xi = 0.2\%$.

Results show that the gust factor is significantly affected by the variability of the parameters γ^* and $\mathcal{T}$. Figure 4a shows that neglecting the role of the background wind (provided by the non-dimensional parameter γ^*), as assumed by other models in the literature [8], may not be conservative. Moreover, Fig. 4b shows that the gust factor increases on increasing the non-dimensional duration of the intense phase of the thunderstorm, provided by the parameter $\mathcal{T}$.

5 Conclusions

The present study aimed to investigate sensitivity of the thunderstorm gust response factor to the analytical model adopted for the modulating function of the slowly-varying mean wind velocity and to the characteristic parameters that shape this function, according to the EPSD model developed by Roncallo and Solari [10] and together with a new formulation of the modulating function that allows an analytical solution of the EFRF.

The analysis of the sensitivity of the gust factor to the modulating function of the wind velocity has shown that very similar trends are obtained employing the different models proposed. In particular, the new model introduced gives almost the same trends of the exponential one from Roncallo and Solari [10].

The sensitivity analysis of the gust factor on the parameters γ^* and T pointed out that the gust factor is sensitive both to the variation of the background wind and to the duration of the intense phase.

In view of the obtained results, the role of the two parameters of the slowly-varying mean wind velocity, along with the transient dynamic effects, cannot be neglected. Therefore, a closed-form solution for the thunderstorm gust response factor, able to account for the role of the parameters and transient effects, is aimed to be developed, furnishing engineers with a handy tool for the structural design against thunderstorm outflows.

Acknowledgements. This research is funded by the European Research Council (ERC) under the European Union's Horizon 2020 research and innovation program (grant agreement No. 741273) for the project THUNDERR - Detection, simulation, modelling and loading of thunderstorm outflows to design wind-safer and cost-efficient structures - supported by an Advanced Grant 2016. The authors are deeply grateful to Prof. Giovanni Solari for his essential input and motivation for this research within the project THUNDERR.

References

1. Caracoglia, L.: Unified stochastic dynamic and damage cost model for the structural analysis of tall buildings in thunderstorm-like winds. ASCE-ASME J. Risk Uncertain. Eng. Syst. Part A Civil Eng. **4**(4), 4018043 (2018)
2. Chen, X.: Analysis of alongwind tall building response to transient nonstationary winds. J. Struct. Eng. **134**, 782–791 (2008)
3. Chen, L., Letchford, C.W.: A deterministic-stochastic hybrid model of downbursts and its impact on a cantilevered structure. Eng. Struct. **26**, 619–629 (2004)
4. Lombardo, F.T., Smith, D.A., Schroeder, J.L., Mehta, K.C.: Thunderstorm characteristics of importance to wind engineering. J. Wind Eng. Ind. Aerodyn. **125**, 121–132 (2014)
5. Kwon, D.-K., Kareem, A.: Gust-front factor: new framework for wind load effects on structures. J. Struct. Eng. **135**(6), 717–732 (2009)
6. Michaelov, G., Lutes, L.D., Sarkani, S.: Extreme value of response to nonstationary excitation. J. Eng. Mech. **127**(4), 352–363 (2001)
7. Kwon, D.-K., Kareem, A.: Generalized gust-front factor: a computational framework for wind load effects. Eng. Struct. **48**, 635–644 (2013)
8. Kwon, D.-K., Kareem, A.: Towards codification of thunderstorm/downburst using gust front factor: model-based and data-driven perspectives. Eng. Struct. **199**, 109608 (2019)
9. Roncallo, L., Solari, G., Muscolino, G., Tubino, F.: Maximum dynamic response of linear elastic SDOF systems based on an evolutionary spectral model for thunderstorm outflows. J. Wind Eng. Ind. Aerodyn. **224**, 104978 (2022)
10. Roncallo, L., Solari, G.: An evolutionary power spectral density model of thunderstorm outflows consistent with real-scale time-history records. J. Wind Eng. Ind. Aerodyn. **203**, 104204 (2020)
11. Solari, G., Piccardo, G.: Probabilistic 3-D turbulence modeling for gust buffeting of structures. Probab. Eng. Mech. **16**, 73–86 (2001)
12. Zhang, S., Solari, G., De Gaetano, P., Burlando, M., Repetto, M.P.: A refined analysis of thunderstorm outflow characteristics relevant to the wind loading of structures. Probab. Eng. Mech. **54**, 9–24 (2018)
13. Davenport, A.G.: Gust loading factors. J. Struct. Division. **93**, 11–34 (1967)
14. Solari, G., et al.: The wind forecast for safety management of port areas. J. Wind Eng. Ind. Aerodyn. **104–106**, 266–277 (2012)
15. Roncallo, L., Tubino, F.: Thunderstorm gust response factor: general tendencies and sensitivity analysis. J. Wind Eng. Ind. Aerodyn. **236**, 105376 (2023)

An Analysis of Truck-Driver System Response to Crosswind in Tunnel Exit Conditions

A. R. Prakash, A. Cioffi(✉), E. Sabbioni, M. Vignati, and P. Schito

Department of Mechanical Engineering, Politecnico di Milano, Milan, Italy
antonio.cioffi@polimi.it

Abstract. Heavy vehicles entering and exiting a tunnel at high speed under strong crosswinds is a particularly critical condition since the aerodynamic load change drastically, greatly affecting the lateral stability of the vehicle. Active control systems (active suspensions, active front steering, etc.) and infrastructure elements (e.g. wind fences) are proposed to reduce the induced risks. To help the design of these devices, the present paper investigates the response of the vehicle-driver system in the case of a high-sided lorry entering and exiting a tunnel under crosswind, by using driver-in-the-loop simulations. The study was performed using the dynamic driving simulator of Politecnico di Milano and 28 test drivers. Vehicle and aerodynamic models have been developed to reproduce the phenomenon in a highly immersive environment. During the tests, several combinations of vehicle and wind speed were considered. The effect of turbulence and vehicle loading condition (empty and full) was also investigated.

Keywords: Vehicle-Driver Interaction · Driver Simulator · Vehicle-Wind Interaction

1 Introduction

According to a 2018 ISTAT [1] survey on road accidents in Italy, it is estimated that about 10% of the total crashes on extra urban roads can be attributed to adverse weather conditions. When considering high sided vehicles, characterized by a big lateral surface, crosswind can be considered dangerous for the driver and road users. The risk is mainly connected to the turbulent nature of the flow, and the dynamic excitation that it can induce on the vehicle and on the driver response. Under this point of view, the presence of infrastructure elements partially shielding the vehicle from crosswind, such as tunnels and bridge pylons, are particularly critical since they induce a sudden variation of the aerodynamic loads acting on the vehicle, which may compromise vehicle lateral stability and/or may induce rollover.

Studies on this topic have been conducted by Rocchi et al. [2]. The authors proposed a numerical-experimental methodology to study the response of a truck when subject to the wake of a bridge tower under crosswind. After characterizing the aerodynamic coefficients of the vehicle through wind tunnel tests, the respective loads are introduced in the dynamic model following a procedure proposed by Charuvisit et al. [3, 4].

P. Schito and A. Zasso (Eds.): IN VENTO 2022, LNCE 461, pp. 31–40, 2024.
https://doi.org/10.1007/978-3-031-53059-3_4

In this study, the vehicle-driver system response to crosswind when exiting a tunnel is investigated. The entire work has been performed on the DiM-400 dynamic driving simulator of Politecnico di Milano, whose main characteristics are reported in Sec. 3.4.

Aim of the paper is to investigate the effect of several parameters (combination of vehicle and wind speed, wind turbulence, vehicle load) on the response of the vehicle-driver couple. This would help the design of the following devices for reducing the risk of rollover/lane invasion:

- Active control systems: active front steering (AFS), active suspensions, electronic stability program (ESP), etc.
- Wind fences which may be added at tunnel enter/exit.

Moreover, the performed analysis may be used to design a driver model to be included into a numerical tool helping engineers in designing the above mentioned devices.

The article is organized as follows. At first, the methodology is presented. Then, the numerical models used are described. Finally, the results and the major conclusions drawn have been reported.

2 Methodology

For this study, a vehicle and aerodynamic model have been coupled to study the vehicle-driver system response when subjected to crosswind at tunnel exit condition.

Everything has then been tested on the DiM-400 dynamic driving simulator facility available at the Politecnico di Milano. After calibrating the model on the driving simulator, an experimental campaign with 28 different drivers was carried out. The drivers were all volunteer university students (average age of 23 years old) with a valid driving license from at least 4 years. Most of them had little to no experience in driving trucks. For these reason, they had a 10 min practice session on the driving simulator prior to conducting the experiment.

The results have been analyzed to assess the relation between vehicle response in terms of yaw, roll and lateral motions and driver control inputs (primarily in terms of steer angles and steer torque application). Other global parameters such as vehicle lane deviation and normalized load transfers have also been evaluated.

3 Numerical Models

In this section the vehicle model, aerodynamic model and the simulation scenario will be described.

3.1 Vehicle Model

The vehicle dynamics model has been developed using the multibody software VI-CarRealTime, from VI-Grade. The relevant vehicle data for the models have been obtained from previous studies carried out by Cheli et al. [5, 6]. A simplified suspension system has been used. The steering system was modelled carefully to reproduce appropriate values of steering torque. The tires have been modeled using the MF-Tyre model including combined slip effects and nonlinear vertical load dependency [7].

3.2 Aerodynamic Model

To reproduce the crosswind excitation on the vehicle, a Simulink model was developed. The loads are calculated using a quasi-static approach. Thus, the relative wind speed is computed at first. Then, the angle of attack is evaluated. This angle is used to evaluate the aerodynamic coefficients through lookup tables. The aerodynamic coefficients used for the current vehicle model are reported in Fig. 1 and have been obtained from the work performed by Cheli et al. [5, 6]. Finally, the loads are obtained. To reproduce the tunnel exit condition, the loads are applied proportionally to the exposed area. The center of pressure has been assumed to be moving proportionally to the exposed area, too. This hypothesis has already been used in the previous work of Walczak [8].

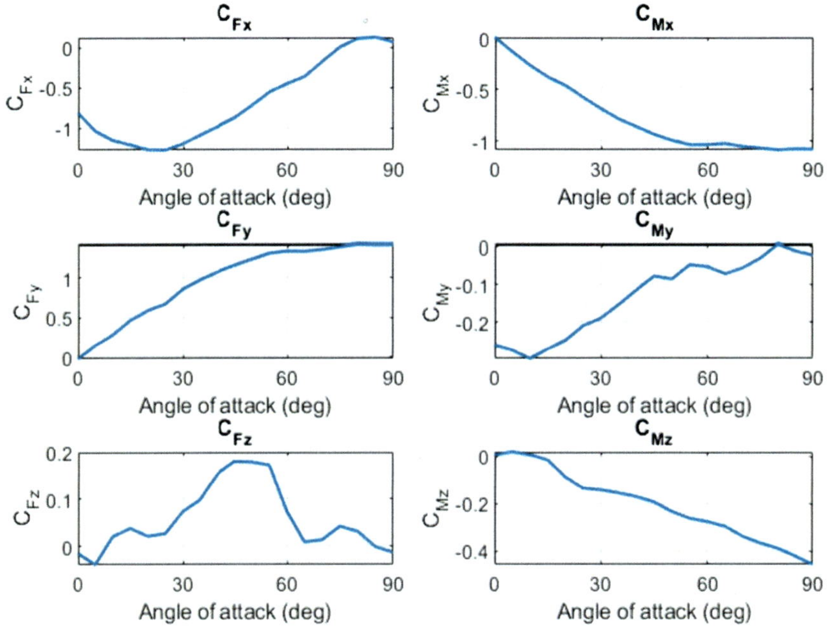

Fig. 1. Aerodynamic coefficients of the VAN vehicle.

In Fig. 2 is reported a time history of the wind profile resulting from the hypothesis above-mentioned. For sake of clarity, wind turbulence is not included into the figure, but it was added during the experiments. The turbulence intensity was set to $I_{wind} = 7\%$ to guarantee a maximum variability of ± 5 km/h with respect to the average wind speed. The turbulent wind time history were generated using the non-dimensional power spectral density function (Von Karman spectrum) with a roughness length $z_0 = 0.05$ m. The grey shaded area represents the tunnel length. Due to the model chosen, the aerodynamic forces show a linear trend at the tunnel exit and entry.

However, the situation is slighter different for the aerodynamic moments. With reference to the yawing moment M_z (the moment that most contributes to the vehicle dynamics), two conflicting trends can be acknowledged. The first trend regards the pure aerodynamic moment (the one evaluated using the $C_{M,z}$ polar that in Eq. (1) is termed $M_{z_{ref}}$) and is linearly increasing due to the increasing lateral area (when in tunnel exit condition). The second trend regards the transport moment due to the aerodynamic force related to the $C_{F,y}$ polar (in Eq. (1) is termed as $F_{y_{ref}}$). In this case, since the moment arm d_1 becomes null when the vehicle is fully outside the tunnel, the moment gradually disappears. The complete equation for the aerodynamic yawing moment M_z is thus reported in Eq. (1).

$$M_z = M_{z_{ref}} + F_{y_{ref}} \cdot d_1 \forall \text{ values of } d_1 \quad (1)$$

The same consideration can be made for the aerodynamic pitching moment M_y due to the lift force F_z.

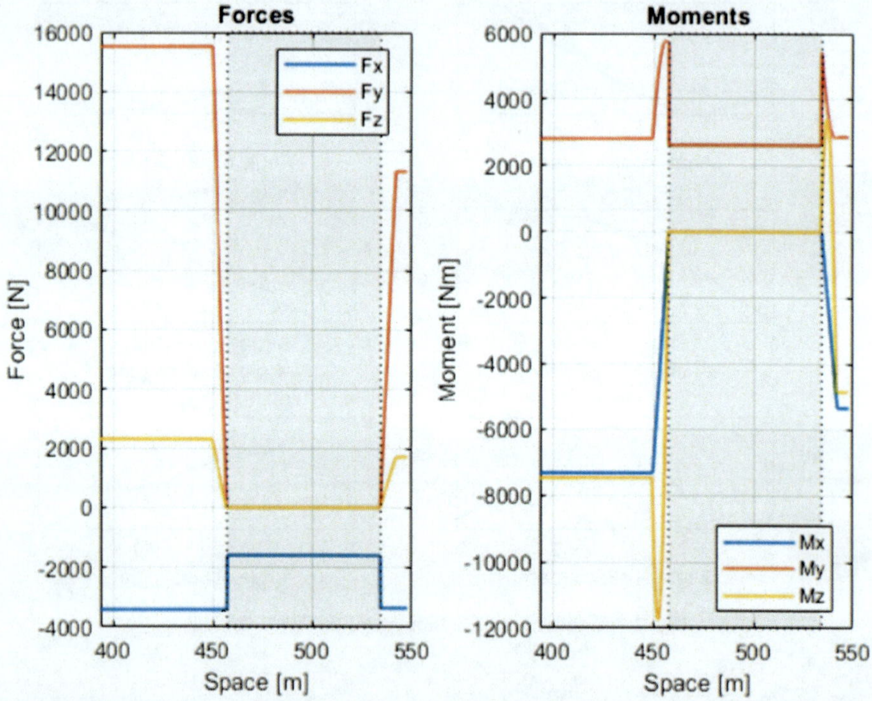

Fig. 2. Aerodynamic forces and moments acting on the VAN vehicle when approaching/exiting the tunnel.

3.3 Scenario

The simulation scenario has been created using Roadrunner. It consists of a straight road with a length of 2 km. On the road, 5 tunnels of length 75 m have been placed. The tunnels are equally spaced. In each section and for each driver, different wind speeds have been selected to conduct the experiment. The wind speed and orientation (wind blowing from left or right) were chosen randomly for each section. The driver was assigned the task to remain at the center of the lane with a 0 rad yaw angle during each test. To maintain a constant speed, a cruise control system was used. The test matrix is reported in Table 1. Both turbulent and non-turbulent wind has been tested.

Table 1. Wind and vehicle conditions tested in the experimental campaign.

Vehicle Configuration	Vehicle Speed	Wind Speed
Empty	65 km/h	±15/20/25 m/s
	80 km/h	±15/20/25 m/s
Laden	65 km/h	±20/25/30 m/s
	80 km/h	±20/25/30 m/s

3.4 Dynamic Driving Simulator

The VI-Grade DiM400 dynamic driving simulator (it is shown in Fig. 3) installed at the DRI.S.MI. Laboratory of Politecnico di Milano is one of the state-of-the-art driving simulators in the world. This simulator is driven by 4 cables capable of controlling 3 degrees of freedom in the plane, namely the longitudinal displacement, lateral displacement and yaw rotation. This leads to large range motion of the diskframe on a working space of dimension 4 m × 4 m. The diskframe houses a moving cabin suspended on six actuators (hexalift system) which control its six degrees of freedom (rotations and translations around the longitudinal, lateral and vertical directions). In addition to these actuators, eight additional shakers are present onboard (with a bandwidth of 200 Hz) to reproduce vibrations from the engine or road irregularities. The workspace of the DiM400 happens to be one of the biggest among the driving simulators of this category with a fixed screen.

The simulator interfaces with VI-CarRealTime and Simulink to model the high sided lorry and with VI-WorldSim that is responsible for reproducing the scenario to be tested (graphic environment). The scenario was created using Roadrunner and Blender and then imported in VI-WorldSim.

Fig. 3. DRI.S.MI. Laboratory of Politecnico di Milano – VI-Grade DiM400 simulator.

4 Results

In this section the results obtained from the experimental campaign will be reported.

Lateral Deviation (LD) and Normalized Load Transfer (NLT) will be analyzed. These two indexes quantify the risk of adjacent lane invasion and rollover. Then, the timeline of events occurring at the tunnel exit will be presented.

The data were postprocessed by cutting each tunnel exit sequence (either up to the end of the scenario or up to the entry of the next tunnel). All the segments related to the same wind and vehicle speed condition were used to obtain the mean and standard deviation of the signals considered in the analysis.

4.1 Lateral Deviation and Normalized Load Transfer

The mean and standard deviation of the maximum Lateral Deviation with respect to the center of the lane and the maximum Normalized Load Transfer (defined in Eq. (2)) is reported in Fig. 4.

$$NLT = \frac{F_{z,load} - F_{z,unload}}{F_{z,load} + F_{z,unload}} \tag{2}$$

The Lateral Deviation signal was obtained measuring the lateral distance between the vehicle center of mass position after the tunnel exit and the vehicle center of mass in the instant where it exits the tunnel ($LD(t) = y_{CG}(t) - y_{CG}(t = TunnelExit)$). The results have been plotted accordingly to the non-dimensional lateral aerodynamic force (divided by the vehicle weight force) F_y/mg since a linear dependence of the two parameters plotted on this non-dimensional force could be acknowledged. On increasing both vehicle and wind speed (and thus increasing the non-dimensional lateral aerodynamic force F_y/mg), there is an increase in both the peaks of the Lateral Deviation and Normalized Load Transfer.

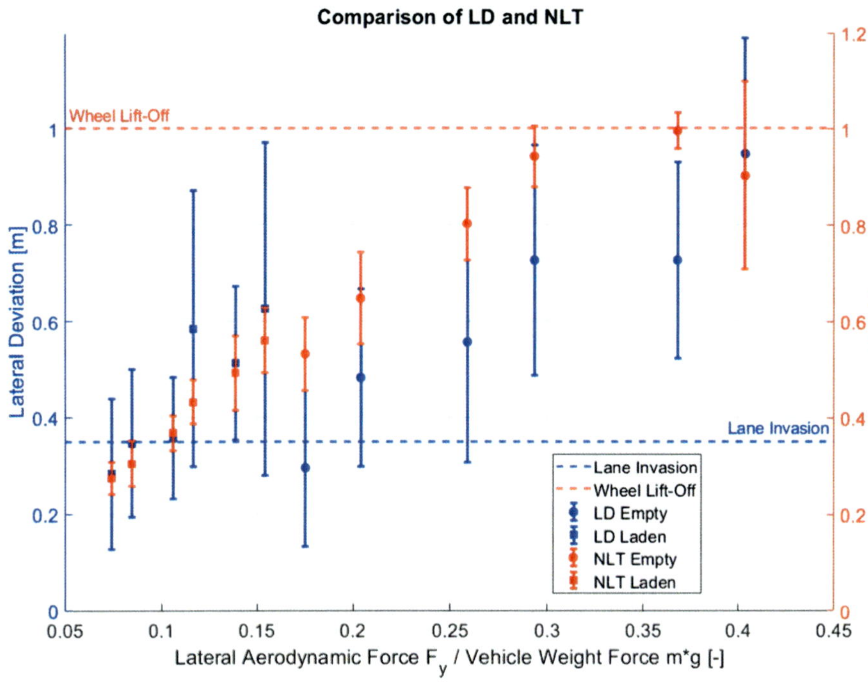

Fig. 4. Mean and standard deviation of the maximum recorded LD and NLT for different vehicle and wind speed conditions.

The NLT defined in Eq. (2) is computed, and the mean and standard deviation of its maximum are plotted in Fig. 4. Again, increasing vehicle and wind speed, there is an increase in NLT.

It is interesting to note that in critical conditions, the empty vehicle is subjected more to the rollover risk than to lane invasion, while in the laden configuration the biggest risk is lane invasion due to the increased inertia of the vehicle, confirming the analysis reported in Sabbioni et al. [9]. Higher dispersion in standard deviation is encountered for the two signal in the case of high non-dimensional lateral aerodynamic force F_y/mg. This happens because the drivers tend to react abruptly in this latter case.

4.2 Timeline of Events

In Fig. 5 is reported the timeline of events happening at the tunnel exit for both vehicle configurations and for both the vehicle velocities tested. On the y-axis it is reported in ascending order the delay (measured in seconds) of the maximum values (peaks) recorded for some of the most important vehicle dynamics and driver's response signals. The vehicle related signals are the Yaw Rate (YR1, YR2), Normalized Load Transfer (NLT1), and Lane Deviation (LD1), while the driver related signals analyzed are the Steering Angle (SA1, SA2) and Steering Torque (ST1). Some of these signals, like the Steering Angle, present two different peaks (SA1 and SA2) while others, like the Steering Torque, only have one (ST1). The SA1 and YR1 peaks are related to the uncontrolled phase of the driver's response (physiological delay in the driver's reaction time to the crosswind stimulus), while all the remaining peaks originate from the control applied by the driver.

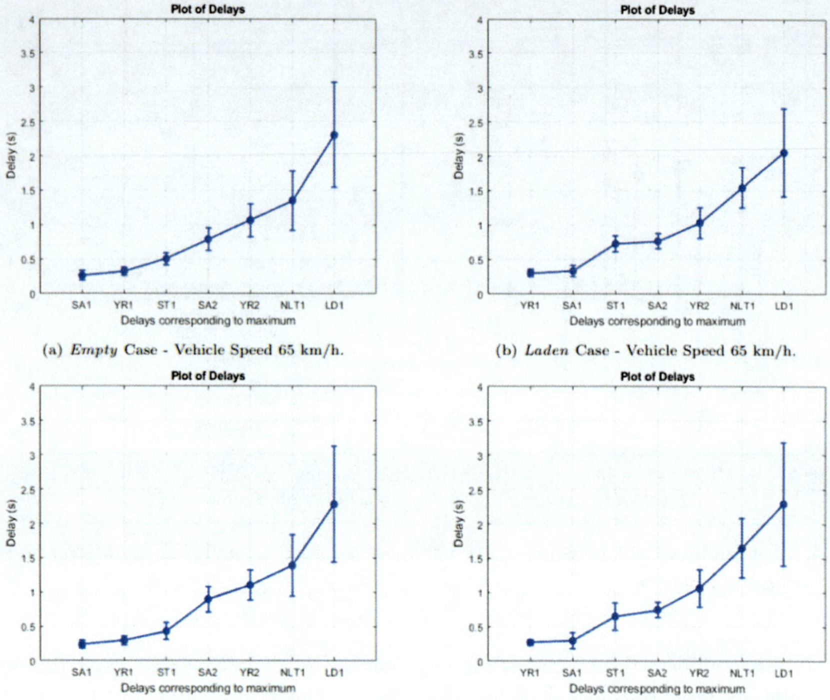

(a) *Empty* Case - Vehicle Speed 65 km/h.

(b) *Laden* Case - Vehicle Speed 65 km/h.

(c) *Empty* Case - Vehicle Speed 80 km/h.

(d) *Laden* Case - Vehicle Speed 80 km/h.

Fig. 5. Timeline of events with mean and standard deviation of the delays.

A typical time history for each of these signals is reported in Fig. 6 in the form of mean and one standard deviation. Figure 6 plots have been obtained using the empty vehicle configuration with a vehicle speed of 65 km/h and a wind speed of 20 m/s (all the data gathered in this configuration has been used to evaluate mean and standard deviation).

SA1 and YR1 refer to the peak due to the uncontrolled vehicle response (driver is not applying any torque on the steering wheel, but nonetheless a non-zero steering angle is found due to the yawing aerodynamic moment M_z).

Briefly after, the driver (that has now acknowledged a change in the vehicle heading and position due to the incident wind) reacts applying a steering torque (ST1) that leads to the second peak (SA2, YR2) in steering angle and yaw rate (these peaks have opposite sign with respect to SA1 and YR1 and a higher magnitude).

Finally, the maximum values of Lane Deviation (LD1) and Normalize Load Transfer (NLT1) come as a result (the delays related to the peaks LD1 and NLT1, together with all the others described in this section, are reported in Fig. 5). The delay recorded for these

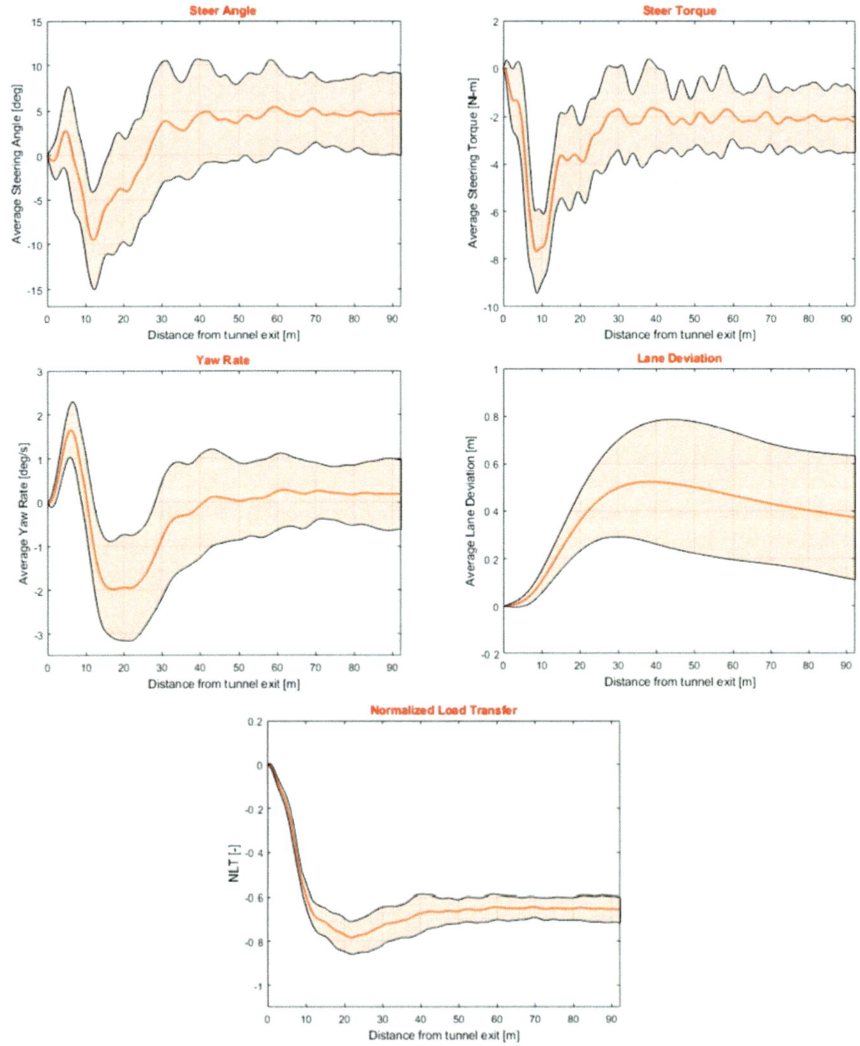

Fig. 6. Mean and standard deviation of the vehicle and driver response signals recorded with an empty vehicle configuration, vehicle speed of 65 km/h and wind speed of 20 m/s.

two last signals are longer due to the slower side-roll dynamics (lateral displacement y and roll φ).

5 Conclusions

A numerical-experimental procedure has been defined to study the response of the vehicle-driver couple at tunnel exit in high crosswind. The methodology relies on an aerodynamic model (which can be easily tuned based on static CFD calculations or static wind tunnel tests) and driver-in-the-loop simulations.

Thanks to these tests, a driver model can be developed or tuned to guarantee the same response shown from the real driver. Developing such a model is necessary to correctly test the effect of devices which may be introduced to mitigate the risk of rollover and/or lane invasion (e.g. vehicle active control systems).

The adopted methodology can be further developed to understand the effect of wind fences, bridge towers, etc. on the vehicle-driver response.

Moreover, the use of a driving simulator allows to further develop and create scenarios (adding other vehicles, low visibility, rain/snow, and crosswind combination, etc.) that can be used to better understand the human driver reaction in such challenging conditions.

References

1. ISTAT: Incidenti Stradali (2019)
2. Rocchi, D., Rosa, L., Sabbioni, E., Sbrosi, M., Belloli, M.: A numerical–experimental methodology for simulating the aerodynamic forces acting on a moving vehicle passing through the wake of a bridge tower under cross wind. J. Wind Eng. Ind. Aerodyn. **104–106**, 256–265 (2012)
3. Charuvisit, S., Kimura, K., Fujino, Y.: Effects of wind barrier on a vehicle passing in the wake of a bridge tower in cross wind and its response. J. Wind Eng. Ind. Aerodyn. **92**, 609–639 (2004)
4. Charuvisit, S., Kimura, K., Fujino, Y.: Experimental and semi-analytical studies on the aerodynamic forces acting on a vehicle passing through the wake of a bridge tower in cross wind. J. Wind Eng. Ind. Aerodyn. **92**, 749–780 (2004)
5. Cheli, F., Corradi, R., Sabbioni, E., Tomasini, G.: Wind tunnel tests on heavy road vehicles: Cross wind induced loads-Part 1. J. Wind Eng. Ind. Aerodyn. **99**, 1000–1010 (2011)
6. Cheli, F., Ripamonti, F., Sabbioni, E., Tomasini, G.: Wind tunnel tests on heavy road vehicles: cross wind induced loads-Part 2. J. Wind Eng. Ind. Aerodyn. **99**, 1011–1024 (2011)
7. Pacejka, H.: Tire and Vehicle Dynamics. Butterworth-Heinemann (2012)
8. Walczak, S.: Analysis of vehicle dynamics under sadden cross wind. IOP Conf. Ser. Mater. Sci. Eng. **148** (2016)
9. Sabbioni, E., Sbrosi, M., Rocchi, D., Galeotti, R.: Dynamic response of vehicle-driver couple to the aerodynamic loads due to the crossing of a bridge tower wake. SAE Int. J. Commer. Veh. **5**(1), 83–93 (2012)

Effects of Downsampling on the Prediction of the Italian Extreme Winds

Andac Akbaba, Vincenzo Picozzi(✉), Alberto M. Avossa, and Francesco Ricciardelli

University of Campania "Luigi Vanvitelli", Aversa, Italy
vincenzo.picozzi@unicampania.it

Abstract. In assessing the design wind speed, downsampling has the effect of underestimating the annual maximum values. This study aims at investigating this effect for the Italian territory, using a correction method for synoptic events recently proposed by the authors. The methodology is briefly recalled and then applied to a dataset consisting of 26 SYNOP and 5 METAR records from Italian weather stations. The observation period varies from station to station, and years lacking 10% of data or more are removed from the datasets. Annual maximum wind speeds were first transformed into standard conditions only considering the effects of anemometer height. Then, the correction for downsampling was made. The Ordinary Least Square fitting method with Gringorten plotting positions was finally applied to both the uncorrected and corrected annual maxima to derive the parameters of the Extreme Value Type I distribution function. The results reveal that there is a significant amount of underestimation when downsampled datasets are used. For SYNOP records, the level of underestimation ranges between 10.6% to 27.6% with a mean of 14% and a coefficient of variation equal to 4.3%. Whereas, for METAR records, the range is obtained between 5.6% to 9.7% with a mean of 7.6% and a coefficient of variation equal to 1.4%.

Keywords: Design Wind Speed · Extreme Value Analysis · Downsampling · Bias Correction · Wind Action Uncertainty

1 Introduction

The evaluation of the design wind speed or velocity pressure is the first step to assess wind actions on structures. Extreme Value (EV) analysis and spectral methods are commonly used for the statistical treatment of historical records, some of which have been reviewed by Palutikof et al. [1]. The available procedures are highly affected by the bias and randomness in the available data, as well as by model and fitting uncertainties. Bias or systematic error can be corrected, whereas random uncertainty can only be accounted for in reliability analysis [11]. Design wind speeds can be either 10-min or 1-h averages, or gust values. For an accurate analysis, averaged wind speeds should be measured continuously. Therefore, if 10-min averages are selected, in compliance with the Eurocode format, the sampling period shall also be 10 min giving rise to 144 readings a day. Yet, the availability of such records is limited, and mostly 10-min averages are sampled three hourly (SYNOP data) or one hourly/half-hourly (METAR data). Within the process of

P. Schito and A. Zasso (Eds.): IN VENTO 2022, LNCE 461, pp. 41–51, 2024.
https://doi.org/10.1007/978-3-031-53059-3_5

the design wind speed evaluation, correction of the available datasets for roughness, orography, anemometer height, and averaging time is commonly carried out. A wide literature exists on the most appropriate procedures to be used for these purposes. On the other hand, correction for downsampling is very seldom made and the effect of it has not been investigated extensively. In fact, just few researchers contributed to this topic, e.g., Larsén and Mann [2], Chiodi and Ricciardelli [3], and Li [4].

Larsén and Mann [2] for first investigated the effects of the sampling rate on the expected annual maximum mean wind speed sampled in non-standard conditions. The time series of 10-min averages at five observation sites in Denmark and three in Gulf of Suez were selected for the study. The observation periods of the sites of interest varied from six to twenty-one years. The longest gap of missing data was reported as 66 days. The theoretical approach of the procedure relies on the assumption that the time series of the mean wind speed are Gaussian Markov chains [5]. The method brings a closed-form equation as a solution to the theory, and a simplified empirical equation was also derived from the measured data. As an outcome, the ratio is given between the average annual maximum wind speed from downsampled data and the corresponding value that one should expect if the wind speeds were contiguously sampled. However, the application of the mathematical solution requires determination of the correlation time scale, possibly introducing further uncertainties to the solution. The simplified empirical equation, which gives the same outcome as the theoretical solution, also seemed to provide too heterogeneous results to be considered as universal. In fact, it brings an underestimation of the expected annual maximum wind speed of 3.1% and 6.0% for $\Delta T = 1$hr and $\Delta T = 3$hrs sampling rates, respectively.

Chiodi and Ricciardelli [3] investigated the effect of downsampling without relying on a mathematical background and provided an empirical correction for the underestimation of the return wind speeds (i.e. the expected value of the maximum wind speed in a time length equal to the return period) employing the dataset provided by the Meteorological Service of the Italian Air Force. Data were used from 119 stations in Italy, covering an observation period of 14 to 38 years, recorded in METAR format and consisting of hourly measurements of the 10-min averaged wind speeds. The dataset was downsampled for sampling periods of 2, 3, 4, 6, 8, 12, and 24 h. For each sampling, 50-years return wind speeds were calculated, which resulted in a large scatter when plotted against the sampling rates. To reduce this scatter, the median values of the estimated 50-years return wind speed were evaluated and plotted against the corresponding sampling period ΔT. It was observed that a logarithmic formula well fits the data. The correction factors were thus derived by taking the ratio of the 50-years return wind speed for $\Delta T = 10$min obtained by extrapolating the fit, and the 50-years return wind speed with a sampling period of ΔT hours. Yet, the downsampling correction factor was later found inaccurate compared to results from a 10-min contiguously sampled dataset.

A third study was performed by Li [4], that advocates the elimination of the sample size effect on the calibration of design wind speeds, which requires the availability of very long records. Hence, employing Monte Carlo simulations, 18,000 years long time series were generated from the Weibull parent of 2-min averaged records at six stations in the US; these were used to calculate 10-min and 1-h averaged wind speeds. As a result, it is concluded that there is an underestimation of the return wind speeds when this is evaluated from disjunct downsampled data, and it is suggested that increasing the length of the records from 10 to 50 years reduces the underestimation of the return wind speeds. The procedure requires high-resolution data, which are generally rare, making it unpractical to apply.

More recently, Picozzi et al. [6] presented a revised and more advanced version of the approach developed by Chiodi and Ricciardelli [3], aimed at eliminating the inaccuracies it contained. In this paper, first the uncertainty in the evaluation of annual maximum wind speed is discussed, accounting for both the systematic (bias) and the random errors. This is done by applying the procedure proposed by Picozzi et al. [6] to derive the bias, while a method for the estimation of the random error is shown. Then, the correction for downsampling calculated for 26 SYNOP and 5 METAR records from Italian meteorological stations. Finally, the impact of downsampling on design wind speeds and velocity pressures is discussed.

2 Methodology

2.1 Uncertainty Due to Downsampling

Picozzi et al. [6] reconsidered the impact of downsampled data on the evaluation of the return wind speeds by employing high-resolution data from the German Weather Service DWD [7]. The homogenous dataset complies with the WMO guidelines [8], in which standard conditions are defined as 10-min averaged values measured at a height of 10m above the ground in open flat terrain conditions. Data are sampled contiguously, that is the dataset it meant to contains 144 daily and 52,560 annual measurements; missing data are rather limited. As such, it was a valuable tool for calibrating and validating the proposed technique. Analyses were performed on 117 representative stations in Germany, selected to have the same observation intervals and a percentage of missing data less than 5%.

To estimate the downsampling correction factor, the following equation was found to well approximate the mean error, i.e. the bias:

$$\rho_{\Delta T}(T_s) = \frac{v_{Ts}}{v_{\Delta T}} = \left[1 - a \cdot \log\left(\frac{\Delta T}{T_s}\right)\right]^{-1/b} \tag{1}$$

where $\log(\cdot)$ denotes the natural logarithm, and a and b are coefficients to be calibrated for each station of interest. For the calibration of Eq. (1), starting from the original time series sampled with period T_s, the annual maximum wind speeds are extracted and their mean value v_{Ts} evaluated. Then, the time series is artificially downsampled with downsampling period ΔT; so doing, $m = \Delta T / T_s$ different time series can be extracted, shifted of T_s from one another; from each of these time series, the mean value of the annual maximum is evaluated, thus obtaining the sample $\boldsymbol{v}_{\Delta T} = \left[v_{\Delta T,1}, \ldots, v_{\Delta T,m}\right]$. The mean value of the sample $\boldsymbol{v}_{\Delta T}$ is assumed to coincide with the expected value of the annual maximum wind speed when sampled with a sampling period ΔT, $E[v_{\Delta T}] = mean\left(v_{\Delta T,i}\right)$. By repeating the procedure for different values of ΔT, the empirical correction factors $\rho_{\Delta T}(T_s) = v_{Ts}/v_{\Delta T}$ are finally estimated and can be fitted by Eq. (1). As an example, in Fig. 1 the calibration procedure is applied to time series originally sampled with $T_s = 3$hrs. In details, in Fig. 1a the samples $\boldsymbol{v}_{\Delta T}$ are shown as dots. Instead, in Fig. 3b the empirical correction factors $\rho_{\Delta T}(T_s)$ are plotted against ΔT (dots) together with the calibrated Eq. (1) (continuous line).

Correcting extreme wind speeds with $\rho_{\Delta T}(T_s)$ allows the bias to be removed. However, a random error remains, which can be evaluated. To this aim we can calculate the corrected value of the annual maximum wind speed for each value $v_{\Delta T,i}$ of the sample $\boldsymbol{v}_{\Delta T}$ as:

$$v_{T_s} = v_{\Delta T,i} \cdot \rho_{\Delta T}(T_s) \cdot e_{\Delta T,i}(T_s) \tag{2}$$

where $e_{\Delta T,i}(T_s)$ represents the residual error when the averaged value $\rho_{\Delta T}(T_s)$ is applied to the element $v_{\Delta T,i}$ of the sample. Referring to Fig. 1a, $e_{\Delta T,i}(T_s)$ is a measure of the distance between the i th black dot and the expected value $v_{\Delta T}$, for a given ΔT. This is also reported in Fig. 1c where an increment of residual error is shown as ΔT increases. For each downsampling period ΔT, the values $e_{\Delta T,i}(T_s)$ give rise to a sample $\boldsymbol{e}_{\Delta T}(T_s)$ of the random error due to downsampling, a measure of which is the coefficient of variation $V[\boldsymbol{e}_{\Delta T}(T_s)]$. For the example of Fig. 1, this is reported in Fig. 1d.

The procedure outlined for calibration of Eq. (1) allows the bias to be evaluated for the artificial downsampled time series. However, what we need is a factor correcting the wind speed v_{T_s} to obtain the corrected values v_{T_a}, i.e. the maximum wind speed that would be obtained if the time series were originally sampled with a period T_a corresponding to the averaging time. This is the value ρ_{T_s} defined as:

$$\rho_{T_s} = \frac{1}{\rho_{T_a}(T_s)} = \frac{v_{T_a}}{v_{T_s}} = \left[1 - a \cdot \log\left(\frac{T_a}{T_s}\right)\right]^{1/b} \tag{3}$$

As for the bias, also the random error can be projected to T_a to evaluate the random error due to downsampling for a sampling period T_s. Evaluating Eq. (2) for $\Delta T = T_a$ and reversing it, the following relationship can be derived:

$$v_{T_a,i} = \rho_{T_s} \cdot v_{T_s} \cdot \varepsilon_i(T_s) = v_{T_a} \cdot \varepsilon_i(T_s) \tag{4}$$

where v_{T_a} is the expected value of the wind speed after correction for downsampling, and $\varepsilon_i(T_s) = 1/e_{10',i}(T_s)$ represent the residual error after elimination of bias. The corrected wind speed is thus represented by the random variable:

$$\boldsymbol{v}_{T_a} = \rho_{T_s} \cdot v_{T_s} \cdot \boldsymbol{\varepsilon}_{T_s} = v_{T_a} \cdot \boldsymbol{\varepsilon}_{T_s} \tag{5}$$

with mean value v_{T_a} and random error $\boldsymbol{\varepsilon}_{T_s}$ having coefficient of variation $V\left[\boldsymbol{\varepsilon}_{T_s}\right]$.

2.2 Available Data and Data Processing

Records from twenty-six SYNOP and five METAR stations of the Meteorological Service of the Italian Air Force are used. Before processing the data, their quality is checked, since it is the primary factor affecting reliability of the results. Data are thus pre-processed eliminating erroneous values from the records with custom algorithms, and a final check is made by eye. As a first step of pre-processing, plausible values are selected as those which wind speeds ranges from 0 to 75 m/s and with wind direction in the range 0° to

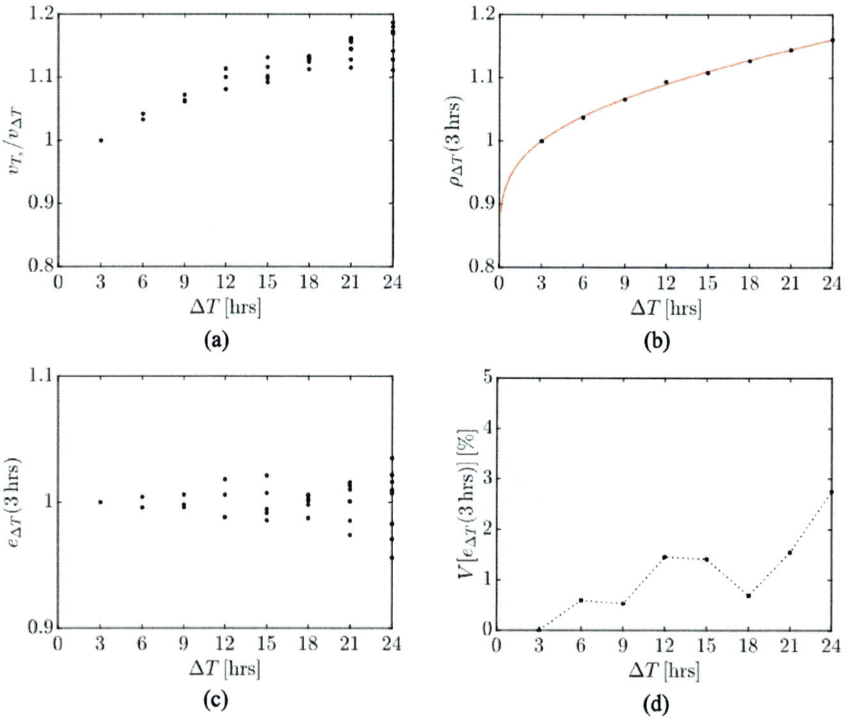

Fig. 1. Example of downsampling uncertainty evaluation: (a) plotting of samples $v_{\Delta T}$; (b) bias; (c) residual error; (d) coefficient of variation of the random error.

360°. Internal consistency is also checked, considering the relationship between different parameters: a zero-wind speed sequence with varying direction can be an internal consistency error, and cases are isolated for a further manual check; abnormally high wind speed values are filtered by moving statistics detection;

changes of more than 10 m/s in consecutive readings are isolated for the manual check.

In Table 1 the information of analyzed stations are reported, including: ICAO code, altitude, latitude and longitude, anemometer height, height correction factors due to the non-standard condition. Moreover, in Fig. 2 the location of the stations is shown.

The application of the procedure outlined in Sect. 2.1 represents the last step in preprocessing the datasets. After that, the annual maximum wind speeds can be analyzed. To derive the 50-yrs return wind speed, Extreme Value (EV) analysis is applied. In details, the parameters of the EV Type I distribution are estimated by using the Ordinary Least Square Method with Gringorten plotting positions.

3 Results and Discussion

The downsampling correction factor $\rho_{3\text{h}}$ resulting from the application of the methodology outlined in Sect. 2 to Italian SYNOP stations are reported in Fig. 2. Overall, factors ranging between 1.11 and 1.28 are found correcting 3 hourly downsampled wind

Table 1. General information of the analysed Italian meteorological stations.

ICAO (Station Name)	Altitude [m]	Latitude [N]	Longitude [E]	Height [m]	Height correction
LIBC (Crotone)	159	38° 59′ 50.01"	17° 04′ 48.61"	10.0	1.00
LIBD (Bari Palese)	54	41° 07′ 59.95"	16° 46′ 00.01"	20.0	0.88
LIBE (Monte S. Angelo)	838	41° 42′ 00.00"	15° 56′ 60.00"	10.0	1.00
LIBP (Pescara)	10	42° 25′ 53.96"	14° 10′ 51.84"	6.5	1.09
LIBQ (Monte Scuro)	1669	39° 19′ 60.00"	16° 23′ 60.00"	10.0	1.00
LIBR (Brindisi)	15	40° 39′ 27.00"	17° 56′ 49.00"	10.0	1.00
LIBS (Campobasso)	793	41° 34′ 00.00"	14° 39′ 00.00"	10.0	1.00
LIBT (Termoli)	16	42° 00′ 00.00"	15° 00′ 00.00"	10.0	1.00
LIBY (S. Maria di Leuca)	104	39° 49′ 00.00"	18° 21′ 00.00"	10.0	1.00
LIBZ (Potenza)	845	40° 37′ 60.00"	15° 48′ 00.00"	27.0	0.84
LICA (Lamezia Terme)	12	38° 54′ 30.00"	16° 14′ 30.00"	10.0	1.00
LICF (Messina)	54	38° 12′ 00.00"	15° 33′ 00.00"	19.0	0.89
LICG (Pantelleria)	198	36° 48′ 46.80"	11° 57′ 39.00"	6.5	1.09
LICJ (Palermo Punta Raisi)	20	38° 10′ 55.20"	13° 05′ 58.20"	10.0	1.00
LICZ (Catania Sigonella)	29	37° 24′ 09.82"	14° 55′ 21.07"	10.5	1.00
LIMC (Milano Malpensa)	234	45° 37′ 01.98"	08° 43′ 42.67"	10.0	1.00
LIQC (Capri)	160	40° 32′ 60.00"	14° 15′ 00.00"	10.0	1.00
LIQK (Capo Palinuro)	184	40° 02′ 27.10"	15° 17′ 16.98"	2.0	0.86
LIQZ (Ponza)	184	40° 53′ 58.24"	12° 57′ 20.98"	8.0	1.04
LIRA (Roma Ciampino)	129	41° 48′ 18.71"	12° 35′ 21.61"	10.5	0.99
LIRK (Monte Terminillo)	1874	42° 28′ 00.00"	12° 58′ 60.00"	6.0	1.11

(*continued*)

speeds to contiguous sampling. On average, a downsampling correction factor equal to

Table 1. *(continued)*

ICAO (Station Name)	Altitude [m]	Latitude [N]	Longitude [E]	Height [m]	Height correction
LIRM (Grazzanise)	9	41° 03′ 37.73"	14° 05′ 14.75"	10.0	1.00
LIRN (Napoli Capodichino)	90	40° 53′ 03.38"	14° 17′ 01.03"	6.5	1.09
LIRP (Pisa S. Giusto)	6	43° 40′ 57.60"	10° 23′ 43.80"	6.5	1.09
LIRT (Trevico)	1085	41° 02′ 60.00"	15° 13′ 60.00"	10.0	1.00
LIRZ (Perugia S. Egidio)	208	43° 06′ 07.05"	12° 30′ 38.55"	10.0	1.00

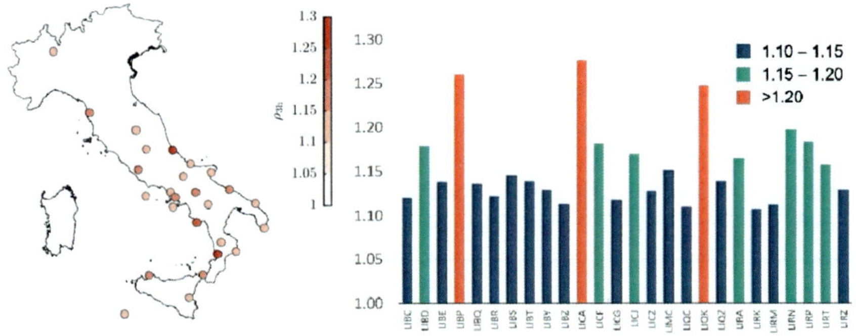

Fig. 2. Location of the analyzed Italian stations and correction factor ρ_{3h}.

1.16 is estimated for the 26 analyzed SYNOP stations, with a coefficient of variation $V[\rho_{3h}] = 4.4\%$. The minimum value $\rho_{3h} = 1.11$ is observed for four of twenty-six stations (LIBZ, LIQC, LIRK, LIRM), while the maximum value of correction factor equal to 1.28 is evaluated for the LICA (Lamezia Terme) station. No correlation is observed with the geographic coordinates.

In Fig. 3, the envelope of the inverse of Eq. (1) calibrated for each single station is shown with an hatched area, together with the envelope of the mean values (continuous line); the dashed vertical line indicates the sampling period of 10 min, corresponding to contiguous sampling. It provides the bird's eye view of the variation of bias and random error associated with the downsampling of the wind speeds.

Implications of wind speed correction are shown in Table 2. The EV Type I scale (α) and location (β) parameters calibrated from the data affected by downsampling are reported in columns 2 and 4, respectively, whereas columns 3 and 5 show the same parameters for the corrected data. Corrected and uncorrected 50-yrs return wind speeds are given in columns 6 and 7, respectively. To compare with, design wind speed v_{CNR} evaluated according to the Guidelines of the Italian National Research Council [9] is

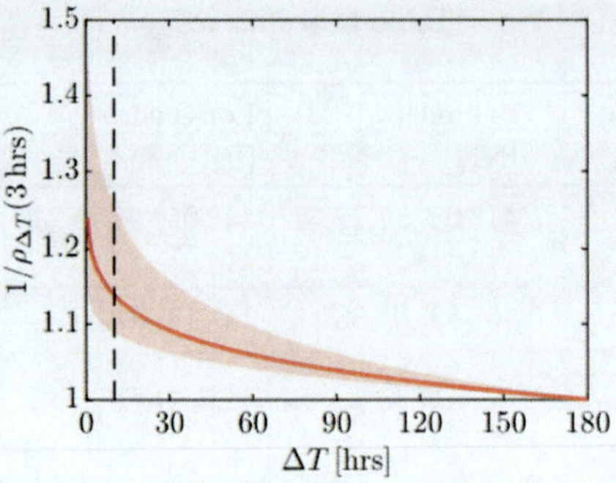

Fig. 3. Results of the downsampling procedure: projection to $T_a = 10$min (dashed line) for the correction factor $\rho_{3\text{h}}$.

shown in column 8; these derive from analyses carried out in the '90s [10], with data from the period 1951–1992. Finally, in column 9 the ratio between the uncorrected q_{50} and corrected q^*_{50} velocity pressure is given. As pointed out by Picozzi et al. [6], the results show that correcting the annual maximum wind speeds for downsampling has the effect of increasing EV Type I parameters. The ratios q_{50}/q^*_{50}, indicate that when downsampling effects are neglected, velocity pressure is underestimated on average by 24.7%, with an underestimation larger than 30% at three stations and a maximum value of 38.6%. This translates into equal underestimation of wind actions.

Comparing the corrected values of wind speed v^*_{50} with the results obtained by application of CNR Guidelines, it is shown that the design wind speeds $v_{50,CNR}$ are underestimated by about 11% on average. The larger underestimation is shown for the stations of LIBP (−46%), LIBE (−34%), LICA (−33%), and LIBS (−30%). Instead, the larger overestimation is shown for the LIRZ station (41%).

Following the same format as in Table 2, results of similar analyses for five METAR stations are reported in Table 3. On average, there results a correction factor $\rho_{1\text{h}}$ for 1 h sampling period equal to 1.08, with a coefficient of variation $V[\rho_{1\text{h}}] = 1.3\%$. The maximum value of the dowsampling correction factor is found to be equal to 1.09 for the LIRN station, while the minimum value of 1.06 is found for the LIRA station. In this case, the ratio between the uncorrected and corrected velocity pressures (last column of Table 3) is on average equal to 0.85, resulting in an underestimation of the wind action of 15%.

A comparison could be made between the results of the analysis on METAR stations and those on the corresponding SYNOP stations, the results of which are shown in Fig. 4. A good agreement is found in comparing the corrected 50-yrs return wind speed with differences in the range of −7% to +8%. However, this uncertainty can be neglected.

when compared with the bias and can be partially attributed to the random error due to downsampling.

As to the random error, for SYNOP stations a coefficient of variation $V[\varepsilon_{3\text{h}}] \cong 4.0\%$ was found for the downsampling correction factor $\rho_{3\text{h}}$, whereas for METAR stations this value reduces to $V[\varepsilon_{1\text{h}}] \cong 1.4\%$. The lower variability of the correction factor in the case of 1-h sampling can potentially be ascribed to the low number of stations for

Table 2. Results of the downsampling correction procedure for Italian SYNOP records.

ICAO	α[m/s]	α^*[m/s]	β[m/s]	β^*[m/s]	v_{50}[m/s]	v_{50}^*[m/s]	$v_{50,CNR}$[m/s]	q_{50}/q_{50}^*[−]
LIBC	2.39	2.67	15.8	17.7	25.1	28.1	27.0	0.800
LIBD	1.67	1.96	13.1	15.4	19.6	23.1	27.0	0.724
LIBE	4.29	4.89	24.6	28.0	41.3	47.0	33.8	0.770
LIBP	3.64	4.56	16.7	21.0	30.8	38.8	27.0	0.633
LIBQ	3.51	3.99	19.6	22.2	33.3	37.8	50.4	0.776
LIBR	2.58	2.90	16.2	18.2	26.3	29.5	27.0	0.793
LIBS	3.11	3.55	19.6	22.4	31.7	36.3	32.9	0.765
LIBT	1.98	2.25	21.8	24.8	29.5	33.6	27.0	0.770
LIBY	2.39	2.70	17.0	19.2	26.4	29.8	27.0	0.784
LIBZ	1.85	2.06	14.9	16.7	22.2	24.7	33.9	0.805
LICA	3.23	4.11	16.5	21.0	29.0	37.0	27.0	0.614
LICF	1.61	1.90	11.4	13.5	17.7	20.9	28.0	0.718
LICG	4.42	4.94	24.5	27.4	41.8	46.7	31.0	0.802
LICJ	3.21	3.76	18.8	21.9	31.3	36.6	28.0	0.735
LICZ	1.84	2.09	15.6	17.6	22.8	25.7	28.0	0.783
LIMC	1.99	2.28	13.5	15.5	21.2	24.4	25.0	0.757
LIQC	4.79	5.33	15.0	16.7	33.8	37.5	31.0	0.806
LIQK	2.03	2.49	17.6	21.6	25.1	31.3	27.0	0.662
LIQZ	2.83	3.22	22.0	25.1	33.1	37.7	31.0	0.771
LIRA	2.20	2.57	15.8	18.4	24.4	28.4	27.0	0.734
LIRK	4.57	5.06	33.5	37.1	51.3	56.8	54.5	0.817
LIRM	1.45	1.61	14.0	15.6	19.6	21.8	27.0	0.810
LIRN	2.91	3.49	16.5	19.8	27.9	33.4	27.0	0.696
LIRP	1.62	1.91	17.2	20.3	23.5	27.8	27.0	0.715
LIRT	3.68	4.27	20.0	23.6	34.4	39.8	38.7	0.744
LIRZ	1.11	1.27	12.7	14.4	17.1	19.3	27.0	0.781

Table 3. Results of the downsampling correction procedure for Italian METAR records.

ICAO	α[m/s]	α^*[m/s]	β[m/s]	β^*[m/s]	v_{50}[m/s]	v_{50}^*[m/s]	q_{50}/q_{50}^*[−]
LIBR	3.18	3.47	16.9	18.1	29.3	31.7	0.853
LIMC	1.80	1.96	16.0	17.4	23.0	25.1	0.842
LIRA	2.19	2.34	16.3	17.1	24.8	26.2	0.896
LIRN	2.92	3.21	18.1	19.8	29.5	32.3	0.830
LIRP	1.95	2.15	16.8	18.1	24.4	26.5	0.845

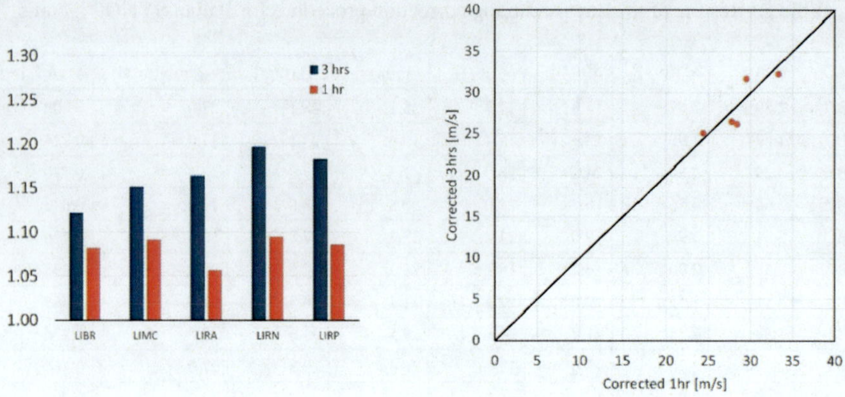

Fig. 4. Comparison between the correction of three-hourly and hourly sampled records for the same stations; correction factors (left) and corrected return wind speeds.

which data were available or to the better quality of these; this effect would need further investigation.

Finally, to explore the generality of the findings, a comparison is made with similar results available for German stations, and already presented in [6]. If the same procedure described above and applied to Italian data is instead applied to 117 German stations [7], values of the downsampling correction factors of 1.16 and 1.08 are found for three-hourly and hourly sampled data, respectively. This result opens a door to the possibility that the bias associated with downsampling is only little dependent on local wind climate. As to the random error, for the German stations values of $V[\varepsilon_{3h}] \cong 3.3\%$ and of $V[\varepsilon_{1h}] \cong 2.0\%$ were found, respectively.

4 Conclusion

METAR and SYNOP records are commonly used in Italy to evaluate return wind speeds; these are sampled with periods in the range of half an hour and three hours, larger than averaging time of 10 min. Consequently, an underestimation is expected of return wind speeds which is larger for longer sampling periods. To quantify the downsampling effect for the Italian wind climate, the application is made to twenty-six SYNOP and five METAR stations. On average, a correction factor is estimated of 1.08 for 1-h sampling period, and of 1.16 for 3-h sampling period. These lead to an underestimation of the return wind speed on average equal to 7% and 14%, respectively. When wind speeds are translated into velocity pressures, this leads to an underestimation of the wind action of 14% and 26% for 1-h and 3-h sampling periods. Agreement between METAR and SYNOP records is good, with differences in the corrected wind speeds in the range of -7% to + 8%, that should be partially attributed to the random error associated with downsampling. Finally comparison of the correction factors evaluated for the Italian stations and those evaluated for German stations shows a very good agreement, suggesting that the bias in return wind speeds deriving from downsampling is little affected by the local wind climate.

References

1. Palutikof, J.P., Brabson, B.B., Lister, D.H., Adcock, S.T.: A review of methods to calculate extreme wind speeds. Meteorol. Appl. **6**, 119–132 (1999)
2. Larsén, X.G., Mann, J.: The effects of disjunct sampling and averaging time on maximum mean wind speeds. J. Wind Eng. Ind. Aerodyn. **94**, 581–602 (2006)
3. Chiodi, R., Ricciardelli, F.: Three issues concerning the statistics of mean and extreme wind speeds J. Wind Eng. Ind. Aerodyn. **125**, 156–167 (2014)
4. Li, S.H.: Effect of disjunct sampling on calibration of design wind speed. J. Wind Eng. Ind. Aerodyn. **183**, 283–294 (2018)
5. Gagniuc, Paul A.: Markov Chains: From Theory to Implementation and Experimentation. USA, NJ: John Wiley & Sons. pp. 159–163 (2017). ISBN 978-1-119-38755-8
6. Picozzi V., Akbaba A., Avossa A.M., Ricciardelli F.: Correction of historical records to improve the reliability of design wind speeds. Eng. Struct. **265**, 114473 (2022)
7. Deutscher Wetterdienst (DWD) Homepage, https://www.dwd.de. Accessed 02 Jun 2022
8. World Meteorological Organization (WMO): Guide to Instruments and Methods of Observation (WMO-No. 8), Geneva, Switzerland (2018)
9. CNR: Istruzioni per la valutazione delle azioni e degli effetti del vento sulle costruzioni, CNR-DT 207/2018 (*in Italian*), Consiglio Nazionale delle Ricerche (2018)
10. Ballio, G., Lagomarsino, S., Piccardo, G., Solari, G.: Probabilistic analysis of Italian extreme winds: reference velocity and return criterion. Wind Struct. **2**, 51–68 (1999)
11. Picozzi, V., Landi, F., Avossa, A.M., Croce, P., Formichi, P., Ricciardelli, F.: The climatic action uncertainty chain. Eng. Struct. **301**, 117357 (2024)

Insights into the Transcritical Reynolds Number Range Based on Field Measurements of a Wind Turbine Tower

Ika Kurniawati[1](✉), Francesca Lupi[1,2], Marc Seidel[3], Rüdiger Höffer[1,2], and Hans-Jürgen Niemann[1,2]

[1] Ruhr-Universität Bochum, Bochum, Germany
ika.kurniawati@ruhr-uni-bochum.de
[2] Niemann Ingenieure GbR, Bochum, Germany
[3] Siemens Gamesa Renewable Energy GmbH & Co. KG, Hamburg, Germany

Abstract. The estimation of vortex shedding load is essential for the prediction of Vortex-Induced Vibration (VIV) on slender bodies. The application of the VIV model to a full-scale slender tower, such as wind turbine towers and chimneys, depends on the estimation of actual vortex shedding load under field conditions. This paper aims to provide insight on the vortex shedding spectrum with a real-life wind and the transcritical flow-regime, which requires high effort to observe in the wind tunnel scale. The vortex shedding spectrum is evaluated from the response measurement is focused on this paper, where challenges from the investigation and an introduction to the wind pressure measurement are then discussed. Global Strouhal number at high Reynolds numbers, spectrum of tower response in lift direction, and the estimation of critical velocity are presented. The influence of ambient condition such as wind profile, wind direction and turbulence intensity are included.

Keywords: Field measurement · Full-Scale · Transcritical Reynolds number range · Vortex shedding

1 Introduction

The development of VIV models began in the 1960s and to this day encompasses a wide range of approaches, mainly distinguished between 1-DOF or 2-DOFs models. In between, the models are also distinguished by their approach, such as wake-oscillator models, aerodynamic damping models, and effective correlation length models. Currently, the Vickery&Basu (V&B) approach (Vickery and Basu, 1983) is used as a basis for the VIV response model in the European building code Eurocode 1 2010 and in the CICIND guidelines 2010 and 2011. In this approach, 1-DOF equation of motion is used and the aeroelastic interaction is represented by the "aerodynamic damping", K_a (See Eq. (1) and (2)). $S_{L,n}$ refers to modal vortex shedding load, ω_n is the natural frequency, M_n is the modal mass and $m_{eq,n}$ is the equivalent mass of mode n. The equivalent term

P. Schito and A. Zasso (Eds.): IN VENTO 2022, LNCE 461, pp. 52–63, 2024.
https://doi.org/10.1007/978-3-031-53059-3_6

of structural damping that can described by Scruton number (Sc) divided by 4π is subtracted by the aerodynamic damping to define the aeroelastic interaction. D refers to the diameter of the tower and ρ is the air density. In Eq. (2), the aerodynamic damping is modelled to depend on the tower's deflection σ_y and a_L is normalized limiting amplitude.

$$\sigma_{y,n}^2 = S_{L,n}(f_n) \cdot \frac{1}{(\omega_n^2 \cdot M_n)^2} \cdot \frac{\omega_n}{8 \cdot \frac{\rho \cdot D^2}{m_{eq,n}} \cdot \left(\frac{Sc}{4\pi} - Ka\right)} \tag{1}$$

$$K_a = K_{a0} \cdot \left[1 - \left(\frac{\sigma_y}{a_L \cdot D}\right)^2\right] \tag{2}$$

This model successfully introduces that the response of a structure when VIV occurs is influenced by the aeroelastic interaction. However, this model often overestimates the structural response, as shown in recent studies on full-scale structures (Lupi et al., 2017). Through a series of wind tunnel experiments, Lupi et al. (Lupi et al., 2018) developed a model called "Damping Modified Spectral Method" (DMSM) based on the V&B approach. This model has a new aerodynamic damping function (See Eq. (3), where a = 0.3475, b = 5.808, and c = 0.3582 are constants) that differs in terms of the shape of K_a curve, however it improves the prediction of structural response compared to the measured full-scale data. Several other studies had supported the proposed model (Guo et al., 2021; Arunachalam et al., 2020). Full-scale measurement series are needed to further evaluate the model, especially for flow conditions that are rarely achieved in wind tunnel experiments. For example, high Reynolds numbers above Re = 10^6 occurred often for the flow around a full-scale wind turbine tower. The applicability of the model to predict VIV in full-scale can then be investigated by comparison to the field measurements.

$$K_a = a \cdot e^{-b \cdot \left(\frac{\sigma_y}{D}\right)} \cdot \left(\frac{\sigma_y}{D}\right)^{-c} \tag{3}$$

The first step in this currently ongoing investigation is to model the vortex shedding load as a narrow band spectral process centered around the Strouhal frequency, e.g. following Vickery's approach using a Gaussian spectrum (Vickery and Clark, 1972). Then, to estimate the modal vortex shedding load (i.e. $S_{L,n}$, see Eq. 4) along the height of the structure. In Eq. (4), Strouhal frequency is defined in terms of Strouhal number St, mean wind speed V_m at respective height z, and tower's diameter D and effective diameter D_r. The spectrum is defined considering the standard deviation of lift coefficient σ_{cL}. The modal load will consider the tower's mode shape Φ and a non-dimensional correlation length λ. This paper focuses on addressing the vortex shedding spectrum directly from full-scale response measurement data and estimating the global Strouhal frequency. Field conditions such as wind direction and turbulence intensity, as well as tower surface roughness, are also considered to provide a realistic interpretation of the Reynolds number effects.

$$S_{L,n}(f_n,z) \cong 2\lambda D_r \cdot \int_0^H \left(\frac{1}{2}\rho V_m(z)^2\right)^2 \cdot D^2(z) \cdot \frac{\sigma_{cL}^2}{\sqrt{\pi} \cdot St(z) \cdot \frac{V_m(z)}{D(z)} \cdot B} \cdot \exp\left(-\left(\frac{1 - \frac{f_1}{St(z) \cdot \frac{V_m(z)}{D(z)}}}{B}\right)^2\right) \cdot \Phi_n^2(z) \cdot dz \tag{4}$$

2 Field Measurement

Two wind turbine towers located in Østerild test field in Denmark are subjected to the field measurement. Measurements for Tower 1 (D8) have been taken for the standalone wind turbine tower without nacelle and measurements for Tower 2 (SG 14-222 DD, shorten to SG14) have been made both with and without nacelle and blades. For measurement on D8 Tower, bending moment at the bottom and acceleration at top of the tower were measured. Period of data of D8 lasts from July until February 2022. The first met mast tower (MET 1) is available 0.61 km west and 1.06 km south from D8 (See Fig. 1). For completeness, SG14 is also shown in Fig. 1. And it is located 0.58 km north from D8. The second met mast tower (MET 2) is located 0.35 km west and 0.58 km south from D8. Surrounding area of the test fields includes flat land with grass and some parts of 10–20 m height forest with dominating wind coming from the west. The wind turbine D8 has the shape of a tapered circular cylinder with effective diameter $D = D_r = 5.05$ m and height $H = 116.46$ m. Based on the measured data, the critical velocity is estimated around 15 m/s. To observe the vortex shedding spectrum, the signal from bending moment at the bottom of D8 is evaluated. The data are processed and classified as a 10-min time history, for both tower response and wind conditions. Numerous wind speed with high transcritical Reynolds number ranges up to $Re \approx 10^7$ are found throughout the data period. The roughness of both D8 and SG14 tower is very smooth with the relative surface roughness $k/D \approx 1.0 \times 10^{-6}$.

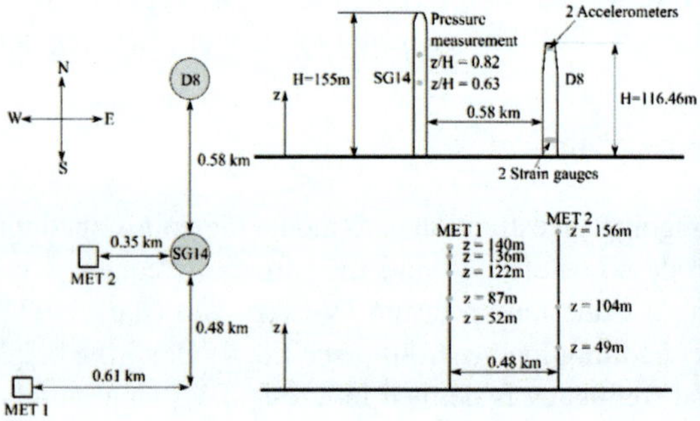

(a)

(b)

Fig. 1. (a) Sketch of D8, SG 14-222 DD and met masts; (b) Satellite picture of the site (©2018 Google-Imagery)

The selection of appropriate datasets to observe the vortex shedding spectrum is important, where a clear and visible Strouhal peak usually occurs at wind speeds that are far from resonance. For the D8 response signal, the wind speed range that is suitable for vortex shedding observation is in the range of 5–8 m/s. At wind speeds closer to the critical value, i.e. where vortex resonance occurs, the resonant oscillation dominates the tower response. This makes it hard to identify explicitly the vortex shedding peak. Figure 2a and Fig. 2b show two different time histories without and with the domination of oscillation in resonance. The Strouhal frequency, which characterizes the vortex

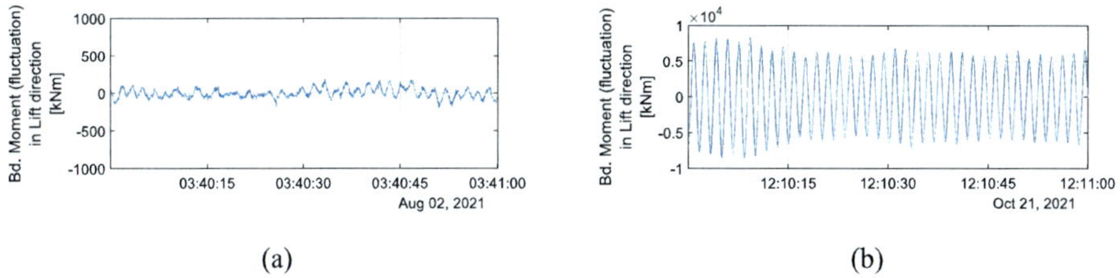

Fig. 2. Time histories of lift bending moment at (a) low wind speed and (b) critical wind speed

shedding spectrum, is calculated in the frequency domain by fitting a Gaussian function (Vickery and Clark, 1972) as shown in Eq. 5. By fitting the equation to the lift coefficient spectral distribution, the variable St (Strouhal number) and B (Bandwidth) are obtained directly from the fitting process. $\sigma^2_{c_L}$ refers to the variance of lift coefficient, which is also represented as the area under the spectral distribution. f_{red} Refers to the non-dimensional reduced frequency, in respect to the diameter D of the tower and mean wind speed $V_m(z)$ at respective height z.

$$\frac{S_{c_L}(f,z)}{\sigma^2_{c_L}} = \frac{1}{\sqrt{\pi} \cdot St(z) \cdot B(z)} \cdot \exp\left(-\left(\frac{1-\frac{f_{red}}{St(z)}}{B(z)}\right)^2\right) \tag{5}$$

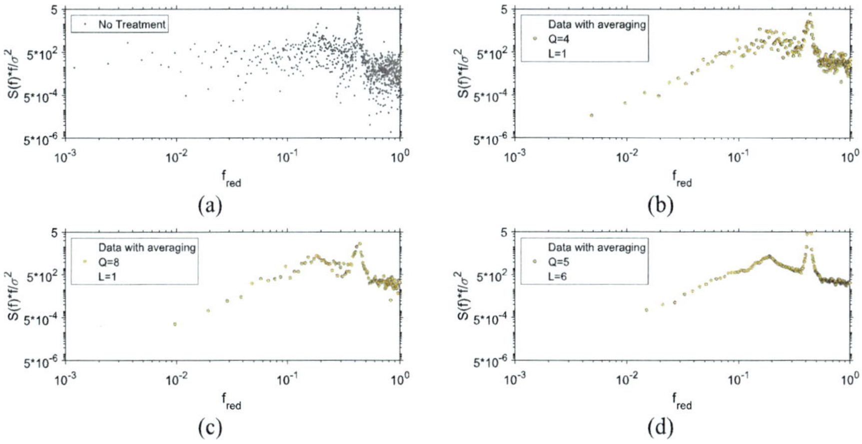

Fig. 3. Block and frequency averaging. (a) Without averaging, (b) Q = 4 Block averaging, L = 1 Frequency averaging, (c) Q = 8, L = 1, (d) Q = 5, L = 6. Where Q is number of blocks and L is number of data points.

Prior to the fitting, block averaging and frequency averaging are applied to the lift coefficient data to enhance the interpretation of spectral distribution. In block averaging technique, the 10-min time history is truncated into Q blocks, where Q refers to the number of blocks. For each block, the spectral distribution is calculated and averaged over all blocks. In frequency averaging, the lift coefficient is averaged in the spectral domain for each L data points. The frequency averaging is done in a trailing manner. Figure 3 shows the effect of block and frequency averaging.

3 Results

3.1 Vortex Shedding Spectrum Observed from Measurement on D8 Tower

Estimation of Strouhal Peak

The response measurement is dominated by the first natural frequency (resonance peak), and the vortex shedding peak is best seen at wind speeds far from resonance, i.e. in the range of 5–8 m/s. Figure 5a shows the prominent vortex shedding peak (Strouhal peak) separated from the resonance peak. As shown in Fig. 5b, the measured spectral distribution of the lift bending moment fluctuations is fitted with a Gaussian function. For the evaluation on the D8 Tower, the reduced frequency f_{red} is calculated using the wind speed at tower height $V_m(H)$. $V_m(H)$ is calculated based on the wind profile, respectively for each 10-min data point. Considering $H = 116.46$m for the D8 Tower, the MET 1 wind speed data is used to estimate the wind profile based on elevations $z = 87$m, $z = 136$m and $z = 140$m. In this way, the wind profile is assumed to have only the power-law shape of the atmospheric boundary layer. Figure 4 shows the example of a 10-min-mean wind profile.

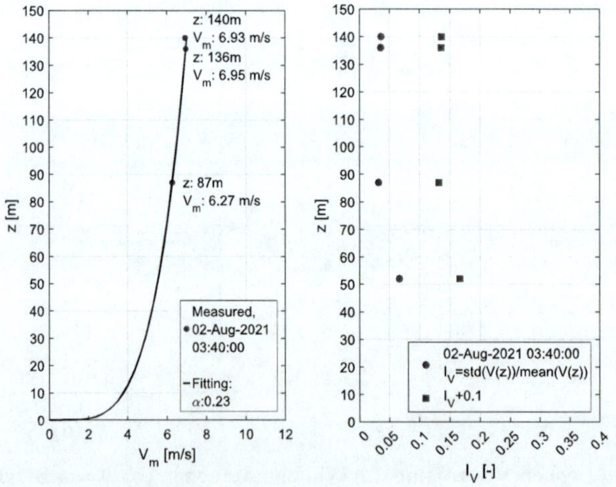

Fig. 4. MET 1 Mean wind profile and turbulence intensity profile along height from one of the 10-min sample

Effect of Turbulence Intensity

The surrounding turbulence intensity and the origin of the wind must be considered. Figure 6 presents four examples of the lift bending moment spectrum for four different

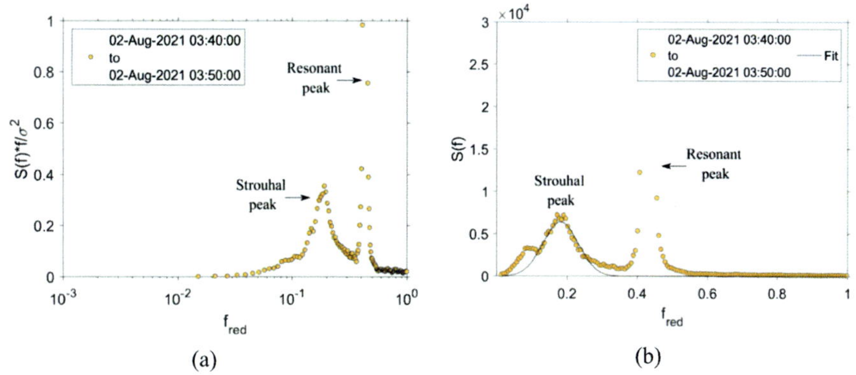

Fig. 5. Resonant (First-natural frequency) peak and Strouhal peak observed in the spectral distribution. (b) Measured spectral distribution of lift bending moment (dot) and a fitted gaussian function (line) to the Strouhal peak.

turbulence intensities $I_V = 0.062$, $I_V = 0.080$, $I_V = 0.094$ and $I_V = 0.152$. I_V is calculated by dividing the standard deviation of the 10-min wind speed time history from MET 1 at z = 136m by the wind speed at tower height $V_m(H)$. It is shown that the parameter B increases with increasing turbulence intensity. B = 0.23 is found from one of the samples with $I_V = 0.062$ (see Fig. 6a) and B = 0.33 is found from one of the samples with $I_V = 0.094$ (Fig. 6c). In Fig. 6d, smaller B is found with higher $I_V = 0.152$ and a higher energy around $f_{red} \approx 0.11$ appears which is guessed to be associated to the higher I_V. Based on these four samples, St is estimated around 0.21–0.22.

To obtain a statistically representative value, four different datasets from four different dates are evaluated as detailed in Table 1. Figure 7 shows the distribution of B in terms of turbulence intensity I_V from each 10-min data points of each dataset. The points are taken from the 24-h data from each date, and they have been curated so that the data contain a noticeable vortex shedding peak. The value B ranges between 0.15–0.4 which describes the turbulence flow. In Fig. 7, black dots plot that describes the expression of $B^2 = B_0^2 + 2 \cdot I_V^2$ is included for comparison purpose, where $B_0 = 0.1$ represents the higher bound of typical bandwidth in smooth flow (Vickery and Basu, 1983). The values of B are mostly larger in full-scale compared to the expression from Vickery and Basu. Yet scatters are obvious that engages further investigation of the exact value of B and its trend in this measurement.

In Fig. 7d, fewer data is available for 24-Nov-2021 with a clear vortex shedding peak. It is suspected that the considerable distance between MET 1 and D8 Tower contributes to some uncertainty in predicting the exact turbulence intensity outside the tower. By comparing Fig. 7a to Fig. 7c, higher turbulence intensities up to $I_V \approx 0.2$ are found with north-westerly wind, while in southerly winds the range of turbulence intensity is usually smaller, i.e. $I_V \lesssim 0.15$. This is because of the rougher terrain due to the forest height on the north-westerly side of the MET 1, while southern side of the MET 1 is a flat land. This is also regarded in Fig. 8, $B \approx 0.3$ has highest occurrence with north-westerly wind found of the dataset 02-Aug-2021 and smaller $B \approx 0.23$ has highest occurrence with southerly wind of dataset 16-Nov-2021.

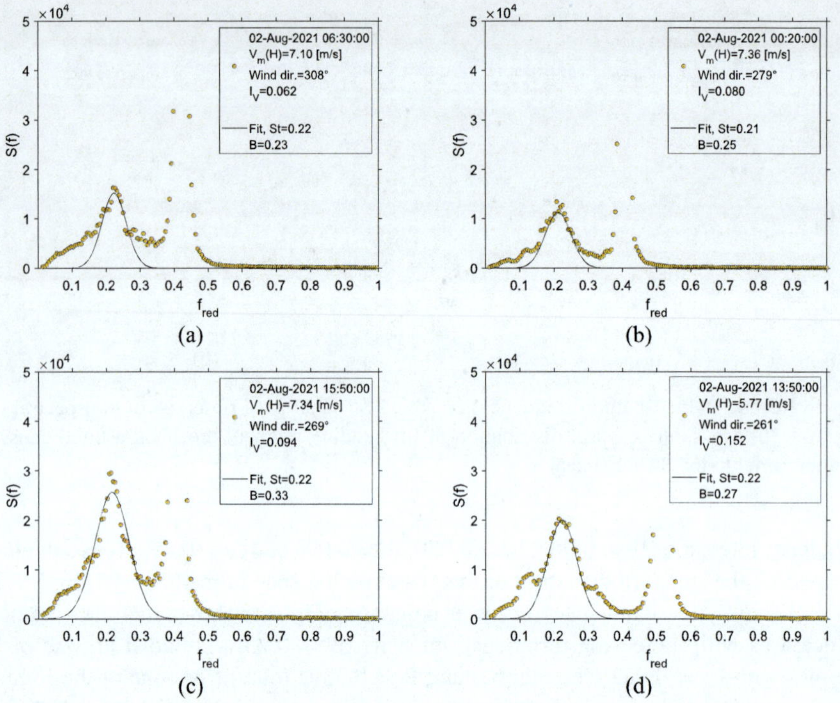

Fig. 6. Spectral distribution of lift bending moment fluctuations from D8 Tower measurement of different turbulence intensity. (a) $I_V = 0.0617$; (b) $I_V = 0.080$; (c) $I_V = 0.094$; (d) $I_V = 0.152$

Table 1. Datasets from D8 Response measurement

Dataset	Min. Wind direction	Max. Wind direction	Orientation
02-Aug-2021	257°	336°	North-West
29-Aug-2021	25°	48°	North-East
16-Nov-2021	184°	208°	South-West
24-Nov-2021	237°	267°	South-West

Strouhal Numbers and Critical Velocity Estimated from the Response Measurement

The Strouhal number corresponds to the vortex shedding frequency is obtained by the fitting for each dataset. In the 02-Aug-2021 dataset (Fig. 9a), St≈0.23 occurs most frequently, whereas in dataset from 29-Aug-2021 (Fig. 9b), there is wider distribution around St≈0.18–0.2. In the 16-Nov-2021 and 24-Nov-2021 dataset (Fig. 9c and Fig. 9d), the highest occurrence of St≈0.2 and lower St≈0.15 values appeared. The large variation of St in the D8 measurement campaign is further investigated, one of the main arguments being that the spectral analysis is calculated from the tower response. Accordingly, the Strouhal number represents the integration of the response along the tower height and

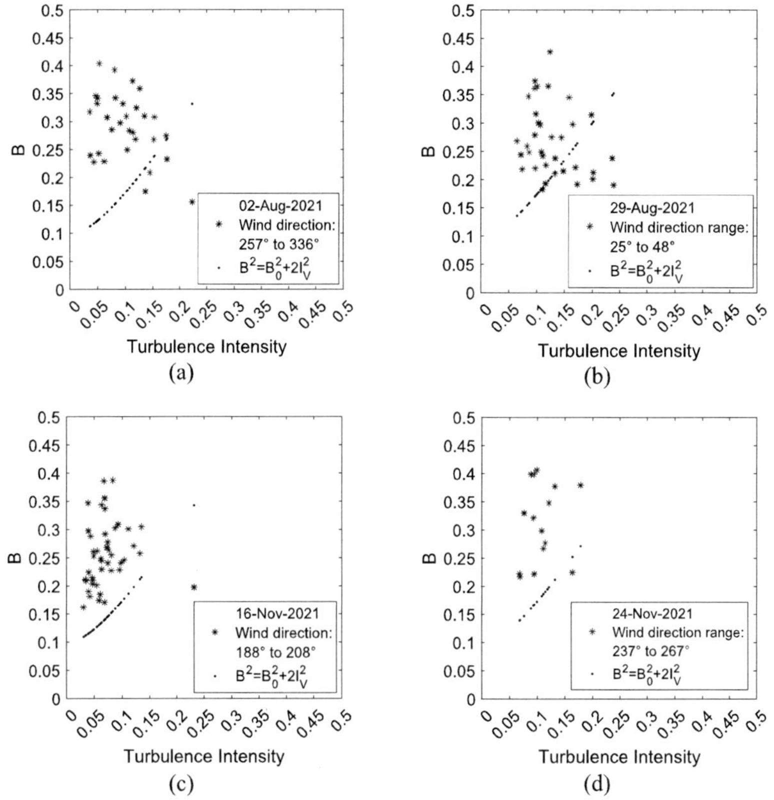

Fig. 7. Distribution of B in respect to the turbulence intensity I_V. From: (a) 02-Aug-2021; (b) 29-Aug-2021; (c) 16-Nov-2021; (d) 24-Nov-2021

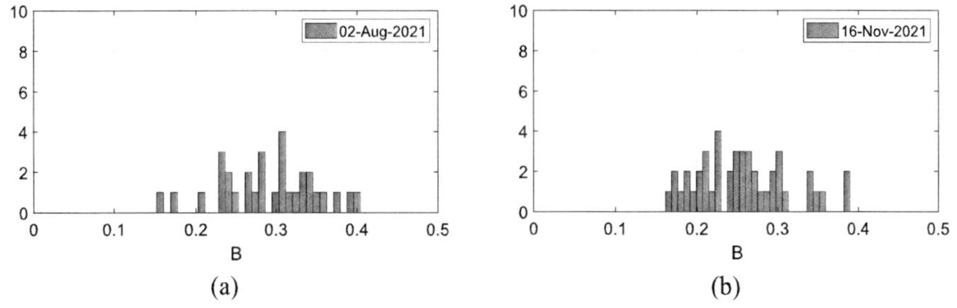

Fig. 8. Distribution (Histogram) of B from dataset of (a) 02-Aug-2021 (north-westerly wind) and (b) 16-Nov-2021 (southerly wind)

the across-wind load was not measured directly. They can thus be regarded as "resultant" St and B, which are a function of the wind profile. The values do not represent sectional St(z) and B(z) values and should be used with care.

In comparison to the previous investigation, Ruscheweyh (Ruscheweyh, 1974) reported St = 0.23 from wind pressure measurement on the Hamburg television tower with rough surface, $k/D = 1.085\times10^{-3}$ at Re range 6×10^{6} to 14.4×10^{6} in a field condition. Referred by Vickery and Clark (Vickery and Clark, 1972), Cincotta et al. (Cincotta et al., 1966) and Schmidt (Schmidt, 1966) reported St = 0.25 to 0.29 for $Re\approx3.5\times10^{6}$ to 2×10^{7} based on experiment on original scale. A higher St in field test seems to be expected, and Ellingsen (Ellingsen, 2021) reports a "higher Strouhal number" in a field test on a custom-made steel chimney with low Scruton number. In his work, the maximum amplitude was found at $V \approx 1.3V_{crit}$, which refers to St = 0.3 at the critical wind speed. Vickery and Basu (Vickery and Basu, 1983) estimated the maximum amplitude at $V \approx 1.1V_{crit}$.

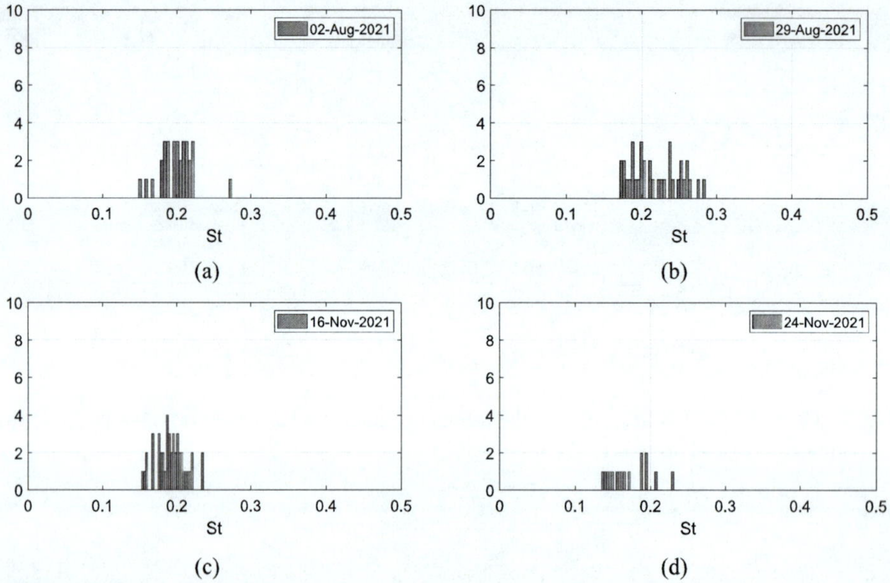

Fig. 9. Distribution (Histogram) of Strouhal number from dataset: (a) 02-Aug-2021; (b) 29-Aug-2021; (c) 16-Nov-2021; (d) 24-Nov-2021

For comparison purpose, the critical wind speed is estimated using the 02-Aug-2021 dataset to represent the westerly wind that is common in the test area (Peña, 2019; Hansen et al., 2014). Figure 10a shows the Strouhal frequency f_s in respect wind speed at the tower top, where a linear fit is included to estimate the gradient of the trend. Considering the first resonance f_1 of the tower, the critical wind speed is estimated at $V_{crit} = 14.2$ m/s. This refers to $St = \frac{f_1 \cdot D_{eff}}{V_{crit}} = 0.21$. This value is then to be compared with the information from first resonance events. Two resonance events with similar westerly winds are found on 22-Sept-2021 and 21-Oct-2021 as shown in Fig. 10b and Fig. 10c. On 22-Sept-2021 (Fig. 10b), highest amplitude is recorded at the wind speed 18.9 m/s (measured) and on 21-Oct-2021 (Fig. 10c) 15.8 m/s. Respectively, the highest amplitude occurred at

$V=1.33V_{crit}$ and $V = 1.1V_{crit}$. This difference can potentially be explained by different wind profiles and will be investigated further.

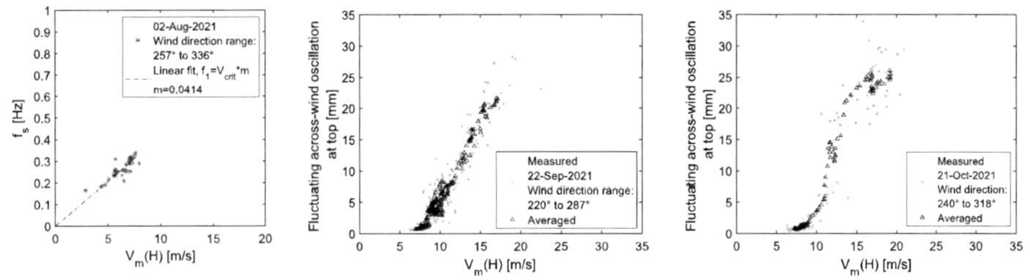

(a) (b) (c)

Fig. 10. (a) Strouhal frequency from dataset 02-Aug-2021 in comparison to the wind speed at D8 tower top. (b) Across-wind oscillation at the top of the tower from 22-Sept-2021; and from (c) 21-Oct-2021

3.2 Short Introduction of Further Investigation on Vortex Shedding Load Through Wind Pressure Measurement

The challenge of accurately estimating the parameters St and B of vortex shedding spectrum under real conditions using the response signal urges the direct measurement of wind pressure. The wind pressure measurement campaign on tower SG 14-222 DD (SG14) aims to provide full-scale wind load data at transcritical Reynolds numbers. Before calculating the lift coefficient and its spectral distribution from the wind pressure measurement, the quality and plausibility of the data must first be assessed. This is done by checking the pressure coefficient distribution. The pressure data is curated to ensure that no rain occurs, and a reasonable pressure distribution curve is obtained. It is also a challenge to obtain the appropriate dynamic pressure q since the reference pressure is not the same for all sensors. The dynamic pressure determined by the met mast cannot be used for quantitative assessment because it is not the exact dynamic pressure that occurs outside the tower surface. On the other hand, knowledge of the dynamic pressure is needed to determine the actual incoming wind speed. In this case, the pressure distribution is calculated by iteration, assuming that the initial dynamic pressure is related to the base pressure coefficient $c_{P,b} = -0.5$.

To obtain the knowledge of the real wind direction outside the tower, the wind direction is then referred to the stagnation point of the pressure distribution. This can be calculated using half the angular distance of the zero crossing ϕ_0 of the pressure coefficient distribution. Figure 11 shows the pressure coefficient distribution of a 10-min average from $z/H = 0.82$.

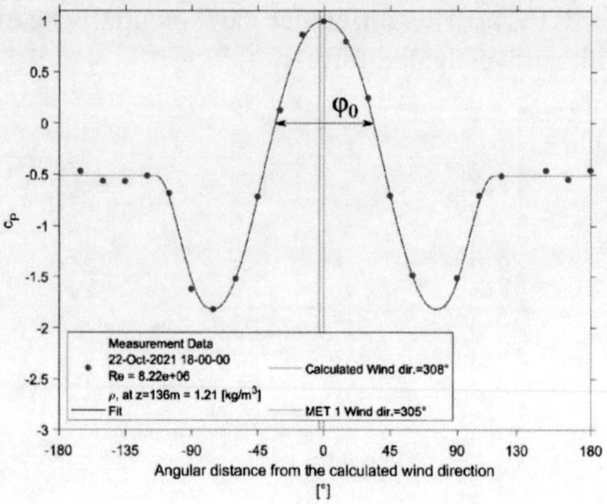

Fig. 11. Pressure distribution of one 10-min sample from measurement on SG14 in a standalone tower condition.

4 Conclusions

Full-scale measurement campaigns are carried out on two wind turbine towers (D8 and SG 14-222 DD) with very smooth surfaces to investigate vortex shedding load under transcritical flow and field condition. Specifically for D8 measurement, spectral evaluation of the lift bending moment fluctuations at non-resonant wind speeds help to observe the vortex shedding peak. A Gaussian function is fitted to estimate the parameters St and B. Knowledge of turbulence intensity and wind profile are considered. The question arises of a robust value of St and B for a field and a full-scale condition. The transcritical flow and the very smooth roughness of the tower are unique characteristics to be considered when comparing with previous studies. The scatter of St and B values based on the evaluation on D8 data still indicates the uncertainty due to the use of the response signal. The considerable distance between the tower and the met mast also introduces additional uncertainty. Direct measurement of wind pressure on the tower surface should provide further opportunities for investigation and is being carried out in this ongoing research. On that note, wind pressure measurement from SG14 should enable the evaluation of the lift coefficient and consequently, an insight regarding the vortex shedding spectrum. Quality and plausibility of the pressure distribution should be considered before analyzing the vortex shedding spectrum from the lift coefficient.

Acknowledgements. This project is funded by the Deutsche Forschungsgemeinschaft (DFG, German Research Foundation) – Project number 493357786 and 426322127. The authors would like to also acknowledge CICIND (International Committee for Industrial Construction) for the support through the CICIND research project "Reynolds number disparity and its effect on vortex excitation – Insight from full-scale tests at wind turbine towers". Last but not least, we are very thankful to Siemens Gamesa Renewable Energy for providing the measurement data and excellent measurement system. The supports are gratefully acknowledged.

References

Arunachalam, S.L.: Non-linear modelling of vortex induced lock-in effects on circular chimneys. J. Wind Eng. Ind. Aerodyn. **202**, 104201 (2020)

Cincotta, J.J., Jones, G.W., Walker, R.W.: Experimental investigation of wind-induced oscillation effects on cylinders in two-dimensional flow at high Reynolds numbers. NASA Langley Research Centre (1966)

Ellingsen, Ø.: Vortex-induced vibrations on industrial chimneys. PhD Dissertation. l'Institut Polytechnique de Paris, Paris, France (2021)

Guo, K., Yang, Q., Liu, M., Li, B.: Aerodynamic damping model for Vortex-induced Vibration of suspended circular cylinder in uniform flow. J. Wind Eng. Ind. Aerodyn. **209**, 104497 (2021)

Hansen, B.O., Courtney, M., Mortensen, N.G.: Wind resource assessment – Østerild National Test Centre for large wind turbines. DTU Wind Energy (2014)

Lupi, F., Niemann, H.-J., Höffer, R.: A novel spectral method for cross-wind vibrations: application to 27 full-scale chimneys. J. Wind Eng. Ind. Aerodyn. **171**, 353–365 (2017)

Lupi, F., Niemann, H.-J., Höffer, R.: Aerodynamic damping model in vortex-induced vibrations for wind engineering applications. J. Wind Eng. Ind. Aerodyn. **174**, 281–295 (2018)

Peña, A.: Østerild: A natural laboratory for atmospheric turbulence. Journal of Renewable and Sustainable Energy **11** (2019)

Ruscheweyh, H.: Beitrag zur Windbelastung hoher kreiszylinderähnlicher schlanker Bauwerke im natürlichen Wind bei Reynoldszahlen bis $Re=1.4\times10^7$. PhD Dissertation. Technical University Aachen, Aachen, Germany (1974)

Schmidt, L.: Fluctuating force measurements on a circular cylinder at reynolds numbers up to 5×10^6. NASA Langley Research Centre (1966)

Vickery, B.J., Clark, A.W.: Lift or across-wind response to tapered stacks. J. Struct. Div. **98**(1), 1–20 (1972)

Vickery, B., Basu, R.: Across-wind vibrations of structures of circular cross-section. Part I. Development of a mathematical model for two-dimensional conditions. J. Wind Eng. Ind. Aerodyn. **12**(1), 49–73 (1983)

Vickery, B., Basu, R.: Simplified approaches to the evaluation of the across-wind response of chimneys. J. Wind Eng. Ind. Aerodyn. **14**, 153–166 (1983)

A Numerical Assessment to Estimate the Dynamic Factor for Wind Excited Façade Elements Having a Natural Frequency Smaller Than 5 Hz

Lori Guido[1(✉)] and Manara Giampiero[1,2]

[1] Permasteelisa Group, Vittorio Veneto, Italy
{g.lori,g.manara}@permasteelisagroup.com
[2] Permasteelisa SpA, Vittorio Veneto, Italy

Summary. Wind is the major action affecting a façade and local effects are governing the behaviour. The European standard (EN1991-1-4, 2005) follows and equivalent static approach, taking in consideration possible resonances by means of a structural factor, which should account for the effect of along wind actions deriving from the non-simultaneous occurrence of peak wind pressure on the surface and the effect of the vibrations of the structure due to turbulence. According to the standard, for façade and roof elements having a natural frequency greater than 5 Hz, the structural factor can be taken as 1. However, the 5 Hz value is not accepted as an adequate limit by the entire scientific community and no justification is provided about the choice of this value. The authors have investigated how the natural frequency influences the response of a Single Degree of Freedom (SDOF) system excited by the Wind Power Spectrum, comparing the outcomes with the structural factor proposed by the standard.

Keywords: facade · wind · vibrations

1 Introduction

The façade is the element of the building that is firstly impinged by the external actions and it definitely resists and transforms the loads, both in magnitude and directions. Moreover, experimental evidences show that the envelopes give a certain contribution to the damping ratio of the building, comparing measurements conducted before and after the façade installation (Meyyappa et al., 1980; Moore, 2016). In addition, more and more academic studies, sometimes in collaborations with façade contractors, are aiming at demonstrating that the mass of the façade could be in the future a significant source of additional structural damping, with effects comparable to the tuned mass dampers mitigation potential. Often the façade is presented in the codes ignoring the relevant mass, its critical need of movement accommodation and the possibility that induced vibrations may downgrade the weather-tightness of the envelope. It is clear that the strict connections between envelope and primary building structure should require in

P. Schito and A. Zasso (Eds.): IN VENTO 2022, LNCE 461, pp. 64–78, 2024.
https://doi.org/10.1007/978-3-031-53059-3_7

general a simultaneous and coupled dynamic method of investigation, but the state of the art shows even a rawer evidence, i.e. that a consistent approach for dynamic effects on a façade in isolation is not existing yet. The example of the wind induced vibrations is quite representative at the scope. European standard (EN1991-1-4, 2005) states that a dynamic factor must be included, within the equivalent static approach, only when the facade fundamental frequency of vibration is smaller than 5 Hz. The American Building code (ASCE7-10, 2010) is, if possible, even less clear for the façade designer, as it introduces only a general limit of 1 Hz, for the critical natural frequency. The reason of this limit is probably because it refers to large size structures, for which the effect of small probability of wind gust spatial correlation, represented in the European standard by the c_s factor, plays a fundamental role in reducing the risk of vibrations.

One of the first literature studies regarding dynamic wind loading on cladding design, by the Canadian Research Council (Allen et al., 1974), reported that in general metal and glazed cladding natural frequencies are in the range between 5 and 50 Hz, which means, for the majority of these elements, several times higher than the significant wind turbulent frequencies. Thus the dynamic amplification can be neglected. Only for unusual, high-frequency wind turbulence, created locally by building corners or by surface irregularities, the dynamic magnification becomes significant. More recently Nagakami (Nagakami, 2003) has studied the problem from the numerical simulation point of view. Starting from DIN 1055 statement: "If the natural frequency of the surface vibration is very low, less than 5 Hz, the vibration effect must be considered", he searched for a confirmation of the statement, by numerous FEM simulations. He took in account 4 edge linear support and 4 points support glazing, varying the dimensions among typical sizes (1400 × 1800 mm, 2000 × 3000 mm, 1000 × 3000 mm) and getting first natural frequencies ranging from 3.6 Hz to 21.8 Hz. Focusing in particular on the lay-out with minimum frequency, under the condition of mean wind velocity of 24 m/s, the difference between quasi-static and dynamic deflection is in the range of 4%, while it is completely negligible for mean wind velocity of 17 m/s. Approximately 10% difference appears when a wind of 30 m/s mean wind velocity is adopted for the simulation.

However, experimental testing conducted during the last years, have shown that relatively small scale elements, like PV systems, can be subjected to significant wind-induced vibrations even if they are characterized by much higher natural frequencies (Moravej et al., 2014) than the <1.0 Hz criterion indicated in (ASCE7-10, 2010) for categorizing dynamically wind sensitive structures.

Limiting the discussion to the along wind effects, the scenario appears anyway much more complex than what a simple threshold limit for occurrence of vibrations could describe. Even if the 5 Hz limit seems for the most part of the applications a reasonable, probably overly-conservative recommendation, some specific systems have shown dynamically sensitive behaviour not in accomplishment with that guideline. It means that the phenomenon should be described making use of additional constraints and parameters. In this sense, for sure the average wind velocity, as well as the turbulence factor, represent an important variable for the recognition of possible wind induced vibrations. On the other side, the natural frequency can't be the only system parameter to govern the dynamic effects: it is expected that a fundamental role is played by the damping ratio of the system and a comprehensive numerical study should determine different thresholds for different damping ratio values.

The extent of the uncertainties involved in the design process of cladding surfaces and other elements due to the dynamic amplification is larger than what can be expected. Fin and shading device first natural frequencies are frequently close to the 5 Hz limit and the issue of risk of failure may arise not just because of ultimate stress limits, but sometimes it involves low fatigue cycle risk, especially at the connections.

In this paper, the authors describe the method proposed by the European Standard (EN1991-1-4, 2005) in order to consider dynamic effects on the response of facades to wind loading. In particular shortcomings of the approach are discussed and challenged, with reference to typical application conditions. In a second part, a methodology of study is proposed, based on frequency domain calculation of the façade response to the wind gust power spectral density and on system analytical model by means of their modal decomposition representation. The methodology is applied for a wide range of variability for the major design parameters, supplying dynamic factor charts and threshold frequencies under specific scenarios. Finally, a comparison with the dynamic factor carried out by the European standard (EN1991-1-4, 2005) is presented for an applicative case, highlighting potential gaps and overestimations when the standard method is applied (or not applied), irrespective of the damping and wind loading specific conditions. Regardless of the SDOF simplification, the proposed method can be applicable without restrictions, identifying the first governing modes of the structure and applying for each of them the SDOF based method.

2 The Dynamic Effects in the European Standard

The current revision of the European standard for the wind actions on structures, (EN1991-1-4, 2005), prescribes a structural factor $c_s c_d$ that should be applied to façade and roof elements that can undergo vibrations under the wind turbulence. The structural factor is formed by two components: the size factor c_s and the dynamic factor $c_{d.}$ The first one takes into account the effect of the non-simultaneous occurrence of peak wind pressure on the surface, while the second accounts for the vibration of the structure due to turbulence. For slender, flexible or low damped structures or components, the increase in the dynamic response dominates over the reduction effects due to the practical correlation of peak actions and the structural factor is generally greater than 1. On the other hand, for stiff and/or high damped structures or components, or in the case in which these have large surfaces, the reduction effects due to the partial correlation of peak actions dominates over the increased dynamic response and the structural factor is generally less than 1. For façade and roof elements having a natural frequency greater than 5 Hz, the value of $c_s c_d$ may be taken as 1.

The calculation of the structural factor is performed by the formula:

$$c_S c_d = \frac{1 + 2k_p I_v(z_s)\sqrt{B^2 + R^2}}{1 + 7I_v(z_s)} \tag{1}$$

where z_s is the reference height for determining the structural factor (1), which can be set as the height of the structure. k_p is the peak factor defined as the ratio of the maximum value of the fluctuating part of the response to its standard deviation. I_v is the turbulence intensity, B^2 is the background factor, allowing for the lack of full correlation of pressure on the structure surface, while R^2 is the resonance response factor, allowing for turbulence in the resonance with the vibration mode. The size factor c_s takes in account the reduction of the wind action due to the non-simultaneity of occurrence of the peak wind pressures on the surface and may be obtained from:

$$c_S = \frac{1 + 7I_v(z_s)\sqrt{B^2}}{1 + 7I_v(z_s)} \tag{2}$$

In the (EN1991-1-4, 2005) it is said that the wind loading on buildings is determined by application of the external pressure coefficients c_{pe} on windward and leeward side of the building simultaneously, the lack of correlation may have to be taken in account. Several National Annexes prescribe the use of a reduction factor 0.85 for this lack of correlation. The dynamic factor c_d takes in account the increasing effect from vibrations due to the turbulence in resonance with the structure and may be obtained by:

$$c_d = \frac{1 + 2k_p I_v(z_s)\sqrt{B^2 + R^2}}{1 + 7I_v(z_s)\sqrt{B^2}} \tag{3}$$

It should be said that the dynamic factor is not, in general, equal to 1 when no resonance occurs. Indeed, the peak factor is not less than 3, reason why, assuming $B = 1$, the limit value of c_d can be given by:

$$c_{d-lim} = \frac{1 + 6I_v(z_s)}{1 + 7I_v(z_s)} \tag{4}$$

For instance, considering a turbulence factor ranging between 0.1 and 0.3, the limit value of the dynamic factor is in between 0.9 and 0.94. This method is similar to the approach proposed by Davenport. In the (EN1991-1-4, 2005) two different methods to derive the dynamic factor are given, in Annex B and Annex C. The method of Annex B is based on empirical studies by (Solari, 1982), while the method of Annex C is a theoretical approach, generally recommended by the national Annexes, which originates from (Drybye and Hansen, 1999). In (Steenbergen et al. 2012), the methods were compared to the theoretically correct values and it was concluded that there are significant differences. The most relevant is that the mode shape in the method of Annex B is taken outside of the integral, while in the method of Annex C it is evaluated inside the integral, more consistent with the theoretical approach. The differences between the methods do not exceed 5% according to the comparison.

3 The Proposed Methodology

The authors propose a numerical approach that is based on the representation of the façade or façade component by one single degree of freedom system, which is a common practice in the analysis of systems under dynamic loading (Biggs, 1964). Both analytical methods for the calculation of the structural factor in the Eurocode (EN1991-1-4, 2005) are also derived by the modal approach theory and it is clearly stated that the designer must pay attention and check that higher modes of vibrations are not contributing significantly to the dynamic response.

The proposed method can be considered a mixed time domain-frequency domain approach, as the most part of the flow is working in frequency domain, but it relies on a reconstruction of one sample of time history in order to transform the Power Spectral Density (PSD) of the turbulent wind in the PSD of the wind Pressure. A sensitivity analysis has been conducted, varying mass, natural frequency, damping ratio from the system side, and mean and standard deviation wind velocity from the side of the excitation.

3.1 Loading Pressure Power Spectral Density

The analysis has been conducted considering the wind spectrum proposed by the European Standard. As described in the code, the spectrum requires the knowledge of the basic parameters of the wind turbulence, in particular the turbulent length scale *L(z)*:

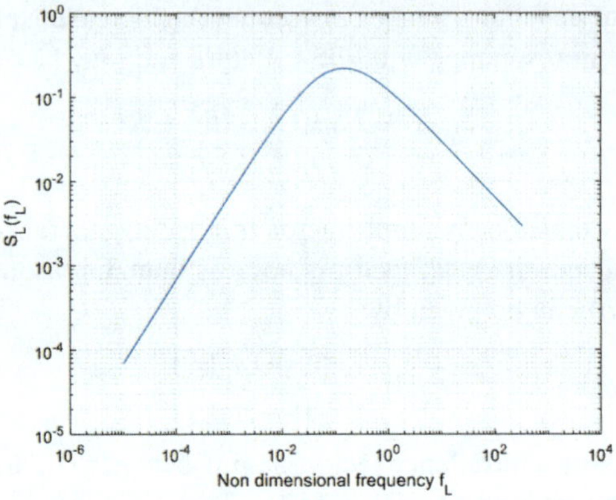

Fig. 1. Wind Power Spectrum of the wind turbulence according to the Eurocode (EN1991-1-4, 2005)

$$L(z) = L_t\left(\frac{z}{z_t}\right)^{\alpha}, z \geq z_{min} \tag{5}$$

$$L(z) = L(z_{min}), z < z_{min} \tag{6}$$

In (5) and (6), the reference height z_t is 200 m and the reference scale length L_t is 300 m, while the roughness length z_0 and the minimum height z_{min} can be defined in function of the terrain category. In this assessment, it has been decided to adopt the terrain category IV, because one of the most relevant in the applications. For the terrain category IV, $z_0 = 1$ m and $z_{min} = 10$ m. The exponent α in (5) can be calculated as:

$$\alpha = 0.67 + 0.05 ln(z_0) \tag{7}$$

Once the average velocity v_m and the standard deviation of the turbulent component σ_V are known, the PSD $S_V(z,f)$ of the turbulent wind can be calculated from the non dimensional PSD function $S_L(z,f)$, shown in Fig. 1:

$$S_L(z,f) = \frac{f \cdot S_V(z,f)}{\sigma_V^2} = \frac{6.8 f_L(z,f)}{(1 + 10.2 f_L(z,f))^{\frac{5}{3}}} \tag{8}$$

The non dimensional frequency $f_L(z,f)$ is expressed in terms of the frequency f, the mean wind velocity v_m and the turbulence length scale $L(z)$ by:

$$f_L(z, f) = \frac{f \cdot L(z)}{v_m(z)} \tag{9}$$

The turbulence intensity $I_v(z)$ is defined as the ratio between standard deviation and mean velocity:

$$I_V(z) = \frac{\sigma_V}{v_m(z)} \tag{10}$$

As discussed, a time history $v(t)$ can be reconstructed from the spectrum $SV(z,f)$, for any specific combination of length scale, wind velocity and turbulence intensity. The PSD should be divided in a certain number N of steps and for each of them a random phase difference is assigned. So, using the Montecarlo method, a possible velocity time history of the full random phenomenon can be reconstructed by superimposing the sine waves at the discrete frequency fi with amplitude given by the square root of the spectral area around fi and random phase:

$$v(t) = v_m(z) + \sum_i \sqrt{2 \bullet S_i(z, f_i) df} \sin(2\pi f_i t + \varphi_i) \tag{11}$$

The wind time history with the right spectral content will be then transformed in pressure time history:

$$p(t) = \frac{1}{2}\rho(v(t))^2 \tag{12}$$

By an operation of Fourier Transform, squaring and dividing for the frequency interval, a PSD of the dynamic pressure from the wind turbulence $S_P(z,f)$ is finally obtained.

3.2 Single Degree of Freedom Frequency Response

A single degree of freedom system is represented in the frequency domain of the angular velocity ω by its Frequency Response Function (FRF), $H(\omega)$:

$$H(\omega) = \frac{X(\omega)}{F(\omega)} \tag{13}$$

The function $H(\omega)$ is a complex function that represents the linear Single Degree of Freedom transformation applied to each sine wave frequency component of the loading input. The specific frequency component of the response will be multiplied by the magnitude of H and a phase difference ϕ.

$$|H(\omega)| = \frac{1}{K}\left[\frac{\omega_n^2}{\sqrt{\left(\omega_n^2 - \omega^2\right)^2 + (2\xi\omega\omega_n)^2}}\right] \tag{14}$$

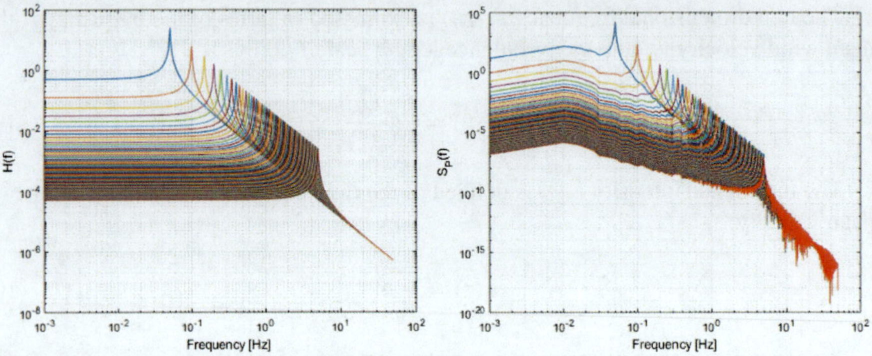

Fig. 2. Series of frequency responses *H(ω)* for same mass and different natural frequencies (left) and corresponding displacement Power Spectral Densities of system response (right)

$$\phi = atan\left[\frac{-\omega_n^2 + \omega^2}{2\xi\omega_n}\right] \tag{15}$$

The magnitude and phase are functions of the lumped parameters of the system, mass M, stiffness K, damping ratio ξ and natural angular velocity ω_n, which in turn are dependent on damping c and the other parameters by:

$$\omega_n = 2\pi f_n = \sqrt{\frac{K}{M}} \tag{16}$$

$$\xi = \frac{C}{C_D} = \frac{c}{2M\omega_n} \tag{17}$$

A set of frequency response functions is shown at the left chart of Fig. 2, by making use of the same mass M and varying the natural frequencies by increasing the stiffness K. Although a continuous system has an infinite number of degrees of freedom and so an infinite number of modes of vibration, under many excitations only the first few natural frequencies have a relevant contribution. A façade system has in general at least two major natural frequencies that contribute to its dynamic response: the first one given by the framing stiffness and the mass of the glazing and of the framing and the second one formed by the glazing mode of vibration on the deformable supports. In general, the first one is also the one with smaller natural frequency and it is relevant to check the framing deflection and stress. For the glass dynamic response, both modes would contribute in general, reason why a Multi Degrees of Freedom (MDOF) is required. However, the second mode governs the glazing behaviour, as both stress and deflections are mostly given by the shape mode relative to the frame. In addition, it should be said like also in the (EN1991-1-4, 2005) the proposed approach for the structural factor calculation is limited to conditions for which only one mode governs the dynamic response. In literature it can be found how to transform the mass, stiffness, damping and applied force to any mechanical system, like for instance a glass panel, to its equivalent SDOF system with equivalent lumped parameters. For this reason, the methodology described into this chapter can be considered of general applicability.

3.3 Displacement Power Spectral Density

The PSD representative of the system displacement $S_D(\omega)$ can be obtained by simply multiplying the PSD of the wind pressure $S_P(\omega)$ (properly calculated in terms of angular velocity), for the square of the magnitude of the specific frequency response function $\boldsymbol{H(\omega)}$:

$$S_D(\omega) = (H(\omega))^2 S_P(\omega) \tag{18}$$

The corresponding series of power spectra to the frequency response function of Fig. 2 at the left are shown at the right chart of the same figure.

4 Results

In order to represent the results of the assessment, a proposed dynamic factor c_{prop} has been defined, by the formulas:

$$c_{prop} = \frac{d_{dyn}}{d_{QS}} \tag{19}$$

$$d_{QS} = \frac{1}{K}(P_{mean} + 3.5\sigma_P) \tag{20}$$

$$d_{dyn} = \frac{1}{K}(P_{mean}) + 3.5\sigma_D) \tag{21}$$

$$\sigma_P = \sqrt{\int S_P(\omega)} \tag{22}$$

$$\sigma_D = \sqrt{\int S_D(\omega)} \tag{23}$$

The dynamic factor is given by the ratio between the expected peak of the quasi-static response of the system, without any influence of the vibration, and the peak of the dynamic response, which includes the calculated effect of the natural frequency excitation. The peak is given by the sum between the average displacement and the expected peak of the turbulent displacement. The substantial difference is that the turbulent displacement component in the quasi-static approach is calculated as the product between the static response *1/K* and the turbulent pressure standard deviation, which corresponds to the square root of the $S_P(\omega)$ spectral area. On the contrary, the turbulent displacement component for the dynamic response is calculated directly as integral of the PSD of the displacement response $S_D(\omega)$. A peak factor 3.5 has been assumed in the calculation. It should be said that this definition requires a refinement, as it ignores that for small frequencies the expected peak of response is lower than 3.5 times the standard deviation σ_D, while for high natural frequencies it can be much larger (effect of the peak factor k_P in (EN1991-1-4, 2005) formulation, Annex B, Figure B.2). In practice the proposed formulation in the current version tends to be conservative for low natural frequencies, but non conservative for high natural frequencies.

4.1 Sensitivity Analysis on Dynamic Factor

The Fig. 3 shows a compact representation for the dynamic factor, plotted in six different charts, varying the damping ratio from 0.01 to 0.05 along the rows and the turbulent length scale along the columns (left plot cases corresponds to 100 m height, giving a turbulent scale length $L(z)$ of about 188 m, while right plot cases are for 10 m of height, so about 40.3 m for $L(z)$. Turbulence factor $I_v(z)$ *and* mass M of the system provides only small or no effect at all, reason why the results are not shown with respect to those parameters.

For each combination of damping ratio ξ and turbulent scale length $L(z)$, the contour plot shows in abscissa the variable mean velocity of the wind v_m and in ordinates the different natural frequencies, obtained by the mass system M varying the stiffness K. An allowable limit for the dynamic factor must be defined. A difference with the dynamic factor of the code is that by the proposed approach, the minimum value can't be less than 1. In addition, it should be considered that in order to compare the outcomes with the structural factor $c_s c_d$ of the code, the effect of the size factor must be taken in account. A rigorous analysis should consider the corresponding size factors for different geometry and turbulent length scale, but in first approximation a size factor in the range [0.9÷0.95] is considered, allowing a dynamic factor 1.1 under all the conditions for this assessment, so by a resulting structural factor marginally larger than 1.

A different representation of the results is given in Fig. 4 (left), in which the frequency that corresponds to the dynamic factor 1.1 is plotted for each combination of damping ratio and mean wind velocity and $L = 188$ m. Considering the hypothesis of the assessment, this frequency is directly comparable with the 5 Hz limit stated in the (EN1991-1-4, 2005). It can be seen like only with a damping ratio larger than 2.5%, the 5 Hz limit can be considered adequate for all the range of mean wind velocity considered. However, it is well known that glass facades are not characterized by large damping ratios (Lenk et al, 2010). The inherent glass damping exceeds 1% only by means of the contribution of interlayers. Even considering the relevant contribution of framing and bracket connections and other source of damping (like aerodynamic), it is not expected that the damping ratio exceeds the value of 2% under the most general conditions. It is interesting to notice that the same minimum frequency calculated by the (EN1991-1-4, 2005) Annex B approach (Fig. 4, right) for certain combinations of mean velocity and damping ratio exceeds the 5 Hz limit. This means that for many applicative conditions it is true that the 5 Hz limit is conservative, but probably some scenarios exist, in particular those ones with a simultaneous occurrence of large wind velocities and small turbulent length scale, for which the designer risks to ignore some dynamic effect because of a not-exhaustive definition of the frequency limit in the code.

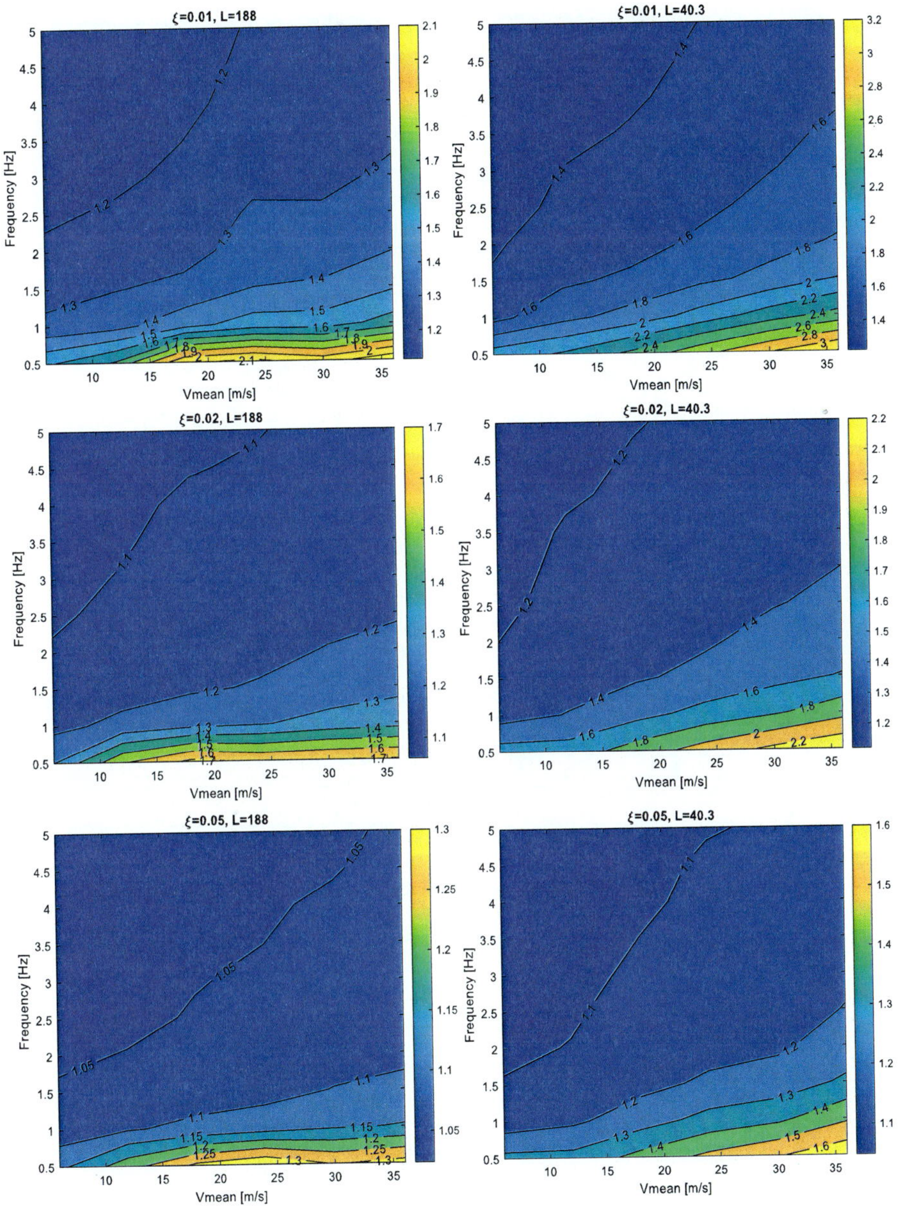

Fig. 3. Contour plot of the proposed dynamic factor c_{prop} for different conditions of f_n, v_{mean}, ξ and L.

4.2 Application Example

A monolithic glass 3000 × 6000 × 12 mm has been modeled by the finite element commercial code Straus®, version 2.4.6. The glass has been initially characterized by an assigned thickness of 12 mm, showing natural frequency of 3.9 Hz.

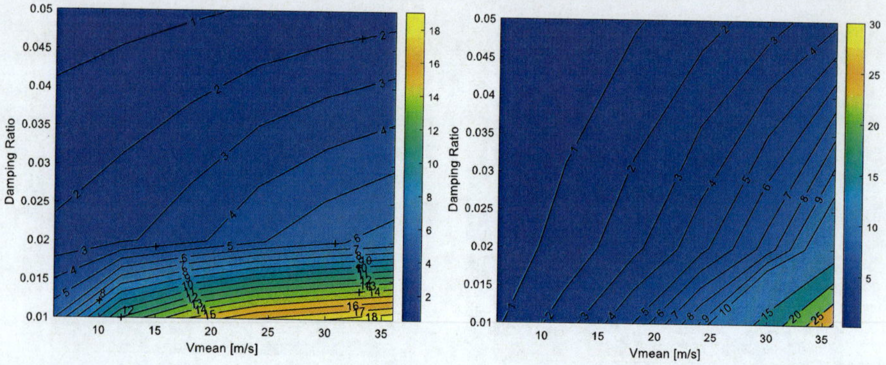

Fig. 4. Minimum frequency without dynamic effects for different conditions of damping and mean wind velocity calculated by the proposed approach by $L = 188$ m (left) and by (EN1991-1-4, 2005) Annex B (right)

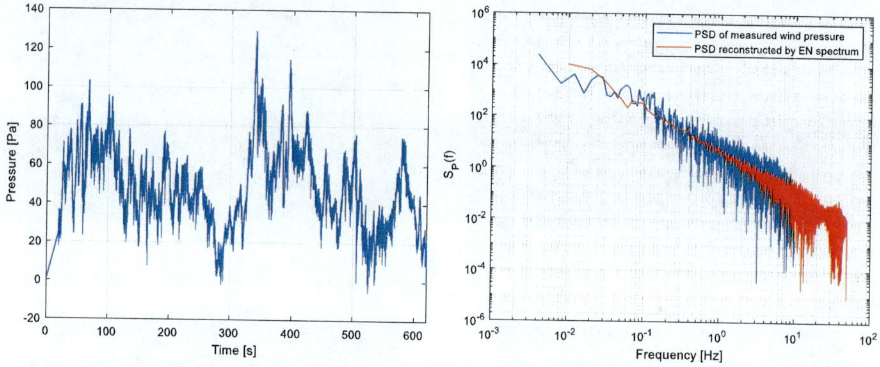

Fig. 5. Pressure time history measured on site and PSD comparison ($v_m = 8$ m/s, $Iv = 0.2$, L = 188 m)

Table 1. Results for damping ratio $\xi = 0.01$

f [Hz]	K[Pa/mm]	$\sigma_{d\text{-}dyn}$[mm]	d_{maxDyn}[mm]	cprop
0.036	1e-2	4.97e3	2.28e4	1,771
0.5	2	14.883	89.923	1.392
1.12	10	2.531	16.111	1.247
2.5	50	0.462	3.155	1.221
3.9	120	0.181	1.191	1.106
5.03	200	0.108	0.688	1.065
6.2	300	0.071	0.493	1.146
8	500	0.042	0.263	1.019

Table 2. Results for damping ratio $\xi = 0.02$

f [Hz]	K [Pa/mm]	$\sigma_{d\text{-}dyn}$ [mm]	d_{maxDyn} [mm]	c_{prop}
0.036	1e-2	4.15e3	2.04e4	1.581
0.5	2	12.741	83.463	1.292
1.12	10	2.281	15.698	1.215
2.5	50	0.431	2.876	1.113
3.9	120	0.173	1.165	1.082
5.03	200	0.103	0.664	1.028
6.2	300	0.068	0.468	1.086
8	500	0.041	0.259	1.004

Table 3. Results for damping ratio $\xi = 0.05$

f [Hz]	K[Pa/mm]	$\sigma_{d\text{-}dyn}$[mm]	d_{maxDyn}[mm]	c_{prop}
0.036	1e-2	3.09e3	1.52e4	1.176
0.5	2	11.191	68.411	1.059
1.12	10	2.105	14.548	1.126
2.5	50	0.409	2.599	1.006
3.9	120	0.168	1.069	0.993
5.03	200	0.101	0.655	1.014
6.2	300	0.067	0.447	1.037
8	500	0.041	0.257	0.994

The glass thickness has been modified covering several options for the natural frequency, but always maintaining an actual mass of 583 kg by means of a fictitious density. The simulated natural frequencies have been limited to 8 Hz, to not exceed half of the sampling rate of the measured time history (20 Hz). The transformation factor for the mass has been set to 0.34, giving an effective mass of 200 kg. An input extracted from real wind pressure measurements (Fig. 5) on a building façade has been applied to the different glass systems. The output in terms of center displacement has been obtained, as shown for one case in Fig. 6. Three different damping ratios has been applied, in agreement with the inputs and results summarized in the Tables 1, 2 and 3. The results, illustrated also in Fig. 7, demonstrates that the FEM example supplies dynamic factors in terms of displacement which are consistent with the outcomes of the proposed approach based on SDOF system behaviour, by differences limited to around 5%. On the plot, the value for the factor 1.1 is included in order to highlight the almost coincident estimation of the minimum frequency which allows to ignore the dynamic effects for the different damping ratios. The applicative example proves that the FRF approach represents in accurate way the dynamic interaction between reconstructed or measured wind pressure time histories and continuous systems, even if represented by SDOF models. This preliminary result seems to confirm that the current code reveals shortcomings, which sometimes results in conservative estimations, but that have also some risks for underestimation of potential resonances. Indeed, Fig. 8 shows the comparison between the code approach and the proposed method with the FEM example results for the case ξ=0.05. It appears that the code formula is not matching the dynamic factor assessed by the numerical example, as the peak factor is underestimated in almost the entire frequency range. It could be concluded that the code approach has some inconsistencies when compared with numerical analysis. However, further investigations must be conducted in order to properly correlate the results with experimental along-wind resonances of mechanical systems and in particular in order to understand how the size factor influences the numerical ideal simulation result.

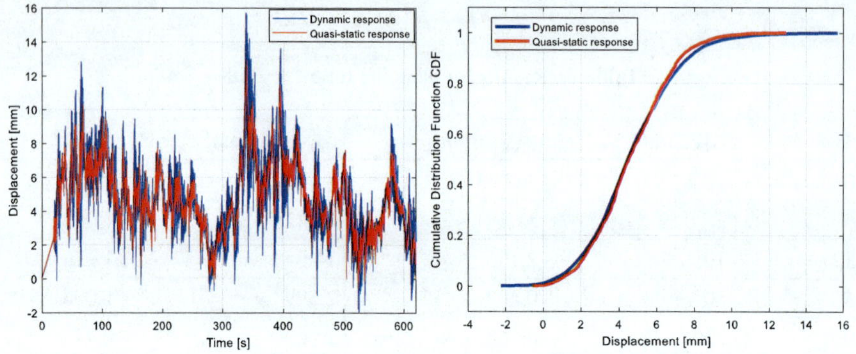

Fig. 6. Example of dynamic response versus quasi-static for the case of $f_n = 1$ Hz and $\xi = 0.02$.

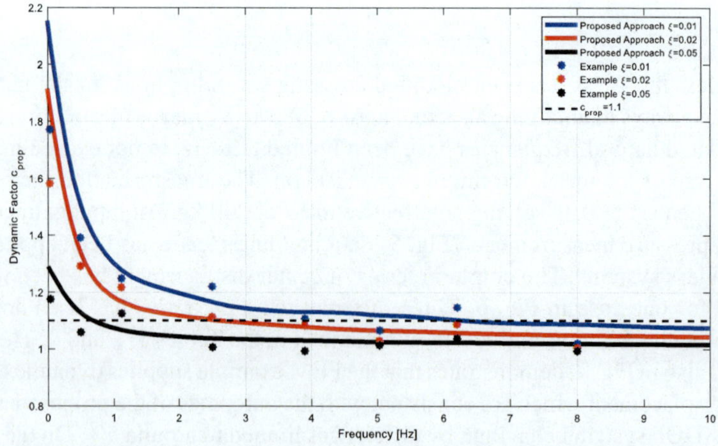

Fig. 7. Result comparison between dynamic factor calculated by the proposed approach and by the application example

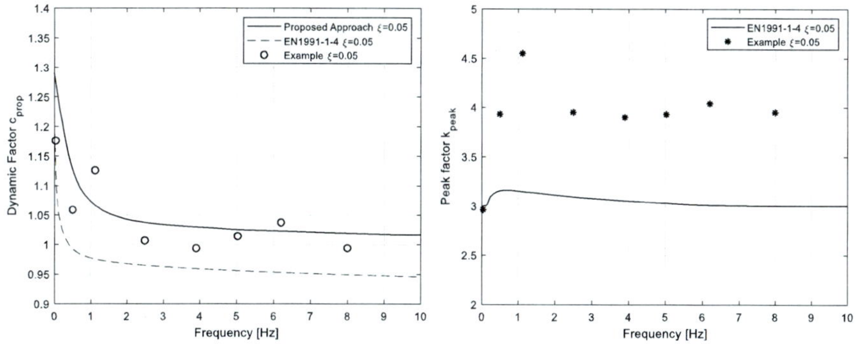

Fig. 8. Dynamic factor comparison between proposed method and (EN1991-1-4, 2005) (left). Difference between peak factor kp estimated by the code approach and calculated by the FEM example

5 Conclusions and Future Work

This paper has been focused on the problem of the dynamic effects on façade elements excited by wind loading. The authors have proposed a numerical assessment based on a Single Degree of Freedom procedure, to verify the adequacy of the 5 Hz limit for the first natural frequency proposed by the European Standard in order to ignore amplifications due to the system resonance. The results of the assessments, validated by means of a finite element model of one large glass panel, loaded by a sample of experimental wind pressure time history, have shown that the limit, although it may appear conservative under many practical applications, it is probably inaccurate under certain critical combinations of the governing parameters. The same conclusion appears also by using the (EN1991-1-4, 2005) analytical approach. The authors believe that a possible improvement in the code could involve a refinement of the frequency limit definition, based on several different scenarios. At this purpose, the authors recognize that two major parameters need a detailed investigation in order to collect relevant information for the method upgrade: the damping ratio, as it provides the major effects in the dynamic response and its extent is not yet fully understood for the overall façade dynamic behaviour and the correlation between the size factor and the dynamic factor. Both the topics will be subject of future work by the authors, together with a refinement based on the integration of a likewise peak factor parameter, which actually has shown some inconsistency when used by the code definition and compared with the numerical results. Another aspect that requires more investigation is the impact of the higher modes, as very slender system can contribute to the dynamic response with more than one mode of vibration: in this circumstances, an extension of the proposed approach to Multi Degrees of Freedom (MDOF) is a possible solution.

References

1. ASCE/SEI 7-10. Minimum Design Loads for Buildings and Other Structures (2010)
2. Allen, D.E., Dalgliesh, W.A.: Dynamic Wind Loads and Cladding Design, NRC Publication Archive (1974)

3. Biggs, J.M.: Introduction to Structural Dynamics. McGraw-Hill Book Company, New York (1964)
4. Dyrbye, C., Hansen, S.O.: Wind Loads on Structures. John Wiley & Sons, New York (1999)
5. Geurts, C.P.W., Van Staalduinen, P.C., De Wit, M.S.: Towards a reliable design of façade and roof elements against wind loading. In: TNO Building and Construction Research, Delft, The Netherlands, Heron, vol. 49, no. 2 (2004)
6. European Committee for Standardization (CEN). EN1991-1-4, Eurocode 1: Actions on Structures, Part 1–4: General Actions – Wind Actions (2005)
7. La Gasse L.C.: Structural reliability assessment of buildings subjected to wind loadings. An assessment of the main bearing structure at foundation level of dynamically sensitive buildings designed within the Eurocode Framework, Master Science in Civil Engineering Thesis, University of Delft (2017)
8. Lenk, P., Coult, G.: Damping of Glass Structures and Components, Challenging Glass 2 – Conference on Architectural and Structural Application of Glass. Bos, Louter, Veer. TU Delft (May 2010)
9. Meyyappa, M., Palsson, H., Craig, J.I.: Modal parameter estimation for high-rise building using ambient response data taken during construction. In: Hart, G.C. (ed.) Dynamic Response of Structures: Experimentation Observation, Prediction and Control, pp. 141–215. ASCE, New York (1980)
10. Moore A.J.: Dynamic characteristics and wind induced response of a tall building. Ph.D. Thesis (2016)
11. Moravej, M., Chowdhury, A., Irwin, P., Zisis, J., Bitusmalak, G.: Dynamic Effects on Wind Loading on Photovotaic Systems (ICWE14) (2014)
12. Nagakami, L.: Probabilistic Dynamics of Wind Excitation on Glass Facade. Dissertation of Master Degree, Darmstadt (2003)
13. National Council Research of Italy. Guide for Assessment of Wind Actions and Effects on Structures, CNR-DT 207/2008, Rime 11th of June 2010
14. Solari, G.: Along wind response estimation: closed for solution. ASCE J. Struct. Divis. **108**(1), 225–244 (1982)
15. Steenbergen, R.D.J.M., Vrouwenvelder, A.C.W.M., Geurts, C.P.W.: The use of Eurocode EN1991-1-4 procedures 1 and 2 for building dynamics, a comparative study. J. Wind Eng. Ind. Aerodyn. **107–108**, 299–306 (2012)

Uncertainty Quantification in the Aeroelastic Response of a Twin-Box Deck Depending on the Inlet Conditions by CFD Studies

Antonio J. Álvarez[1], Giuseppe G. Lobriglio[2], Félix Nieto[1](✉), José Á. Jurado[1], and Santiago Hernández[1]

[1] University of A Coruña, A Coruña, Spain
{antonio.jose.alvarez,felix.nieto,jose.angel.jurado, santiago.hernandez}@udc.es
[2] University of Pavia, Pavia, Italy
giuseppegrazia.lobriglio01@universitadipavia.it

Abstract. The stochastic collocation method is applied to assess the impact of uncertainties in the angle of attack and the turbulence intensity in the force coefficients and flutter derivatives of a short-gap twin-box deck. A level-3 sparse grid has been adopted, and the deterministic realizations have been obtained by means of 2D URANS simulations. It has been found that the uncertainty in the aerodynamic and aeroelastic outputs is low to moderate, and in general, is higher for the flutter derivatives at larger reduced velocities. Furthermore, Sobol indices are reported, which provides a metric for the sensitivity of the outputs with the considered random variables. The adoption of nested quadrature points has enabled an efficient use of the computational resources, by requiring a reasonable number of CFD simulations for the deterministic realizations.

Keywords: Uncertainty quantification · stochastic collocation · twin-box deck

1 Introduction

The identification of aerodynamic and aeroelastic characteristics of interest in bridge engineering, such as force coefficients and flutter derivatives, is one of the most important tasks in long-span bridge design. However, it is important to note that the values obtained are sensitive to uncertainties in a number of parameters such as inaccuracies in the geometric details in sectional models, measurement errors or insufficient characterization of the incoming flow in wind tunnel testing, among others. Several authors have addressed this topic in bridge engineering applications, such as Bruno and Fransos (2011) who applied the Multi-element generalized Polynomial Chaos method to address the probabilistic evaluation of the integral parameters of a trapezoidal deck. Similarly, Mannini and Bartoli (2015) studied the uncertainty propagation in the critical flutter velocity considering the probability distribution of the experimental flutter derivatives of a bridge deck; or Fang et al. (2020), who identified the probabilistic critical flutter speed of the Pasco-Kennewick Bridge considering the uncertainty in the experimental flutter derivatives.

P. Schito and A. Zasso (Eds.): IN VENTO 2022, LNCE 461, pp. 79–89, 2024.
https://doi.org/10.1007/978-3-031-53059-3_8

Twin-box decks are considered as a feasible alternative for long-span bridges as the central slot has proved to have a positive impact in flutter performance. Nevertheless, it is important to assess how uncertainties in the incoming flow may affect the aerodynamic and aeroelastic parameters. This work addresses this issue by conducting CFD simulations and applying probabilistic propagation of the uncertainty in the chosen random variables with assumed probability density functions. Previous work by the authors (Lobriglio et al., 2022), addressed this problem considering as random variable only the angle of attack of the incoming flow. In this work, the aforementioned research is extended by including the turbulence intensity in the incoming flow as an additional random variable and obtaining the Sobol indices to assess the impact caused by the random variables in the force coefficients and flutter derivatives of the study case.

The paper is organized as follows: first the formulation adopted for the uncertainty quantification (UQ) assessment is introduced; and afterwards the addressed problem is briefly presented along with some comments regarding the applied methodology. In Sect. 4, the numerical model adopted for the deterministic evaluation of the responses of interest is described. In the following section, the results obtained for the stochastic mean and standard deviations of the force coefficients and the flutter derivatives are reported and discussed. Finally, in the last section, the main conclusions are summarized.

2 Formulation

2.1 Force Coefficients and Aeroelastic Forces

In order to obtain the forces coefficients and the flutter derivatives of the studied long-span bridge deck, numerical simulations were carried out using the $k-\omega$ SST turbulence model, for incompressible flow, implemented in the open source software OpenFOAM (Menter and Esch, 2001).

The non-dimensional time-averaged force components (drag (C_d), lift (C_l) and moment (C_m)) are calculated as:

$$C_D = \frac{D}{\frac{1}{2}\rho U^2 B}, C_L = \frac{L}{\frac{1}{2}\rho U^2 B}, C_M = \frac{M}{\frac{1}{2}\rho U^2 B^2} \tag{1}$$

where B stands for total width of the twin-box deck, and D, L and M are the drag and lift forces and moment per unit of span length respectively, which were calculated considering both the pressure and viscous forces acting on the deck.

Flutter is a self-excited phenomenon, where wind-induced forces and deck motions couple. Considering two degrees of freedom, heave h and pitch α, the aeroelastic lift force and moment per unit of span length are L_{ae} and M_{ae}, and might be written adopting the semi-empirical approach proposed in Scanlan and Tomko (1971) as:

$$L_{ae} = \frac{1}{2}\rho U^2 B \left[KH_1^* \frac{\dot{h}}{U} + KH_2^* \frac{\dot{\alpha}}{U} + K^2 H_3^* \alpha + K^2 H_4^* \frac{h}{B} \right]$$

$$M_{ae} = \frac{1}{2}\rho U^2 B \left[KA_1^* \frac{\dot{h}}{U} + KA_2^* \frac{\dot{\alpha}}{U} + K^2 A_3^* \alpha + K^2 A_4^* \frac{h}{B} \right]. \tag{2}$$

In the above equation, the $\dot{}$ symbol represents the time-derivative, and H_i^* and A_i^* ($i = 1, \ldots, 4$) are the so-called flutter derivatives, which are empirical parameters identified by means of wind tunnel tests or numerical simulations, as in this UQ application. The term $K = (B\omega)/U$ is the reduced frequency, being ω the circular frequency of oscillation.

2.2 Stochastic Collocation

For the Stochastic collocation method adopted in this work, we have followed the work in Mariotti et al. (2016) and Smith (2014), thus the interested reader is referred to the former references for a more detailed explanation.

The Clenshaw-Curtis points in the range $[-1, 1]$ are adopted as interpolation points, taking into account that these might be transformed afterwards to match the range of interest for the considered random variable $[a, b]$. For increasing levels in the quadrature approximation, the number of points (M_l) range from $M_l = 1$ for level 1, up to $M_l = 2^l + 1$ for any other higher level. These points define the samples in the design domain whose responses must be obtained. In our work, the corresponding CFD simulations were carried out for obtaining the outputs of interest for the considered set of samples. These values over the studied range, were interpolated by means of Lagrange polynomials, therefore the values of the simulations are preserved, generating a multidimensional interpolation surface. Finally, these multidimensional surfaces were used for the calculation of the mean and variance of the variables of interest with respect the parameters over which the uncertainty is going to be calculated. In this piece of research these are the angle of attack and the turbulence intensity of the incoming wind flow, expanding the scope of the work reported in Lobriglio et al. (2022).

In the results section, the stochastic mean values and standard deviations of the force coefficients and the flutter derivatives for the studied application case are provided aiming at assessing the impact of the uncertainty of the selected random variables in the outputs of interest in bridge engineering. Furthermore, the Sobol indices were obtained to shed additional light on the impact that each random variable has on the aerodynamic and aeroelastic responses.

3 Problem Description and Methodology

The uncertainty parameters chosen have been the turbulence intensity (I_t) and the angle of attack (AoA), as the main goal has been studying the impact of uncertainties in the inflow conditions upon force coefficients and flutter derivatives. These random variables are susceptible of playing an important role in the uncertainties associated with wind tunnel testing campaigns. Previous research by others has shown the limited impact of the inflow turbulence length scale (Mariotti et al., 2016), when compared with the contributions coming from turbulence intensity and angle of attack. On the other hand, Fransos and Bruno (2011) reported important sensitivity in the integral parameters for a non-streamlined trapezoidal-shaped deck with curved lower corners. Taking into account the streamlined features and the sharp edges in this case study, along with the need to consider a small number of random variables to avoid a large number of collocation points making computationally unfeasible this study, the turbulence intensity and the angle of attack have been selected for the UQ study. The ranges of variation considered are $[-1°, +1°]$ for the flow angle of incidence and $[0.001, 0.03]$ for the turbulence intensity, which correspond to realistic ranges in wind tunnel testing, assuming a uniform probability density function for both. In the 2D URANS simulations, the value of the turbulence intensity level has been imposed at the inlet as a Dirichlet boundary condition.

The stochastic collocation method (the interested reader may refer to Xiu (2010) for a general overview) has been applied adopting the Smolyak sparse grid extension of level-3 Clenshaw-Curtis quadrature points, following the procedure described in Mariotti et al. (2016). The Clenshaw-Curtis nodes assure the "nestedness" of the one-dimensional nodes for increasing levels of the sparse grid. The procedure is applied increasing the level of the Clenshaw-Curtis nodes by one, at each stage, seeking to avoid the risk of oversampling. For the level-3 adopted herein, this defines 13 pairs of values for the angle of attack and turbulence intensity within the considered ranges in the random variables requiring a deterministic realization. In Fig. 1, schemes of the two-dimensional Smolyak sparse grids of levels 2 and 3 are provided.

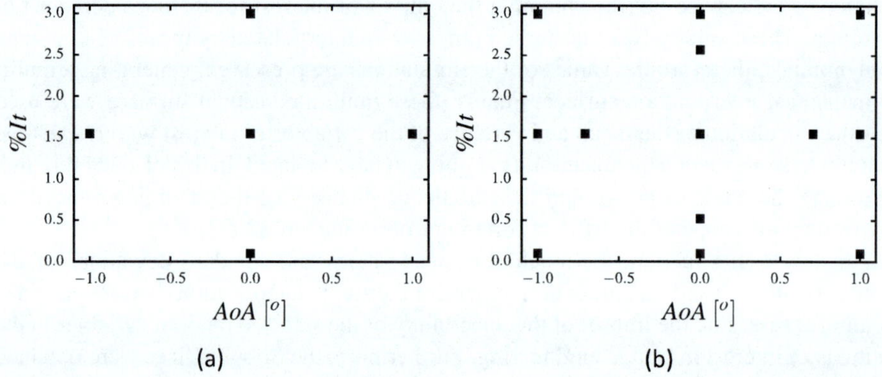

Fig. 1. Smolyak sparse grids identifying the deterministic values for the angle of attack and turbulence intensity in the 2D URANS simulations, (a) Level 2 quadrature points; (b) level 3 quadrature points.

The suitability of the chosen approach relies on the low dimensionality of the problem being considered. Alternative approaches based on uniformly distributed interpolation points, or non-nested quadrature points, may result in high computational demands, as each collocation point requires one CFD simulation for the evaluation of the force coefficients and several for the identification of the flutter derivatives at different reduced velocities. The use of non-uniformly distributed points decreased the interpolation error of the polynomial approximation in the vicinity of the boundaries. Due to the number of points used in the level 3 approximation, the maximum order of the Lagrange polynomials is four. The deterministic realizations for the quantities of interest (force coefficients and flutter derivatives) have been obtained by means of 2D URANS simulations of the so-called g22 twin-box cross-section in Nieto et al. (2020), whose verification and validation studies were reported in that last reference. An image of the deck and surrounding mesh is presented in Fig. 2. In Table 1 a summary of the deck and mesh characteristics is provided along with the y$^+$ values at $Re_C = 1.14 \times 10^5$. A brief description of the numerical model is provided next, in Sect. 4.

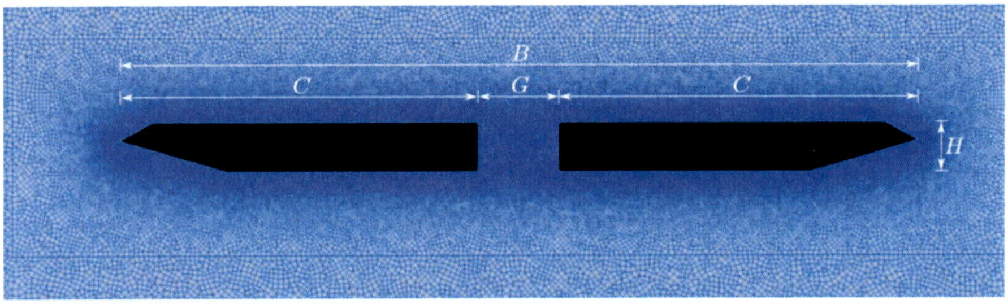

Fig. 2. Detail of the geometry under study and the mesh surrounding it.

Table 1. Main characteristics of the deck and mesh. (δ_1 is the height of the first element of the boundary layer, δ_{bl} is the total height of the boundary layer, UB refers to upwind box and DB for downwind box).

B/C	*H/C*	*G/C*	δ_1/C	δ_{bl}/C	y^+_{mean} UB	y^+_{mean} DB
2.225	0.134	0.224	$3.328 \cdot 10^{-4}$	$7.327 \cdot 10^{-3}$	2.1	1.8

4 2D URANS Modeling

2D URANS simulations are employed to obtain the results for the deterministic realizations of the force coefficients and flutter derivatives that are subsequently used in the application of the Stochastic Collocation method. The selected application case is a short-gap twin-box deck geometry studied previously by the authors in the frame of shape optimization studies (Nieto et al. 2020) that was named as g22. Interested readers may refer to the last reference for information about the verification and validation studies for that case.

The simulations reported herein have been conducted at $Re_C = 1.14 \times 10^5$, keeping the same value of $0.1C$ (being C the width of one single box) for the turbulence length scale. The corners of the twin-box deck are modeled as sharp. The total width of the deck is $B = 2.225C$ and the depth is $H = 0.134C$.

The main numerical settings for the simulations are the following: the Euler scheme was adopted for the time schemes, while for the Laplacian terms the Gauss discretization was adopted, using linear interpolations along with limited interpolations for the surface normal scheme. For the divergence terms, Gaussian discretization was selected, considering linear, linear upwind or upwind interpolation schemes depending on the particular variables being considered. The gradient terms have been discretized using a second order cell limited scheme, and the interpolation from cell centers to face centers was conducted by means of a central differencing scheme. In all the simulations, the maximum allowed Courant number has been 1. In Fig. 3, a snapshot of the instantaneous vorticity fields for the studied geometry under pitch forced rotation are provided. The vortices detached from the upwind box might be identified along with their impingement on the downwind box and the rolling vortices traveling along its upper and lower surfaces. This shows the inherent complexity in the aerodynamics of twin-box decks and the importance of uncertainty propagation studies for this deck arrangement typology. Detailed information on twin-box bridge deck aerodynamics may be found also in Alvarez et al. (2021).

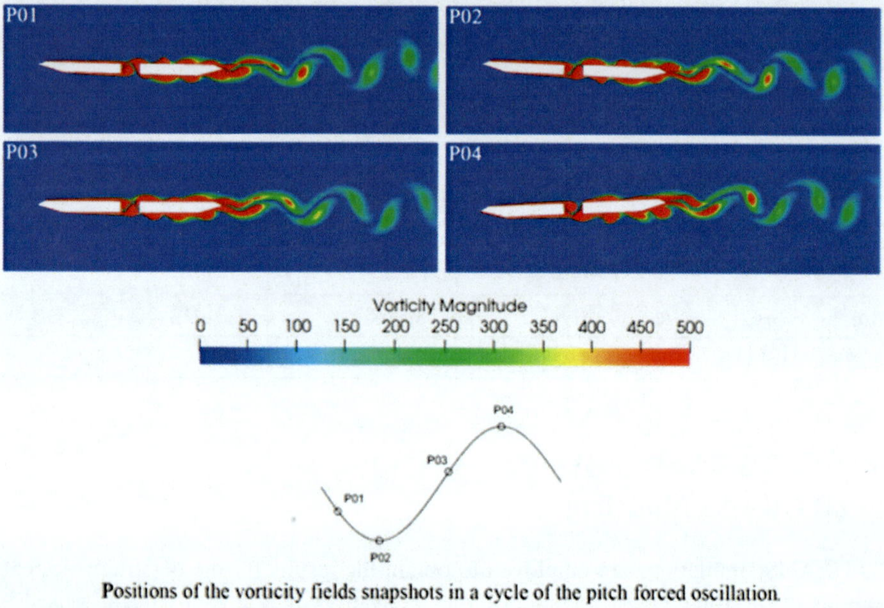

Fig. 3. Snapshot of the vorticity magnitude field for the g22 cross-section under forced pitch vibration.

The evaluation of the flutter derivatives has been done by means of forced oscillation simulations, independently for the heave and pitch degrees of freedom, following the procedure described in Nieto et al. (2015). The frequency of the deck oscillation is modified in order to cover the targeted range of reduced velocities (0, 20). The amplitude of oscillation for the pitch degree of freedom has been 3°, while for the heave degree of freedom the amplitude of the displacement has been $0.141C$, which are similar values to the ones adopted in standard wind tunnel testing. In Fig. 4, the time histories of the drag, lift and moment coefficients along with the forced pitch oscillations are presented for a reduced velocity U/fB=10. In the chart, the high frequency fluctuations in the force coefficients associated with vortex shedding can be identified along with the lower frequency oscillations associated with pitch forcing.

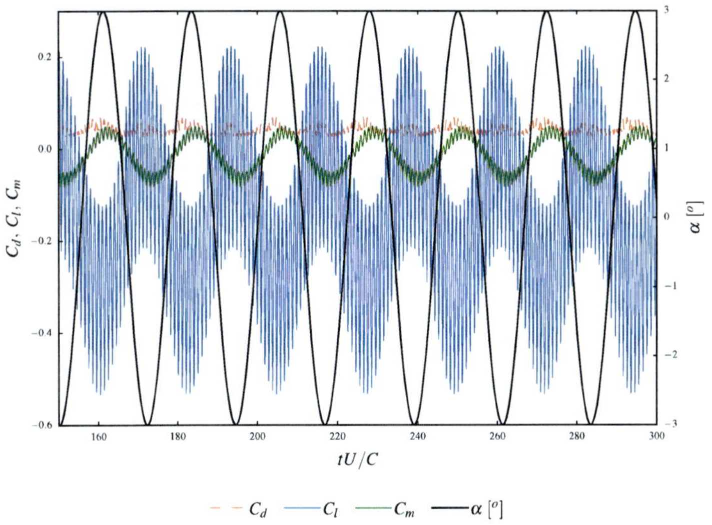

Fig. 4. Time histories of the drag, lift and moment coefficients of the g22 twin-box geometry under forced oscillations in the pitch degree of freedom.

In the next section, the stochastic mean and standard deviation of the force coefficients and flutter derivatives considering as random variables the angle of wind incidence and the turbulence intensity are reported.

5 Results

5.1 Force Coefficients

Based on the deterministic results for the integral parameters of the 13 cases considered, the stochastic mean and standard deviation values have been obtained for level-3 Clenshaw-Curtis quadrature points, and they are reported in the Table 2 below. All the obtained values yielded a slightly lower mean value and larger standard deviation, than in Lobriglio et al. (2022), where only the AoA was considered for the stochastic calculations. This effect resulted in a slight improvement when comparing with experimental values.

Table 2. Stochastic mean and standard deviation for integral parameters.

	C_D	C_L	C_m	$std(C_D)$	$std(C_L)$	$std(C_m)$
Stochastic mean	0.0314	−0.1415	0.0139	0.0028	0.0773	0.0055
Stochastic Std. dev.	0.0078	0.0333	0.0120	0.0032	0.0847	0.0059

In Fig. 5, the stochastic mean ± the stochastic standard deviation obtained for the force coefficients are reported along with the experimental wind tunnel data.

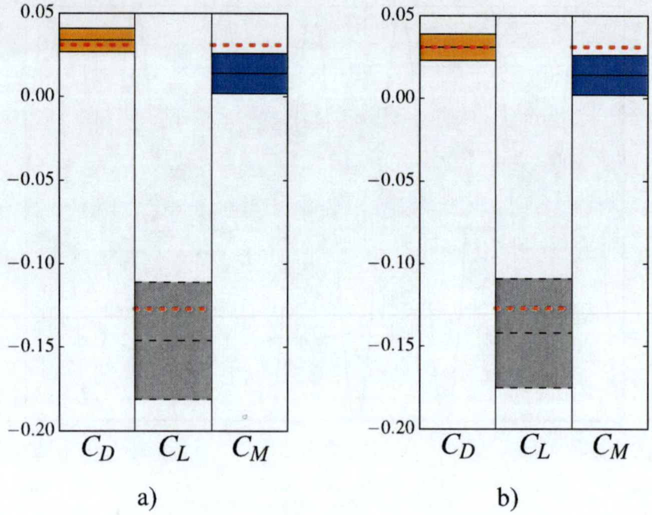

Fig. 5. Stochastic mean ± stochastic standard deviation for the force coefficients including experimental values as dotted red lines. (a) Level 2 quadrature points; (b) level 3 quadrature points.

In Table 3, the results obtained for the variance-based sensitivity analysis conducted for the force coefficients are reported. The Sobol indices show the share of the variance of the output due to each considered random variable and the interaction between them. In this case, the variance in the force coefficients is mainly due to the angle of attack, which explains the very limited difference in the stochastic mean values for the force coefficients when the turbulence intensity was considered as a random variable along with the angle of attack. Furthermore, for the standard deviations of the force coefficients, both variables contribute to the variance similarly.

Table 3. Sobol indexes for the mean and standard deviation of the integral parameters.

	C_D	C_L	C_m	$std(C_D)$	$std(C_L)$	$std(C_m)$
AoA	0.518	0.946	0.999	0.441	0.472	0.458
I_t	0.433	0.048	0.000	0.509	0.477	0.494
AoA & I_t	0.049	0.006	0.000	0.050	0.051	0.048

5.2 Flutter Derivatives

The stochastic values for the flutter derivatives of the g22 short-gap twin-box deck have been also obtained, based on the level-3 sparse grid CFD forced oscillation simulations in heave and pitch degrees of freedom. In Fig. 6, for selected flutter derivatives, the stochastic mean ± stochastic standard deviation are provided along with the wind tunnel values for comparison. The agreement between the numerical flutter derivatives and the wind tunnel test is reasonable. Furthermore, the uncertainty propagation in the flutter derivatives is low to moderate, depending on the particular flutter derivative considered, although in general terms, the stochastic standard deviation increases with the reduced velocity. As for the case of the integral parameters, almost no difference in the mean value reported in Lobriglio et al. (2022) has been obtained after introducing the turbulence intensity as an additional random variable in the uncertainty study, however the uncertainty level has slightly increased for all flutter derivatives when the two random variables are considered simultaneously.

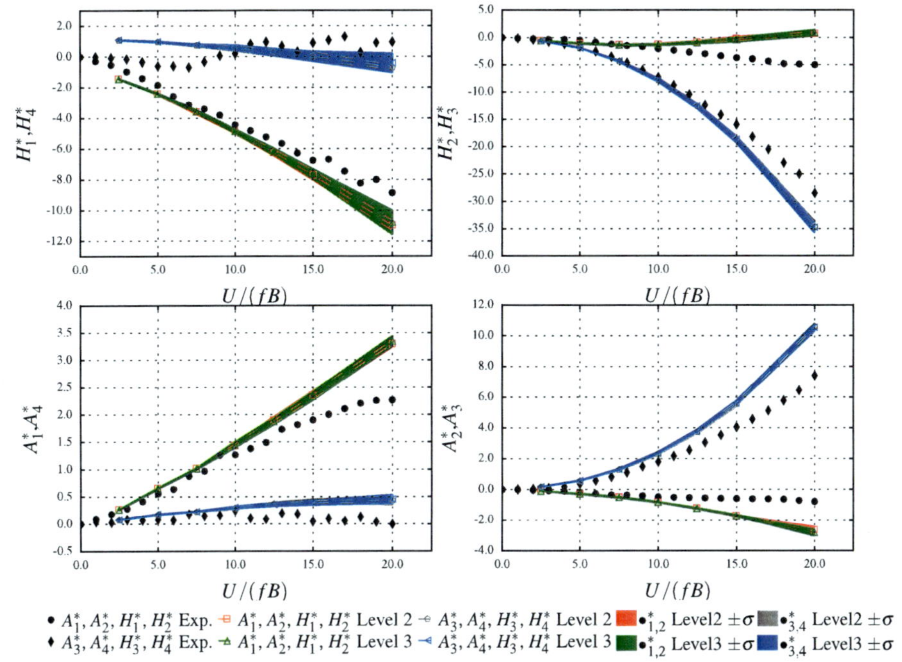

Fig. 6. Stochastic mean ± stochastic standard deviation and experimental values for the flutter derivatives of the g22 geometry.

In Tables 4 and 5, the Sobol indices are provided for the flutter derivatives H_i^* and A_i^* $(i = 1, \ldots, 4)$, for reduced velocities $U/fB = 10$ and 20. For the H_i^* $(i = 1, \ldots, 4)$, is the angle of attack the random variable mainly responsible for the variance in the flutter derivatives, as the turbulence intensity plays a minor role, with the exception of the H_2^* at $U/fB = 10$. As in the case of the force coefficients, this explains the minor impact of considering the turbulence intensity as a random variable in the stochastic mean values of the flutter derivatives. Interestingly, for the A_i^* $(i = 1, \ldots, 4)$ flutter derivatives set, both random variables play a non-negligible role for the considered reduced velocities and the Sobol index value might change significantly for certain values of the reduced velocity.

This shows that the flutter derivatives related with the aeroelastic moment are more sensitive to uncertainty in the turbulence intensity than the ones related to aeroelastic lift.

Table 4. Sobol indices for the H_i^* flutter derivatives at U/fB=10 and 20.

	H_1^*		H_2^*		H_3^*		H_4^*	
U/fB	10	20	10	20	10	20	10	20
AoA	0.858	0.921	0.156	0.919	0.987	0.997	0.796	0.859
I_t	0.138	0.045	0.779	0.079	0.011	0.000	0.110	0.046
AoA & I_t	0.004	0.034	0.065	0.002	0.002	0.003	0.094	0.095

Table 5. Sobol indices for the A_i^* flutter derivatives at U/fB=10 and 20.

	A_1^*		A_2^*		A_3^*		A_4^*	
U/fB	10	20	10	20	10	20	10	20
AoA	0.632	0.678	0.670	0.374	0.221	0.976	0.245	0.377
I_t	0.323	0.268	0.274	0.586	0.739	0.020	0.665	0.594
AoA & I_t	0.045	0.054	0.056	0.040	0.04	0.004	0.090	0.029

6 Conclusions

This work reports the uncertainty propagation of the flow angle of attack and turbulence intensity considered as random variables in the integral parameters and flutter derivatives of a twin-box deck.

The stochastic collocation method has enabled the assessment of the sensitivity of the quantities of interest to the uncertainty in the considered flow parameters. Furthermore, the adoption of nested quadrature points has enable an efficient use of the computational resources by limiting the number of required 2D URANS simulations.

It has been found that for the flutter derivatives, this sensitivity is not negligible, depending on the reduced velocity and the particular derivative considered. Furthermore, the adoption of nested quadrature points has enable an efficient use of the computational resources by limiting the number of required 2D URANS simulations. The computation of the Sobol indices has permitted the assessment of the contribution of the considered random variables to the variance of the outputs of interest, finding that is the angle of attack the dominant random variable within the considered set. However, for certain flutter derivatives at some reduced velocities the turbulence intensity also plays clear role in the uncertainty of the aeroelastic output.

Acknowledgements. This research has been funded by the Spanish Ministry for Science and Innovation (project PID2019-110786GB-I00), and the Xunta de Galicia (Galician Regional Government), including FEDER funding, (ED431C 2021/33). The Erasmus+ Higher Education Traineeship program has funded Giuseppe G. Lobriglio.

References

Alvarez, A.J., Nieto, F., Kwok, K.C.S., Hernandez, S.: A computational study on the Aerodynamics of a twin-box bridge with a focus on the spanwise features. J. Wind Eng. Ind. Aerodyn. **209**, 104465 (2021)

Bruno, L., Fransos, D.: Probabilistic evaluation of the aerodynamic properties of a bridge deck. J. Wind Eng. Ind. Aerodyn. **99**, 718–728 (2011)

Fang, G., Cao, J., Yang, Y., Zhao, L., Cao, S., Ge, Y.: Experimental uncertainty quantification of flutter derivatives for a PK section girder and its application on probabilistic flutter analysis. J. Brige Eng. ASCE **25**(7), 04020034 (2020)

Lobriglio, G.G., Alvarez, A.J., Nieto, F., Herandez, S., Jurado, J.A.: Uncertainty quantification of aerodynamic and aeroelastic responses of a short-gap twin-box deck depending on the Wind angle of attack. Int. J. Comput. Methods Exp. Measur. **10**(1), 74–84 (2022)

Mannini, C., Bartoli, G.: Aerodynamic uncertainty propagation in bridge flutter analysis. Struct. Saf. **52**, 29–39 (2015)

Mariotti, A., Salvetti, M.V., Shoeibi Omrani, P., Witteveen, J.A.S.: Stochastic analysis of the impact of free stream conditions on the aerodynamics of a rectangular 5:1 cylinder. Comput. Fluids **136**, 170–192 (2016)

Menter, F., Esch, T.: Elements of industrial heat transfer prediction. In: Proceedings of the 16th Brazilian Congress of Mechanical Engineering, vol. 20, pp. 117–127 (2001)

Nieto, F., et al.: Aerodynamic and aeroelastic responses of short gap twin-box decks: box geometry and gap distance dependent surrogate based design. J. Wind Eng. Ind. Aerodyn. **201**, 104147 (2020)

Nieto, F., Owen, J.S., Hargreaves, D.M., Hernandez, S.: Bridge deck flutter derivatives: Efficient numerical evaluation exploiting their interdepence. J. Wind Eng. Ind. Aerodyn. **136**, 138–150 (2015)

Scanlan, R.H., Tomko, J.J.: Airfoil and bridge deck flutter derivatives. J .Eng. Mech. Div. **97**(6), 1717–1737 (1971)

Smith, R.C.: Uncertainty Quantification. Theory, Implementation and Applications. SIAM (2014)

Wilcox, D.C.: Turbulence Modelling for CFD, DCW Industries: La Cañada (2006)

Xiu, D.: Numerical Methods for Stochastic Computations: A Spectral Method Approach. Princeton University Press, New Jersey (2010)

Check for updates

Calibration of Wind Action Combination Factors from Experimental Data

Vincenzo Picozzi[1](✉), Francesco Ricciardelli[1], and John D. Sørensen[2]

[1] University of Campania "Luigi Vanvitelli", Aversa, Italy
vincenzo.picozzi@unicampania.it
[2] Aalborg University, Aalborg, Denmark

Abstract. The semi-probabilistic method is currently used worldwide for the assessment of structural reliability. In European Countries, both partial factors and combination factors recommended in EN1990 are adopted. However, in the case of climatic actions, this choice can lead to overestimation or underestimation of design values of the actions, and calibration at the National level would be appropriate. Based on wind climate data available in the literature, an attempt is made to calibrate combination factors for wind action combined with imposed loads for Italy. Two methods are considered: the classical simplified method based on the Ferry-Borges and Castanheta's approach and Turkstra's rule, and a method based on the numerical simulation of the combined load effects. First, the variability of combination factors over Italy is shown. Then, the influence of uncertainties and of the imposed load model is evaluated. Finally, the differences between the two methods are discussed and a comparison with current combination factors suggested by Eurocode 0 is made. Overall, it is shown that: local combination factors change over Italy without any correlation with the geographic coordinates; the simplified method overestimates combination factors; current values suggested by Eurocode 0 overestimate the combined load effects.

Keywords: Wind Action · Design Criteria · Combination Factors · Eurocodes

1 Introduction

In modelling wind action, the Alan G. Davenport Wind Loading Chain is commonly used, describing how wind is translated into wind action and effects on structures. Once the wind action or its effect are evaluated, then design criteria are required. Following the Eurocode format, the Italian building Code [1] is based on the semi-probabilistic method. Accordingly, evaluation of the load combinations and of their effects on structures are based on a set of load partial factors γ and combination factors ψ. The first account for the uncertainty associated with the evaluation of the characteristic value of the actions and of their effects, the latter account for the non-simultaneous occurrence of extreme events represented by the characteristic value of the action, as well as the beneficial contribution of permanent actions in maximizing load effects. In Italy, Limit State analyses have become mandatory in 2008, and most of the load and combination factors have been

P. Schito and A. Zasso (Eds.): IN VENTO 2022, LNCE 461, pp. 90–101, 2024.
https://doi.org/10.1007/978-3-031-53059-3_9

adapted from Eurocode 0 [2]. On the other hand, it is known that their calibration requires the use of reliability methods involving appropriate stochastic models for both loads and resistances [3]. In particular, in the case of climatic actions, e.g. wind, a regional climate analysis would be required for calibration [13]. When two or more variable loads have to be combined, the stochastic models are not enough to calibrate load factors. In fact, the sum of individual maxima can lead to extremely conservative estimates. A complete description of variable loads in terms of stochastic processes should be adopted, providing information on the variation in time of the load. However, for design purposes it is complicated to model and then combine the stochastic load processes, thus evaluating their maximum effects. For this reason, through the years simplifications have been introduced in modelling stochastic processes [4, 5] leading to the combinations rules to be used within Codes [6, 7].

In this paper, combination factors for the wind action are calibrated based on available Italian climatic data, with the purpose of validating the values currently in use. First, an overview on methods for the calibration of load combination factors is presented. Then, historical wind speed measurements are used to calibrate the probabilistic models for the Italian wind climate. Finally, calibrations are presented for the combination factor for wind action to be combined with imposed loads, and the results are discussed.

2 Combination of Time-Variant Loads

2.1 Limit State Function

In the assessment of structural reliability, the probability of failure is defined as the probability of a structure (or a structural member) to be in a failure state, i.e. to reach a Limit State. A Limit State Function (LSF) is defined as:

$$g(R, E) = R - E \tag{1}$$

where R and E are the resistance and the effect of loads acting on the structure, respectively. It follows that the probability of failure is defined by:

$$P_f = P(R - E \leq 0) \tag{2}$$

In the design of structures, the load effect E derives from a combination of the dead load, G, and variable load, Q. Introducing uncertainties in modelling of both the resistance and the load effect, then Eq. (1) can be written:

$$g(\mathbf{x}, z) = zX_R R - X_E\{(1 - a_Q)G + a_Q Q\} \tag{3}$$

where z is a design parameter, X_R and X_E are model uncertainties for resistance and loads effect, respectively, and a_Q is the relative weight between dead and variable loads. The combination $G = a_G G_s + (1 - a_G)G_p$ between the self weight G_s and the permanent load G_p is commonly used to express the dead load, a_G being the relative weight between them. Associated with Eq. (3), the boundary between the safe and the unsafe domains is defined by the failure surface $g(\boldsymbol{x}, z) = 0$. Within the Partial Safety Factors (PSFs) format, a design equation is defined fulfilling the condition $g(\boldsymbol{x}_d, z_d) = 0$ at the design point $\boldsymbol{x}_d$. Corresponding to Eq. (3), the following design equation can be formulated:

$$pR_d - \{(1 - a_Q)[a_G G_{sd} + (1 - a_G) G_{pd}] + a_Q Q_d\} = 0 \tag{4}$$

where $R_d = R_k/\gamma_R$ is the design value for resistance and $G_{sd} = \gamma_G G_{sk}$, $G_{pd} = \gamma_G G_{pk}$, and $Q_d = \gamma_q Q_k$ are the design values for self-weight, permanent load, and variable load effects, respectively, γ_i being the partial factors.

When Eq. (3) is associated to an Ultimate Limit State, the term Q represents the random variable describing the maximum value of the load effect $Q(t)$ in the reference period T. In the case of combination of two variable loads $Q_1(t)$ and $Q_2(t)$, $Q(t)$ is replaced by the combined load effect $S(t)$ expressed by:

$$S(t) = S_1(t) + S_2(t) = a_s Q_1(t) + (1 - a_s) Q_2(t) \tag{5}$$

where a_s is the relative weight between the two load effects and a linear relationship between loads and load effects is assumed. The maximum value of $S(t)$ in the reference period T is given by:

$$\max_T S(t) = \max_T [S_1(t) + S_2(t)] \tag{6}$$

that it is likely to occur at a time different from the times t_1 and t_2 at which the maximum values of $S_1(t)$ and $S_2(t)$ occur. To obtain the distribution of the variable S of the maximum values of $S(t)$ in the reference period T, time series of the two processes $S_1(t)$ and $S_2(t)$ can be generated, and then combined.

2.2 Simplified Approach

Simulating and combining two variable loads is not an easy task, since the physics of the two processes are to be modelled. As a simplification, approximate approaches can be used such as that proposed by Ferry-Borges and Castanheta [5] and by Turkstra [6]. Turkstra's rule provides a good approximation of Eq. (6). Accordingly, the maximum value of the combined load is given by:

$$\max_T S(t) \approx \max \left\{ \begin{array}{c} \max_T S_1(t) + S_2^* \\ S_1^* + \max_T S_2(t) \end{array} \right\} \tag{7}$$

where the maximum value of the effect $S_1(t)$ combines with a random-point-in-time value S_2^* of the second load, and the maximum value of the effect $S_2(t)$ combines with a random-point-in-time value S_1^* of the first load. The simplification introduced by Turkstra's rule lends itself to be used within the PSF format, within which the design value S_d of the combined load is given by:

$$S_d = \max \left\{ \begin{array}{c} S_{1d} + S_{2c} \\ S_{1c} + S_{2d} \end{array} \right\} \tag{8}$$

where $S_{i,d} = \gamma_{Q_i} S_{ik}$ is the design value of S_i defined as the product between its characteristic value S_{ik} and the partial factor γ_{Q_i}, and $S_{ic} = \Psi_{0i} S_{id}$ is the combination value of S_i, Ψ_{0i} being the load combination factor.

A further approximation can be derived by the Ferry-Borges and Castanheta (FBC) model. Accordingly, the load effects $S_1(t)$ and $S_2(t)$ are modelled as rectangular processes whose intensities change after equal elementary intervals τ_1 and τ_2, respectively, with $\tau_1 > \tau_2$. The elementary interval τ_i must be chosen such that in the reference period T the $n_i = T/\tau_i$ repetitions are independent and identically distributed, with distribution function $F_{Q_i,\tau_i}(q_i)$. So doing, the combination effect S_{1c} of the load Q_1 is equal to the arbitrary point-in-time load effect S_1^* with distribution function $F_{S_{1,\tau_1}}(s) = \left[F_{S_1,T}(s)\right]^{1/n_1}$, $F_{S_1,T}(s)$ being the distribution function for extremes of S_1 in the reference period T. On the other hand, the combination effect S_{2c} of the load Q_2 is equal to the maximum value occurring in the interval τ_1. Consequently, it must be derived from the distribution function $F_{S_2,\tau_1}(s) = \left[F_{S_2,T}(s)\right]^{1/n_1}$, $F_{S_2,T}(s_2)$ being the distribution function for extremes of S_2 in the reference time T.

As a consequence of the adoption of the FBC model, the following equations can be used to evaluate the load combination factors:

$$\psi_{0i} = \frac{F^{-1}_{S_i,\tau_1}(P_c)}{F^{-1}_{S_i,T}(P_d)} = \frac{F^{-1}_{S_i,T}\left(P_c^{n_1}\right)}{F^{-1}_{S_i,T}(P_d)} \tag{9}$$

where P_c and P_d are the probabilities associated with the design value and the combination value, respectively. To calibrate combination factors by using Eq. (9), values for both P_c and P_d must be assumed. To this aim, the Design Value Format Method (DVM) is commonly used within the FBC model, and this will be discussed in the next section.

2.3 Design Value Method

Load factors, including partial factors, γ, and combination factors, ψ, are meant to lead to a desired safety level for the specified limit state. In the case of design for structural safety, Eurocode 0 [2] prescribes a target reliability index $\beta_t = 3.8$ for 50-yrs return period, corresponding to a probability of failure $P_f = 7.23 \cdot 10^{-5}$. However, evaluation of the reliability index requires a full probabilistic approach. As an alternative, according to the Design Value Format Method, the design values R_d for resistance and E_d for load effect can be evaluated as:

$$R_d = F_R^{-1}[\Phi(-\alpha_R \beta_t)]$$

$$E_d = F_E^{-1}[\Phi(\alpha_E \beta_t)] \tag{10}$$

where α_R and α_E are the sensitivity factors, and $\Phi(\cdot)$ is the Cumulative Density Function (CDF) of the standard normal distribution. Based on the work of König and Hosser [9], the values $\alpha_R = 0.8$ and $\alpha_E = 0.7$ are suggested in Eurocode 0 for dominant variables. Instead, when two variable loads act simultaneously, it is suggested that for the non-dominant load effect the value $\alpha_E \beta_t$ is reduced by a factor equal to 0.4 [9]. This leads to a value $P_c = \Phi(0.4\alpha_E\beta_t)$. Notice that when Eq. (9) is used to evaluate ψ_{02}, the factor 0.4 applies to $\beta_\tau = \Phi^{-1}\left[\Phi(\alpha_E\beta_t)^{1/n_1}\right]$ and the probability $P_c = \Phi(0.4\beta_\tau)$ is to be used.

3 Calibration of Load Combination Factors

3.1 Load Models

The probabilistic models used for calibration of combination factors are mainly based on the Probabilistic Model Code [10]. Variable loads are represented as the product of a time-variant load $Q(t)$ and a time-invariant component θ_Q. The distribution function of the maximum values in the reference period T is given by the distribution of the product $Q = \theta_Q Q_T$, where $Q_T = \max Q(t)$ in the reference period T.

The time-variant uncertainty of imposed load is given by the sum of a sustained component and an intermittent component. The sustained load accounts for the weight of furniture and normal people occupancy; its changes usually corresponds either to changes in use or of users in building. Instead, the intermittent load accounts for exceptional events like crowding or stacking of furniture during refurbishment. In simulating imposed load, the sustained part is modelled as a Poisson process whose intensity slowly changes with a mean rate $\lambda_q = 1/\tau_{1,q}$. The intermittent part is modelled as a Poisson process occurring with a mean rate λ_p and a relative-short duration $\tau_{1,p}$. Four categories of use are considered in this study: residence, office, school classroom, and industrial. The additional parameters to model the imposed load are shown in Table 1 in which $V\left[Q_1\right]$ represents the coefficient of variation of the 50-yrs maximum values of Q_1 modelled by a Gumbel distribution. The model uncertainty θ_{Q1} associated with the imposed load are modelled with a log-normal distribution with coefficient of variation equal to 0.10.

In the case of wind action, the time-variant component is representative of the extreme wind climate, i.e. of maximum values of the wind speed or velocity pressure in the reference period T. It is site-related and regional climatic data are needed for its modelling. For Italy, the probabilistic model is derived from the work of Ballio et al. [11] in which the parameters of the Gumbel distribution are given for 40 Italian stations. It is shown that the coefficient of variation $V\left[Q_{T,2}\right]$ of the velocity pressure for a return period of 50 years is in the range of 0.16 to 0.32, with a mean value of 0.24, which is larger than those reported for other European countries, e.g. 0.14 for Germany. In simulating the wind action, the assumption is made that the daily maximum values of wind speed are independent and identically distributed with the distribution function:

$$F_{V,\tau_2}(v) = [F_V(v)]^{\tau_2/T} \tag{11}$$

where $\tau_2 = 1$day. The time-invariant part θ_{Q_2} of wind load is modelled with a log-normal distribution with coefficient of variation equal to 0.26.

3.2 Calibration Methods

Combination factors account for the non-simultaneous occurrence of extreme events related to two different variable loads. Thus, it is expected that they are mainly influenced by the duration of the events, i.e. by the time-variant component of the load.

Table 1. Parameters for imposed load [10].

	Sustained Load			Intermittent Load				$V[Q_1]$
	μ_q $[kN/m^2]$	σ_q $[kN/m^2]$	$\tau_{1,q}$ [yrs]	μ_p $[kN/m^2]$	σ_p $[kN/m^2]$	$1/\lambda_p$ [yrs]	$\tau_{1,p}$ [days]	
Residence	0.3	0.38	7	0.3	0.46	1.0	3	0.33
Office	0.5	0.48	5	0.2	0.25	0.3	3	0.26
School Class	0.6	0.58	10	0.5	1.98	0.3	5	0.47
Industrial	3.0	5.99	10	–	–	–	–	0.98

However, the random nature of time-invariant component can also affect calibration of the combination factors. In combining the wind load with the imposed load, the role of the duration of events can be highlighted by including or excluding the Intermittent Imposed Load (IIL). On the other hand, the role of the Time-Invariant Uncertainty (TIU) can be quantified by including it or excluding it from the analyses. Thus, combination of the above scenarios make it possible to consider four cases in the calibration process. According to Sect. 2, three methods are available for calibration of combination factors: (1) application of the Design Value Method within the Ferry-Borges and Castanheta approach, hereafter simply referred to as FBC method; (2) application of the Design Value Method using the distribution function of maximum vales of the combined load $S(t)$ obtained by simulation, hereafter simply referred to as simulation method; (3) use of a full probabilistic method with a representative Limit State Equation. In this paper, the first two methods are applied.

Within FBC method, Eq. (9) is used for a first, quick calibration of the combination factors, consistent with the recommendations in Eurocode 0. The worst-case scenario leading to the maximum value of the combination factor is considered. This is represented by modelling only the sustained imposed load for which $\tau_1 = \tau_{1,q} > \tau_2$. Within the simulation method, to model the distribution function of 50-yrs maximum values of the combined load $S(t)$, simulation of time histories of the loads is carried out. The following steps are required within Monte Carlo method:

1. Simulation of 50 years of the intermittent and sustained parts of the imposed load, and their sum $Q_1(t)$ (see Fig. 1a). The time-invariant component is included as constant value in 50 years.
2. Simulation of 50 years of the wind speed and evaluation of the wind action $Q_2(t)$ (see Fig. 1b). The time-invariant component is included as constant value in 50 years.
3. Evaluation of 50 years of the combined effect $S(t) = S_1(t) + S_2(t)$.
4. Evaluation of the maximum value of the combined effect in the simulated 50 years.
5. Repetition of Steps 1 to 4 for N simulations.

The number of simulations N shall be chosen based on the minimum probability value to reach, i.e. on the expected maximum reliability index β. In this work, $N = 5 \cdot 10^6$ is used, allowing the evaluation of probability values of the order $P = 8 \cdot 10^{-5}$ ($\beta = 3.8$) with a coefficient of variation of 5% [12].

Once the Empirical Cumulative Distribution Function (ECDF) $F_S(S)$ of the combined load effect is derived, then the DVM can be applied. First, the design values S_{1d} and S_{2d} are obtained from the ECDF of the maximum load effects S_1 and S_2 in the 50-yrs reference period. Then, the value S_d of the combined load effect must correspond to the probability P_d, the same as that chosen for the design values S_{1d} and S_{2d}. Thus, the combination factors ψ_{01} and ψ_{02} shall meet the requirement:

$$S_d = S_{1d} + \psi_{02} \cdot S_{2d} = \psi_{01} \cdot S_{1d} + S_{2d} \tag{12}$$

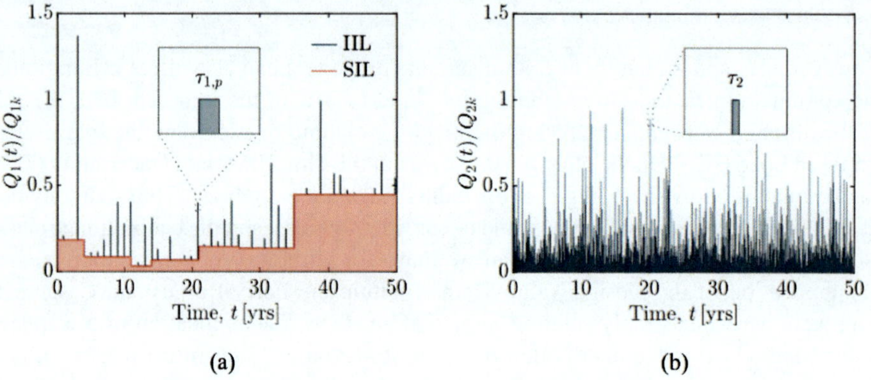

Fig. 1. Example of 50-yrs simulation of process for (a) imposed load and (b) wind action.

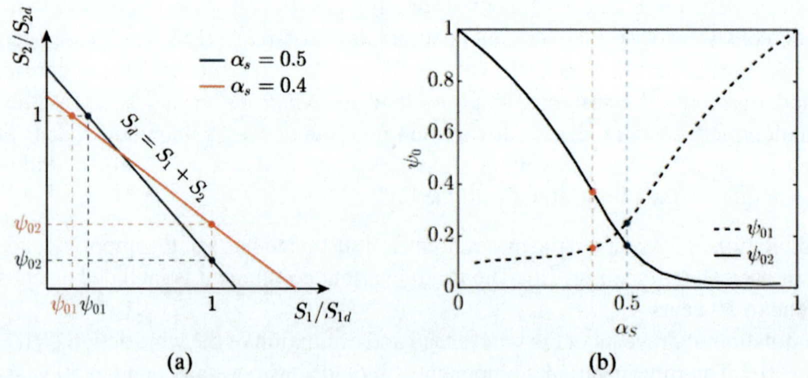

Fig. 2. Example of (a) graphical solution of the Design Value Method and (b) variation of combination factors as function of the weighting factor a_s.

However, when a linear relationship between loads and their effects is considered, then S_d is a function of the relative weighting factor a_S and the same applies to the ψ_0 factors. As an example, in Fig. 2a the meaning of Eq. (12) is illustrated from a geometrical perspective in the space of load effects S_1 and S_2 normalized with respect to the design values S_{1d} and S_{2d}. The straight lines represent pairs of values (S_1, S_2) which sum gives S_d. Between these, only two value pairs satisfy Eq. (12), which are plotted as dots on the straight lines. Moreover, when the weighting factor a_s changes, then the combination factors ψ_{01} and ψ_{02} also change. This relationship is shown in Fig. 2b where the combination factor ψ_0 is plotted against a_s.

The combination factors must apply to all the design situations represented by the a_s ratios. To solve the issue, an optimization problem is formulated. First, given a set of combination factors (ψ_{01}, ψ_{02}), based on Eq. (8) the design value of the combined load effect is defined as:

$$S_d(\psi_{01}, \psi_{02}) = \max\left\{ \begin{matrix} S_{1d} + \psi_{02} S_{2d} \\ \psi_{01} S_{1d} + S_{2d} \end{matrix} \right\} \tag{13}$$

where the DVM is used to determine S_{1d} and S_{2d}. Then, the probability $P[S_d(\psi_{01}, \psi_{02})]$ is evaluated from $F_S(S)$, and the corresponding reliability index $\beta(\psi_{01}, \psi_{02})$ estimated. Finally, the set of factors (ψ_{01}, ψ_{02}) should be calibrated to have (on average) a uniform reliability level for each a_s ratio. The following minimization problem is solved:

$$(\psi_{01}, \psi_{02}) = \text{argmin} \sum_i^{n_s} [\beta_i(\psi_{01}, \psi_{02}) - \alpha_E \beta_t]^2 \tag{14}$$

where n_s is associated with the number of weighting factors a_s considered, i.e. the number of design situations. The application of Eq. (14) to single sites makes it possible to calibrate the combination factors at the local scale. Instead, the calibration at regional scale is made by extending the minimization problem to the number n_{st} of sites considered and Eq. (14) turns to be:

$$(\psi_{01,opt}, \psi_{02,opt}) = \text{argmin} \sum_j^{n_{st}} \sum_i^{n_s} \left[\beta_{ij}(\psi_{01}, \psi_{02}) - \alpha_E \beta_t\right]^2 \tag{15}$$

4 Results and Discussion

The results of the combination factors calibration are summarized in Table 2. Within the FBC method, the combination factor ψ_{01} for imposed load does not depend on the wind action, i.e. on the relative ratio n_2/n_1, but only on the statistics of the imposed load, i.e. on the coefficient of variation of the maximum values characterizing aleatory uncertainty. Calibration results are summarized in column 5 of Table 2 where values of 0.27, 0.38, 0.26, and 0.31 are shown for residence, office, school classroom, and industrial areas, respectively. On the other hand, the combination factor ψ_{02} for wind action resulting from calibration within the FBC method varies for the 40 sites from 0.52 to 0.58, with a mean value of 0.55 for residence use; from 0.55 to 0.67 with a mean value of 0.60 for office use; from 0.50 to 0.62 with a mean value of 0.56 for school classroom and industrial use. Therefore, the category of building use influences the calibration results. In fact, what differs within different categories is the number of repetitions n_1 of imposed load in the reference period T: it is equal to 10 for office use, to 7 for residence use, and to 5 for school classroom and industrial use.

The variation of combination factor for wind action calibrated for single sites within the FBC method is shown in Fig. 3a in terms of bias with respect to their mean value show similar results. A coefficient of variation equal to 4.9% is evaluated for residence use, equal to 5.4% for office use, and equal to 5.4% for school classroom and industrial use. In Fig. 3b the location of the forty analysed sites is reported, together with the corresponding value of ψ_{02} calibrated for wind action combining with imposed load for office use. No correlation seems to exist with geographic coordinates.

Table 2. Results of the calibration of the combination factors.

Use	Case			FBC method		Simulation method	
	#	TIU	IIL	ψ_{01}	ψ_{02}	$\psi_{01,\text{opt}}(E[\psi_{01}])$	$\psi_{02,\text{opt}}(E[\psi_{02}])$
Residence	1	Yes	Yes	–	–	0.11 (0.10)	0.00 (0.02)
	2	Yes	No	0.27	0.58	0.19 (0.20)	0.42 (0.41)
	3	No	Yes	–	–	0.12 (0.11)	0.01 (0.03)
	4	No	No	0.27	0.58	0.25 (0.22)	0.48 (0.49)
Office	1	Yes	Yes	–	–	0.21 (0.21)	0.08 (0.07)
	2	Yes	No	0.38	0.60	0.24 (0.24)	0.40 (0.39)
	3	No	Yes	–	–	0.24 (0.24)	0.13 (0.13)
	4	No	No	0.38	0.60	0.23 (0.23)	0.49 (0.47)
School Classroom	1	Yes	Yes	–	–	0.04 (0.04)	0.01 (0.02)
	2	Yes	No	0.26	0.56	0.29 (0.27)	0.43 (0.43)
	3	No	Yes	–	–	0.03 (0.03)	0.01 (0.02)
	4	No	No	0.26	0.56	0.26 (0.26)	0.49 (0.47)
Industrial	2	Yes	No	0.13	0.56	0.51 (0.55)	0.39 (0.37)
	4	No	No	0.13	0.56	0.50 (0.50)	0.48 (0.48)

The results of application of the simulation method are summarized in columns 7 and 8 of Table 2 in terms of optimum value $\psi_{0,opt}$ of combination factor derived by solving the optimization problem of Eq. (15), and of mean value E[ψ_0] of combination factors derived for single sites according to Eq. (14). Overall, $\psi_{0,opt}$ and E[ψ_0] values are close to each other for both imposed load and wind action. In details, it is observed that combination factors result to be lower when the intermittent imposed load is considered in the analyses. Such effect is larger for wind action, while it is almost negligible for imposed load in the case of office use. On the other hand, time-invariant uncertainty results to affect less the calibration of combination factors. Including TIU in the analyses corresponds to a reduction of ψ_{02} factor for wind action, with values of about 0.8 and 0.86 times the combination factors obtained when neglecting time-independent uncertainties. This is most likely to occur due to the bias in θ_{Q_2} rather than to its variability.

The combination factor $\psi_{01,opt}$ for imposed load takes values ranging from 0.03 (school classroom) to 0.24 (residence) when IIL are considered, i.e. in cases n. 1 and n.

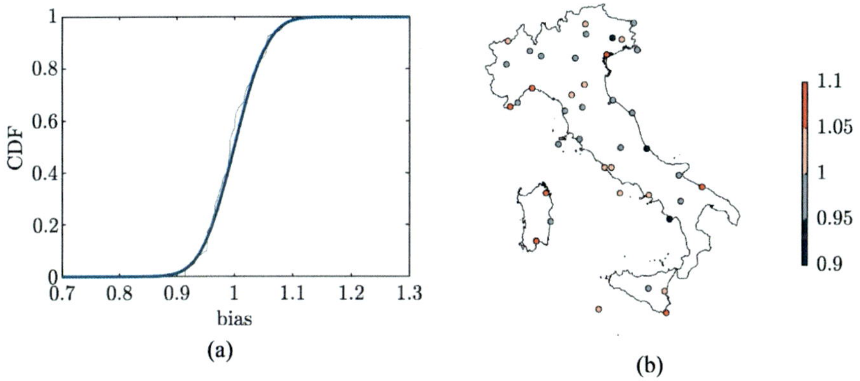

Fig. 3. Combination factors ψ_{02} for wind action resulting from the application of FBC method to Italy for office use: (a) CDF and (b) values at local scale of bias with respect to the mean value.

3. Instead, it ranges between 0.19 (residence) to 0.51 (industrial) when only the sustained part of imposed load is simulated. As for wind action, it is observed that $\psi_{02,opt}$ factor is almost zero for residence and school classroom use, when the intermittent imposed load is accounted for in the calibration, and reaches a value of 0.13 for office category of use. Instead, it assumes values ranging from 0.39 (industrial) to 0.49 (office and school classroom) when wind action combines with only the sustained part of imposed load.

Comparing the results of the two methods, smaller values of ψ_{01} for imposed load are observed for the simulation method with respect to those of the FBC method when analysing residence and office use. Instead, results int better agreement are obtained for school classrooms, while the simulation method provides larger ψ_{01} values for industrial category of use. On the other hand, for wind action smaller ψ_{02} values are observed in all the analysed categories of use. The explanation of this is to be found in the differences between the two methods. The combination factor ψ_{0i} for the load effect S_i is defined as the ratio between the combination value S_{ic} and the design value S_{id}. Both methods refer to the same design value defined within the DVM as the value corresponding to the probability $P_d = \Phi(\alpha_E \beta_t)$. However, in FBC method the approximation is made that S_{ic} corresponds to the value having probability P_c defined as in Sect. 2.3. Hence, the combination factor ψ_{0i} depends on both the reduced sensitivity factor $0.4\alpha_E$ and the ratio n, whose correlation and influence require further investigation.

The calibrated combination factors allow comparison with the values suggested by Eurocode 0 [2]. Within the DVM, the target reliability level $\beta(S_d)$ for the design value S_d of the combined load effect is equal to $\beta(S_d) = 0.7\beta_t = 2.66$, where the target reliability index β_t is equal to 3.8 for the reference period $T = 50$yrs. In Fig. 4 an example of the reliability index $\beta(S_d)$ of the combined load effect S_d as function of weighting factor a_s is shown for wind load combining with imposed load for office use. The green line in the figure corresponds to the value obtained by applying the combination factors $\psi_{01} = 0.7$ and $\psi_{02} = 0.6$, as recommended in the Eurocode. Instead, the red lines show the results obtained from the calibration of Cases 1 (Fig. 4a) and 2 (Fig. 4b). Moreover, hatched areas correspond to the envelope of the results for the 40 sites. Overall, it is observed that Eurocode 0 overestimates the design value of the combined effect S_d.

Instead, the reliability values obtained by optimization approach are much closed to the target reliability level.

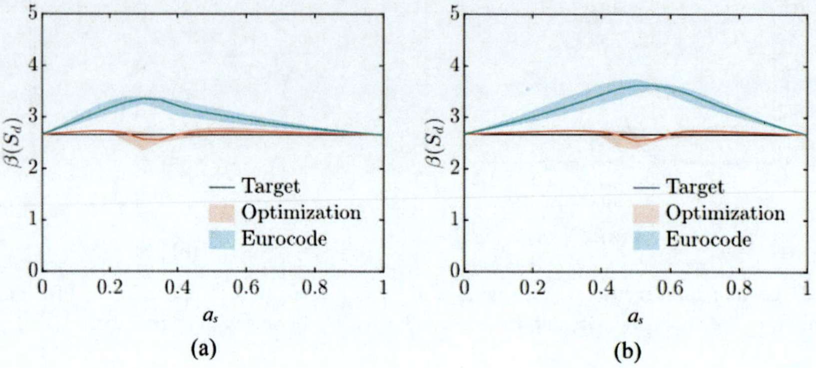

Fig. 4. Reliability index $\beta(S_d)$ for wind action combining with imposed load for office use: (a) Case 1 and (b) Case 2 defined as in Table 2.

5 Conclusions

Results concerning the calibration of the combination factors for wind action combined with imposed load for the Italian wind climate have been shown. Two methods are considered in this paper: (1) application of the DVM based on the FBC approach; and (2) simulations of the stochastic process of the combined load effects to derive the distribution function of the 50-yrs maximum values. Forty sites distributed were investigated. First, it is observed that combination factors change over Italy without correlation with the geographic location. Then, applying the FBC method the combination factors range between 0.13 and 0.38 for the imposed load, and are (on average) in the range of 0.56 to 0.60 for wind action. Within the FBC method, the combination factors for wind action result to be larger than those calibrated when simulating the combined load. In fact, within the simulation method, when only the sustained imposed load is simulated then optimum values are obtained in the range of 0.19 to 0.51 for the imposed load, and in the range of 0.40 to 0.49 for wind action. When also the intermittent part of imposed load is accounted for, then values in the range of 0.03 to 0.51 for the imposed load, and in the range 0.00 to 0.49 for the wind action are derived. Finally, it is shown that current values of combination factors suggested by Eurocode 0 lead to overestimation of design values up to 20% for wind action effects when combined with imposed load effects.

References

1. MIT: Aggiornamento delle "Norme Tecniche per le Costruzioni", D.M. 17/01/2018 (*in Italian*). Ministero delle Infrastrutture e dei Trasporti, Roma (2018)
2. CEN: EN1990:2002. Eurocode - Basis of structural design, Comité Européen de Normalisation (2002)

3. Picozzi, V.: Uncertainties in Wind Action: Modelling and Code Calibration. Ph.D. thesis. University of Campania "L. Vanvitelli", Caserta, Italy (2023)
4. Wen, Y.K.: Structural Load Modeling and Combination for Performance and Safety Evaluation. Elsevier, Amsterdam (1990)
5. Ferry-Borges, J., Castanheta, M.: Structural Safety, 2nd edn. Laboratorio Nacional de Engenhaira Civil, Lisbon (1972)
6. Turkstra, C.J., Madsen, H.O.: Load combinations in codified structural design. J. Struct. Div. ASCE **106**(12), 2527–2543 (1980)
7. Wilson, E.L., Der Kiureghian, A., Bayo, E.P.: A replacement for the SRSS method in seismic analysis. Earthquake Eng. Struct. Dynam. **9**, 419–435 (1981)
8. Hasofer, A.M., Lind, N.C.: Exact and invariant second-moment code format. J .Eng. Mech. Div. **100**(1), 111–121 (1974)
9. König, G., Hosser, D.: The simplified Level II method and its application on the derivation of safety elements for Level I. CEB Bulletin no. 147 (1982)
10. JCSS: Probabilistic Model Code, Joint Committee on Structural Safety (2001)
11. Ballio, G., Lagomarsino, S., Piccardo, G., Solari, G.: A first step towards a map of Italian extreme winds. Part 2. Costruzioni Metalliche **4**, 209–243 (1991)
12. Schneider, J.: Introduction to Safety and Reliability of Structures. International Association for Bridge and Structural Engineering (IABSE), Zürich (1997)
13. Picozzi, V., Landi, F., Avossa, A.M., Croce, P., Formichi, P., Ricciardelli, F.: The climatic action uncertainty chain. Eng. Struct. **301**, 117357 (2024)

Mitigation of Interference-Induced Vibrations for Towers in Group Arrangements

Francesca Lupi[1,2](✉), Marc Seidel[3], Rüdiger Höffer[1,2], Norbert Hölscher[2], and Hans-Jürgen Niemann[1,2]

[1] Ruhr-Universität Bochum, Bochum, Germany
francesca.lupi@rub.de
[2] Niemann Ingenieure GbR, Bochum, Germany
[3] Siemens Gamesa Renewable Energy GmbH & Co. KG, Hamburg, Germany

Abstract. The paper studies tower vibrations in closely spaced group arrangements due to vortex excitation and interference galloping. The ever-growing development of the wind energy industry has recently led to an in-depth examination of the subject, particularly with regard to the transport of towers for offshore wind turbines. Due to the low natural frequencies typical of steel towers and the need of assembling, transporting and installing towers in all possible weather scenarios, the range of verification for aeroelastic instabilities is extended until very high reduced wind velocities. The paper presents a synthesis of aeroelastic wind tunnel tests of towers in several group arrangements for spacings a/D = 1.25, 1.50, 1.875, 2.5 (being "a" the axis-to-axis distance). The Scruton number is varied in the experiments between 13 and 77. The paper shows that Scruton number effects on vortex-induced vibrations, on interference galloping and on their mechanisms of interaction are considerable. Increasing the Scruton number not only shifts the onset velocity of galloping towards higher values (in accordance with Eurocode), but also changes the behaviour of the galloping branch and reduces the range of galloping-prone wind directions. In all cases where vortex-induced oscillations act as a trigger for interference galloping, increasing the Scruton number can be used to effectively eliminate the occurrence of galloping instabilities.

Keywords: Offshore Wind Turbines · Vortex-Induced Vibrations · Interference Galloping · Towers in Group Arrangements

1 Introduction

Interference effects in tower groups involve various aeroelastic phenomena, ranging from predominant vortex induced vibrations (VIV) for larger spacings to interference galloping (IG) for smaller spacings. Interaction between vortex induced vibrations and galloping-like vibrations may occur for intermediate distances. Reviews on the topic can be found e.g. in [1, 5, 6].

The amplification of the response due to interference effects in vortex induced vibrations is considered by the Eurocode EN 1991-1-4 through the reduction of the Strouhal

P. Schito and A. Zasso (Eds.): IN VENTO 2022, LNCE 461, pp. 102–113, 2024.
https://doi.org/10.1007/978-3-031-53059-3_10

number and the increase of the lift coefficient. With regard to interference galloping, the Eurocode contains an equation to calculate the onset velocity based on the distance between towers, the Scruton number and an instability parameter (see EN 1991-1-4 [2], E.3). This approach goes back to Ruscheweyh [3]. However, vortex induced vibrations and interference galloping are treated by the Eurocode as two separate and independent phenomena. Instead, for tower spacings (axis-to-axis) smaller than 2.5D (being D the diameter), the two phenomena strongly interact for small until moderately high Scruton numbers. The interaction further amplifies the response. Additionally, the interaction is made more complex by multiple vortex shedding frequencies that may occur from each tower in the group arrangement.

Whereas a practical solution to avoid interference galloping for chimneys in group is to couple them at the top, this option is not practicable for towers of offshore wind turbines, due to difficulties to disconnect the towers at sea.

Although the Eurocode approach generally provides a good estimate of the onset velocity for interference galloping, simply extending this approach to a range of Scruton numbers, reduced velocities, group configurations and relative tower spacings that are largely different from the original range of application of steel chimneys - for which the Eurocode approach was intended - can lead to over-conservative design requirements, particularly with regard to additional damping and/or additional stiffness for 'soft' towers. The behaviour of the galloping branch and the role of vortex-induced vibrations in triggering the interference galloping are explored in this paper through wind tunnel tests for a wide range of Scruton numbers and group configurations.

2 Methods

2.1 Literature and Code Approach

The Eurocode approach for IG was developed in [3] and [4] based on the following equation of motion:

$$\mathrm{M} \cdot \ddot{\mathrm{y}}(t) + \mathrm{C} \cdot \dot{y}(t) + \mathrm{K} \cdot \mathrm{y}(t) = c_y(t) \cdot \frac{1}{2}\rho V^2 \cdot D \cdot H \tag{1}$$

M is the generalized mass in kg, C and K are the generalized damping and stiffness. $F_y = c_y(t) \cdot 0.5\rho V^2 \cdot D \cdot H$ is the cross-wind force. D and H are reference diameter and length. Unlike classical galloping, where quasi-stationary theory applies and the onset of galloping instability can be estimated from wind tunnel tests on stationary models, the aeroelastic interaction for IG includes a phase shift Θ between the crosswind force $\mathrm{F_y}$(t) and the crosswind oscillation y(t), i.e.:

$$\mathrm{y}(t) = \hat{\mathrm{y}} \cdot \sin(\omega t) \text{ and } c_y(t) = \widehat{c_y} \cdot \sin(\omega t + \Theta) \tag{2}$$

Based on geometrical considerations from Fig. 1 and assuming that the first cylinder is fixed, $\widehat{c_y}$ can be linearized for small angles β_{cr}.

$$\widehat{c_y} = \left.\frac{\partial c_y}{\partial \beta}\right|_{\beta} \cdot \Delta\beta \approx \left.\frac{\partial c_y}{\partial \beta}\right|_{\beta} \cdot \frac{\hat{\mathrm{y}}}{a} \tag{3}$$

β_{cr} is the critical direction, which marks the transition between two flow states: in the first state, the second cylinder is completely immersed in the wake of the first, and in the second state there is a gap flow between the staggered cylinders, which causes a higher velocity on one side and thus a crosswind force on the second cylinder.

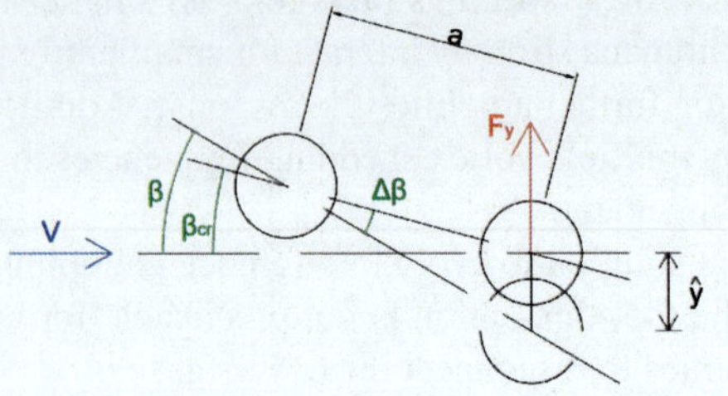

Fig. 1. Reference coordinate system for IG (after [3]).

By combining Eqs. (1) to (3) and assuming that the generalized mass is uniformly distributed on the reference length H, the onset velocity for IG can be calculated by imposing the zero-damping condition (see [2–4]). It results:

$$V_{CIG,red} = \frac{V_{CIG}}{f_1 \cdot d} = \frac{2\pi}{\sqrt{\pi}} \cdot \sqrt{\frac{Sc \cdot \frac{a}{D}}{a_{IG}}} \tag{4}$$

a_{IG} is the combined instability parameter for IG:

$$a_{IG} = \left.\frac{\partial c_y}{\partial \beta}\right|_{\beta} \cdot \sin(\Theta) \tag{5}$$

The phase angle Θ is negative. Therefore, the condition for IG instability is $\frac{\partial c_y}{\partial \beta} > 0$.

2.2 Wind Tunnel Testing

Aeroelastic wind tunnel experiments on vibrating models are performed at the WISt Boundary Layer Wind Tunnel of the Ruhr University Bochum (Fig. 2). The models are cantilevered elastic circular cylinders connected to a force balance at the base. All models are designed as aeroelastic models and are therefore able to oscillate. This is because the interference effects depend on the oscillation of each model with respect to the others. This is an improved setup compared to earlier investigations on the subject, where often the first cylinder was stationary. The dynamic properties of each model (natural frequency and structural damping) are measured to ensure that all models have the same dynamic behaviour within the desired range of accuracy.

During the experiments, the forces and bending moments at the base of each model are measured. The oscillations (displacements) at the top are calculated by considering the distribution of inertial forces in resonance. This procedure has proved to be more accurate than simply applying the static stiffness to relate dynamic reactions to oscillations.

Fig. 2. Towers in group arrangement in full-scale (left) and wind tunnel experiments at the Ruhr University Bochum (right).

An important aspect for the interpretation of the aeroelastic test results is the accurate knowledge of the structural damping. This is measured through snap-back tests in still air. The contribution of still air damping in the free-decay tests is expected to be small and is therefore not separated from pure structural damping in the analysis. Depending on the model properties, the structural damping may deviate from the assumption of linear viscous damping. This results in an amplitude dependent structural damping. The appropriate values for the equivalent linear viscous structural damping are calculated in relation to the range of amplitudes that are of interest for the scope of investigation. The corresponding Scruton numbers are listed in Table 1. The Scruton number of each cylinder in the group does not exceed a 5% variation from the mean value for the group that is reported in Table 1, with the exception for the case Sc = 13. In this case, the Scruton number of the cylinders in the group varies in the range ±20%.

To allow a large variability of the Scruton number (from 13 to 77), the wind tunnel models include a changeable damping element. Further variation of the Scruton number is possible by adding a mass at the top of the models, filling the interior with grains of sand and varying the material of the cylinder from carbon fiber to steel. In all these cases, the stiffness of the models can be adjusted to ensure that the natural frequencies and thus the critical velocity remain in the desired range.

Reynolds number effects are investigated by using uniformly distributed grain roughness. The experiments are performed with k/D = 0.03, k being the averaged roughness height.

The wind tunnel tests include a wide range of parameter variations: the spacing between cylinders varies from a/D = 1.25 and 1.5 (small spacing), to a/D = 1.875 (medium spacing) and a/D = 2.5 (large spacing). The group configurations include 2 cylinders (2x1), 4 cylinders in line (4x1), 4 cylinders placed in two adjacent rows of 2 cylinders each (2x2) with the associated subset 2 + 1 (one cylinder at the side). The wind direction varies between in-line (0°), perpendicular to the main axis of the group

(90°), slightly inclined in the range 0° ÷ 15° with respect to the main axis and other relevant diagonal wind directions.

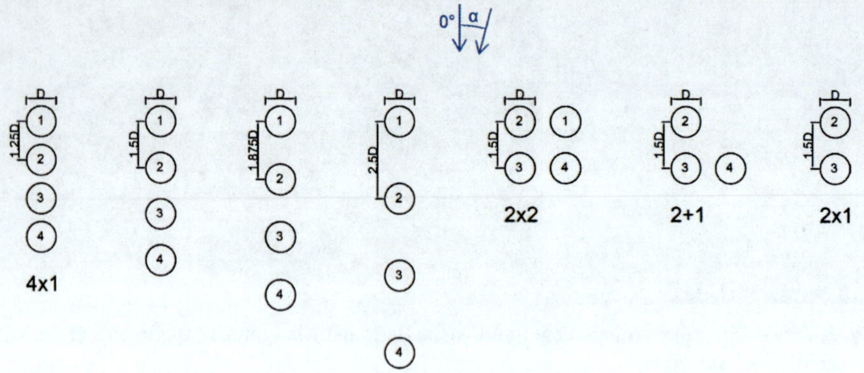

Fig. 3. Group arrangements and definition of wind direction.

Figure 3 shows the configurations investigated and the definition of the wind direction. Table 1 reports the Scruton numbers related to each configuration and the corresponding reduced onset velocity of interference galloping calculated according to EN 1991-1-4 [2]. Further wind tunnel tests are performed for a/D = 1.875 spacing (2x2, 2+1 and 2x1), but are not covered in this paper. Similarly, wind tunnel tests are also performed on rigid models for some 4x1 configurations (a/D = 1.25, 1.875, 2.5), but they are not included in this paper.

Table 1. Group arrangements, Scruton numbers and onset reduced velocities of interference galloping [2]

Spacing	Configuration	Sc	$V_{CIG,red}$ acc. EN 1991-1-4
a/D = 2.50	4x1	13	11.5
a/D = 1.875	4x1	13	10.0
		35	16.4
		47	19.0
		77	24.3
a/D = 1.50	4x1, 2x1, 2+1, 2x2	18	10.5
		35	14.6
		47	17.0
		77	21.7
a/D = 1.25	4x1	13	8.10

3 Results and Discussions

3.1 Spacing a/D = 1.25, Group Arrangement 4x1

The aeroelastic response of the towers in the narrowest arrangement (a/D = 1.25, Fig. 4) is strongly dominated by interference galloping. At $V_{red} \approx 7$ for the downstream cylinders 3 and 4 and at $V_{red} \approx 9$ for the upstream cylinders 1, 2 and 3, lock-in ranges are still visible at wind direction 0°. At small angles of attack (e.g. $\alpha = 10°$) the onset of interference galloping is completely ahead of the critical velocities of VIV. Despite this complex situation, the Eurocode prediction for Sc = 13 ($V_{CIG,red} = 8.1$, see Table 1) matches well with the experiment.

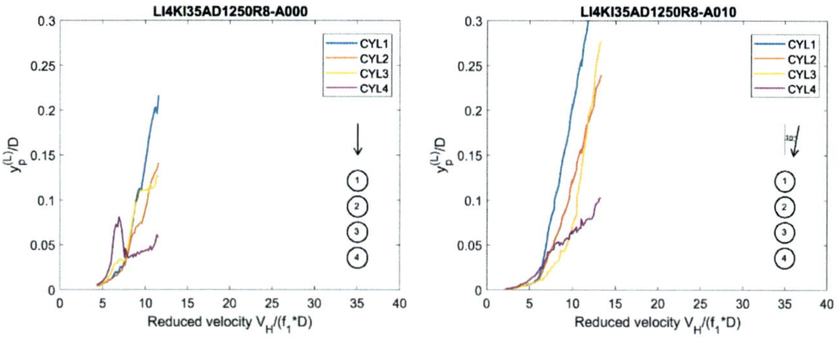

Fig. 4. Four cylinders in line (4x1), a/D = 1.25, wind direction $\alpha = 0°$ (left) and $\alpha = 10°$ (right), Sc = 13.

3.2 Spacing a/D = 1.5, Group Arrangement 2x1

The aeroelastic response of two towers in line at the small spacing a/D = 1.5 presents a peculiar feature for Sc < 20 (see Fig. 5). Differently from the approach discussed in Sect. 2.1, a strong vibration with galloping character is observed also for the upstream cylinder. Two vortex shedding peaks can be identified, called here VIV-category 1 and VIV-category 2. The VIV-category 1 refers to the first lock-in range that is associated with the largest Strouhal number (lowest critical wind velocity), which usually corresponds to the Eurocode [2], (Eq. E.11). In this case, St is about 0.13 ($V_{cr,red} = 7.5$) for the first VIV category. The second VIV category is associated with a smaller Strouhal number (St about 0.06 in this case) and thus a higher critical velocity. This could be associated with a tip effect, since it is more pronounced in the spectral analysis of the base bending moment compared to the base shear force. The galloping vibration of the upstream cylinder for closely spaced arrangements is discussed in [1]. It is related to an interaction between shear layers from upstream cylinder and the downstream cylinder. Multiple VIV ranges can also be observed in [3], Fig. 9.

As the Scruton number almost doubles from 18 (Fig. 5) to 35 (Fig. 6), VIV is greatly reduced and the onset of galloping is shifted to a reduced velocity of about 20 (Fig. 6, left). Further increasing the Scruton number (Fig. 7) not only shifts the initial velocity to higher values, but also flattens the galloping branch to oscillations below 0.03D, which are comparable to typical buffeting oscillations in strong winds.

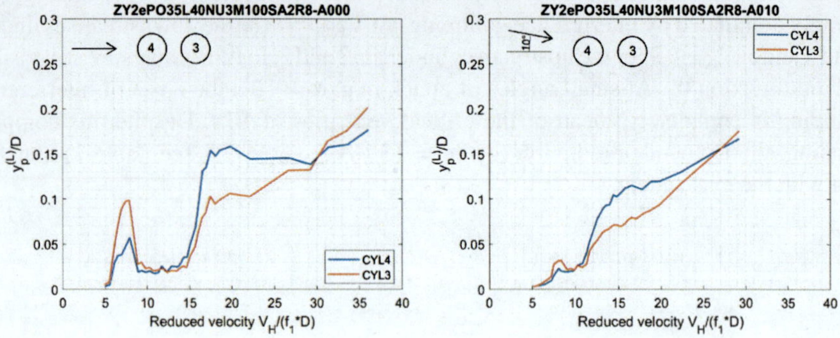

Fig. 5. Two cylinders in line (2x1), a/D = 1.50, wind direction α = 0° (left) and α = 10° (right), Sc = 18. Differently from Fig. 3 and only in this case: cylinder 4 is upstream, cylinder 3 is downstream.

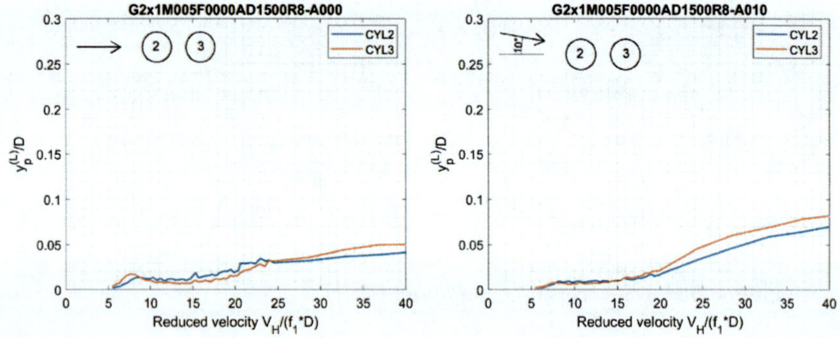

Fig. 6. Two cylinders in line (2x1), a/D = 1.5, wind direction α = 0° (left) and α = 10° (right), Sc = 35.

3.3 Spacing a/D = 1.5, Group Arrangements 2x2 and 2+1

The aeroelastic response of the group of four towers arranged in 2x2 configuration is particularly severe until moderately high Scruton numbers (see Fig. 8, Sc = 18). However, predominant oscillations for reduced velocities below 20 are mainly associated to VIV (St ca. 0.06, $V_{cr,red}$ ca. 17). The wide lock-in range is then followed by the galloping branch. Because of the strong interaction between the two phenomena, a clear onset velocity for interference galloping cannot be identified in Fig. 8. Equation (4) predicts $V_{CIG,red} = 10.5$ (see Table 1). Figure 8 (left) confirms that this value can be used as an estimation in the design. It marks the onset of strong VIV oscillations and also corresponds to the linear extrapolation of the galloping branch until it crosses the horizontal axis.

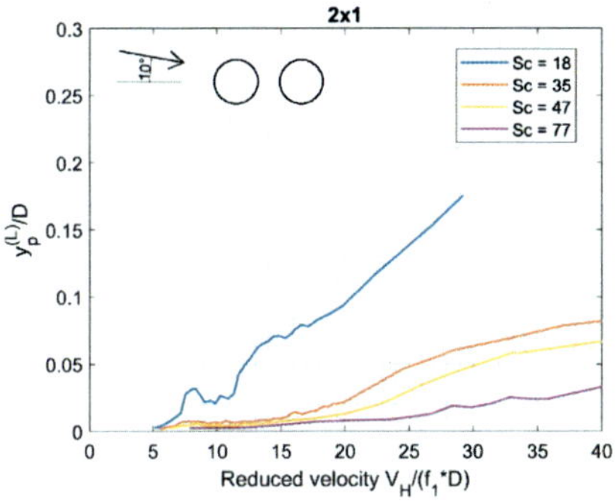

Fig. 7. Two cylinders in line (2x1), a/D = 1.5, wind direction α = 10°, varying Sc, downstream cylinder.

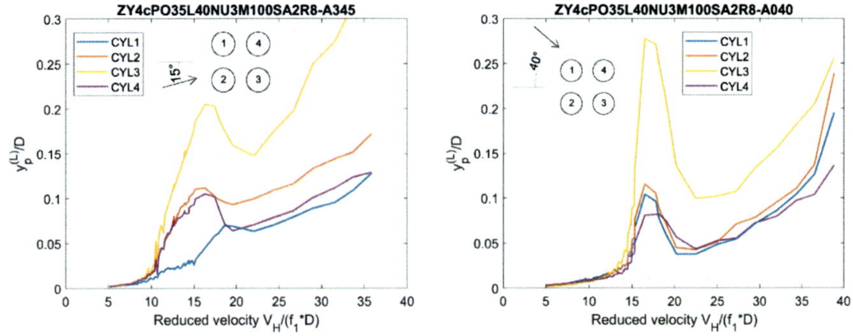

Fig. 8. Four cylinders in two rows (2x2), a/D = 1.5, wind direction α = 345° (left) and α = 40° (right), Sc = 18.

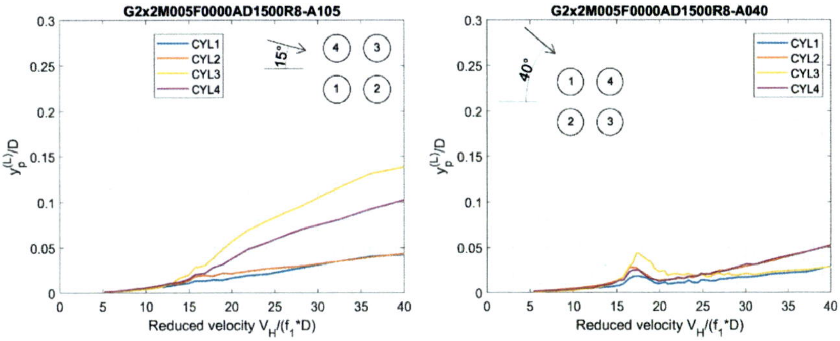

Fig. 9. Four cylinders in two rows (2x2), a/D = 1.5, wind direction α = 105° (left) and α = 40° (right), Sc = 35.

The experiments in Fig. 9 (Sc = 35) shed some further light on the interaction between VIV and IG. The right side of Fig. 8 and Fig. 9 (2x2, α = 40°) is a good example where IG is triggered by VIV. In this case, increasing the Scruton number not only shifts V_{CIG} towards higher velocities (Eq. (4)), but also reduces VIV in such a way that the triggering of the aeroelastic interaction is inhibited. Then, the cross-wind oscillations at high reduced velocity are mainly buffeting oscillations. Consequently, the range of galloping prone wind directions becomes smaller as the Scruton number increases.

With regard to that, the 2x2 and 2+1 configurations show some analogies. For some wind directions, such as 40° for 2x2 (previously mentioned) and 260° for 2+1 (not shown here), the strong vortex-induced vibrations are followed by a steep galloping branch if Sc is less than 18, while for Sc larger than 35 the vortex-induced vibrations are sufficiently damped and are unable to trigger the galloping instability. For both configurations, the most critical wind direction has an inclination of 15° to the axis of the assembly (Fig. 10). The absence of a cylinder in the first row for the 2+1 case has no relevant effect.

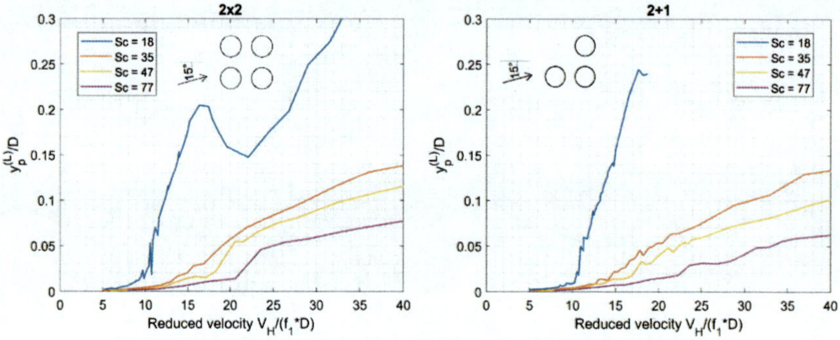

Fig. 10. Four cylinders 2x2, a/D = 1.5, (left) and three cylinders 2+1, a/D = 1.5 (right), α = 345°.

3.4 Spacing a/D = 1.5, Group Arrangement 4x1

In this configuration, the third cylinder has the strongest vibration. The largest oscillations occur for α = 0° (see Fig. 11, left) and α = 5°. However, on the whole, this configuration with four cylinders is less critical than 2x2, as shown by Fig. 11 compared to Fig. 10.

3.5 Spacing a/D = 1.875, Group Arrangement 4x1

The wind tunnel tests for Sc ca. 13 (see Fig. 12) are dominated by strong vortex-induced oscillations for $V_{red} < 10$. These are caused by St ≈ 0.11. This lock-in region is followed by a branch of increasingly strong oscillations and the wind tunnel tests were terminated. The wind tunnel tests at higher Scruton number (Sc = 35, Fig. 13) evidence the existence of two lock-in ranges, one that corresponds to St ≈ 0.115 and one that corresponds to St ≈ 0.06. This is confirmed by the spectra of the base bending moment (not shown), where the smaller Strouhal number is especially pronounced on the first cylinder. Previous wind tunnel experiments with rigid cylinders, where pressure measurements were distributed over the height of each cylinder (not shown in this paper), had shown that the smaller

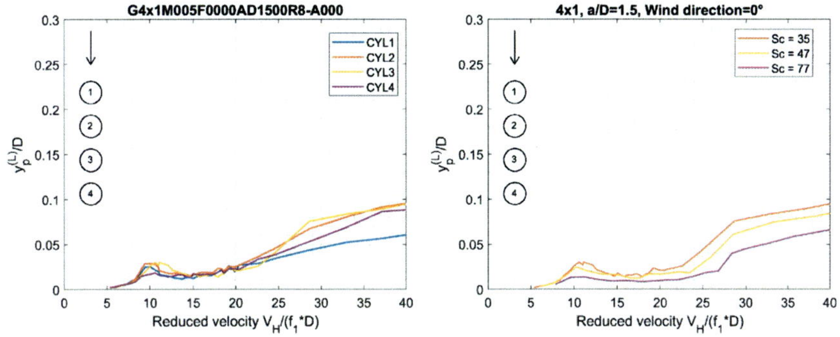

Fig. 11. Four cylinders in line (4x1) at the spacing a/D = 1.5, wind direction α = 0°. Left: oscillations of the four cylinders for Sc = 35. Right: oscillations of the third cylinder for varying Sc.

Strouhal number was related to the tip effect of the upstream cylinder, while the tip effect was suppressed for the downstream cylinders.

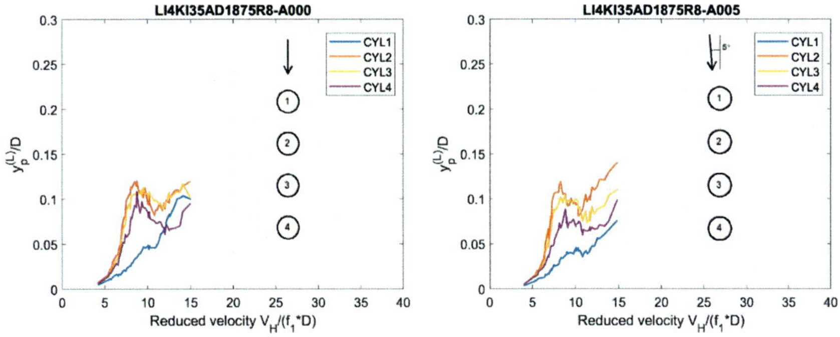

Fig. 12. Four cylinders in line (4x1), a/D = 1.875, wind direction α = 0° (left) and α = 5° (right), Sc = 13.

For this configuration, the Eurocode prediction for V_{CIG} in Table 1 is on the safe side (V_{CIG} = 16.4). Figure 13 (left) shows that the initial velocity of the interference galloping is about 20.

3.6 Spacing a/D = 2.5, Group Arrangement 4x1

For wind turbine towers during all assembly and transport phases, the distance a/D = 2.5 can be considered the upper limit. In comparison with Fig. 12, Fig. 14 shows that VIV becomes stronger and all cylinders, including the first, are in the co-shedding regime. In this configuration, the wind tunnel tests focused on the VIV category 1, as this is the predominant effect up to reduced velocities of about 15. Therefore, the wind tunnel tests were ended after the first lock-in interval. However, Fig. 14 (right) shows an increasing branch for higher wind velocities. This could be a galloping branch or a second vortex resonance, similar to Fig. 12 (right). Further investigation would be necessary.

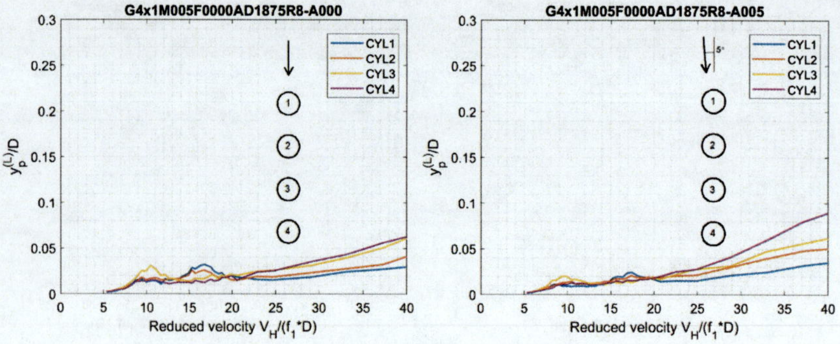

Fig. 13. Four cylinders in line (4x1), a/D = 1.875, wind direction α = 0° (left) and α = 5° (right), Sc = 35.

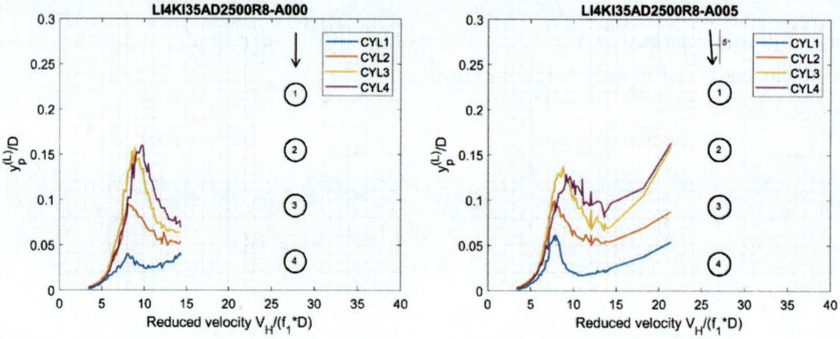

Fig. 14. Four cylinders in line (4x1), a/D = 2.5, wind direction α = 0° (left) and α = 5° (right), Sc = 13.

4 Conclusion

The paper presents wind tunnel experiments on oscillating cylinders in group arrangements for a wide range of reduced velocities, wind directions and Scruton numbers. Despite complex flow fields, the Eurocode approach for interference galloping generally provides a good estimate of the onset velocity. However, the code's model includes neither the interaction with vortex-induced vibrations nor the post-critical behaviour of galloping. In fact, experiments show that for small and moderate Scruton numbers (Sc < 20), large oscillations at low reduced velocities are primarily caused by vortex-induced vibrations. Multiple lock-in ranges are generally followed by a steep galloping branch. Due to the interaction, it is not possible to identify a clear onset velocity for the interference galloping. On the other hand, by increasing the Scruton number, vortex-induced vibrations are significantly reduced. For some angles of attack, this reduction is sufficient to completely avoid the trigger of interference galloping. For other wind directions, despite the onset of the interference galloping at a critical speed that increases with the square root of the increase in the Scruton number (as per Eurocode), the character of the galloping branch changes profoundly and the aeroelastic oscillation, while existing, remains below acceptable levels, comparable to those typical of buffeting or vortex resonance design. For these reasons, the design of wind turbine towers based solely on the estimation of onset velocity for interference galloping can lead to extremely conservative minimum damping requirements. Specific criteria based on the experimental determination of allowable oscillations for certain velocity ranges are needed. These will be developed in future research.

Acknowledgements. This project is partly funded by the German Research Foundation (DFG – Deutsche Forschungsgemeinschaft), Project number 426322127.

References

1. Alam, M.M., Meyer, J.P.: Global aerodynamic instability of twin cylinders in cross flow. J. Fluids Struct. **41**, 135–145 (2013)
2. DIN EN 1991-1-4:2010-12 Eurocode 1: Einwirkungen auf Tragwerke – Teil 1-4: Allgemeine Einwirkungen, Windlasten, Deutsche Fassung EN 199114:2010, Deutsches Institut für Normung e.V., Dez. 2010
3. Ruscheweyh, H.: Aeroelastic interference effects between slender structures. J. Wind Eng. Ind. Aerodyn. **14**(1–3), 129–140 (1983)
4. Dielen, B., Ruscheweyh, H.: Mechanism of interference galloping of two identical circular cylinders in cross flow. J. Wind Eng. Ind. Aerodyn. **54–55**, 289–300 (1995)
5. Sumner, D.: Two circular cylinders in cross-flow: a review. J. Fluids Struct. **26**(6), 849–899 (2010)
6. Zdravkovich, M.M.: Review of interference-induced oscillations in flow past two parallel circular cylinders in various arrangements. J. Wind Eng. Ind. Aerodyn. **28**(1–3), 183–199 (1998)

Evaluation of Wind Comfort Over the Balconies of an Irregular Tall Building by Using Computational Wind Engineering

E. Amaya-Gallardo, M. A. Velázquez-Santiago, and A. Pozos-Estrada(✉)

Instituto de Ingeniería, UNAM, Mexico City, Mexico
{eamayag,apozose}@iingen.unam.mx, marco.vs@comunidad.unam.mx

Abstract. In the last decade, the construction of tall buildings in Mexico has greatly increased, which has raised the need for open spaces, e.g., building balconies, to provide comfort to people. The comfort of people in balconies can be impaired due to adverse wind effects. Although current Mexican standards contemplate service condition assessments from the structural approach, there are still no guidelines to establish criteria due to wind direct perception. In this work, numerical simulations are performed by means of Computational Wind Engineering to predict wind velocities in the balconies of a tall building. These wind velocities were used to evaluate wind comfort with the Beaufort modified scale, which associates wind speeds with physical sensations in people. Based on this, the less comfortable building balconies were identified.

Keywords: wind comfort · wind direct perception · computational wind engineering

1 Introduction

In the last decade, the construction of tall buildings in Mexico has greatly increased, which has raised the need for open spaces, e.g., building balconies, to provide comfort to people. The comfort of people in balconies can be impaired due to adverse wind effects. The structural design approach usually focuses only on structural safety; however, numerous research has shown the need not only to guarantee structural performance but also to improve service and comfort conditions (Pozos-Estrada 2010; Montazeri et al. 2013; Omrani et al. 2017; Zheng et al. 2020).

Different Wind Engineering groups have developed criteria to evaluate the forces produced by the wind, safety, and comfort in people. People can experience two ways of perceiving wind effects, indirectly and directly. The indirect form could be identified by building accelerations associated to wind gusts, which can cause discomfort in the occupants or even trigger some psychological condition while the direct form could be perceived due to loss of balance, cold sensation by winds sustained for long periods or displacement of objects. Open spaces in office and residential buildings can increase or improve the performance of employees or the well-being of the people who inhabit

P. Schito and A. Zasso (Eds.): IN VENTO 2022, LNCE 461, pp. 114–136, 2024.
https://doi.org/10.1007/978-3-031-53059-3_11

them by mitigating the adverse effects of wind (Pozos-Estrada 2018). One of the most relevant studies on balcony comfort in recent years is that of Montazeri et al. (2013). They studied comfort on balconies for a 78 m tall building according to the Dutch standard NEN8100. In this study it was found that the wind can cause discomfort and danger to people on the balconies, in addition, a second skin or second façade concept is proposed with which local speeds are reduced to improve comfort on the balconies. In their study, comprehensive information about the wind climate of the location of the building under study was gather and the probabilistic approach described in NEN8100 was applied.

Since wind effects on structures depend to a great extent on their geometry, national building codes sometimes do not have enough information to design some structures as well as measures to mitigate the discomfort of the users. In order to compensate this lack of information, different methodologies are often used, such as full-scale experimental tests, experimental boundary layer wind tunnel tests (BLWT), aerodynamic databases, or computational fluid dynamics (CFD). In this work, numerical simulations are performed by means of computational wind engineering (CWE) using OpenFOAM® to predict wind velocities in the balconies of a tall building with irregular plan geometry and 158 m height for predefined wind directions. Finally, the comfort condition for each of the balconies on the height of the building are evaluated for three wind velocities and two wind directions.

2 Numerical Model and Wind Directionality

The structure under study is a tall building with balconies for residential use and offices, with irregular geometry in plan and dimensions of 55.4 m (L) × 23.2 m (B) × 158 m (H) (Fig. 1). The building has 40 stories with balconies with semicircular shapes of variable dimensions. The balconies have a maximum width of about 2.65 m with parapets or railings based on glass plates of 1.2 m high. In order to minimize the computational cost of the simulation, as well as the discretization process, the railings thickness were set equal to 15 cm, similar to the thickness of the slab (Fig. 1).

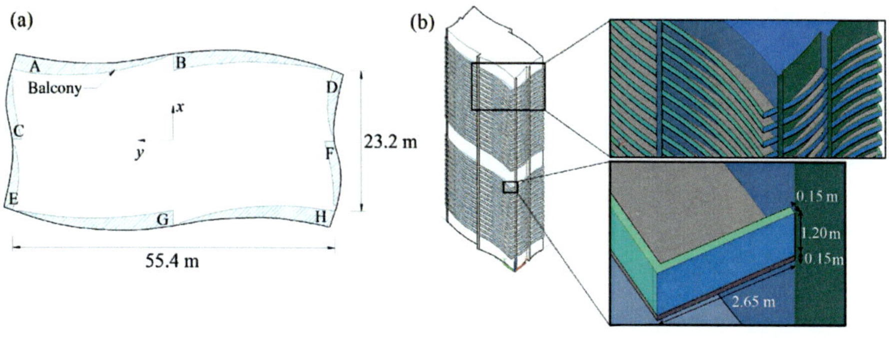

Fig. 1. Building under study: (a) Plan view and dimensions; (b) Isometric and details of the balconies.

To reduce the computational cost, the dimensions to size the computational domain are shorter than those recommended by Franke et al. (2004) due to the thickness of the balconies greatly increases the number of control volumes, and this in turn, the computational analysis time.

The base mesh of this model was structured mainly by hexahedrons of about 1.4 m, with a first cell height of 1.4 m. Likewise, to reduce the number of elements and the computational cost, it was decided to use subdomains, which allow the use of different levels of refinement around the model (Fig. 2); furthermore, to increase refinement in the balconies, it was decided to model the main structure and the balconies separately; that is, modeling two surfaces that are connected with the "snappyHexMesh" utility, which allows to increase the mesh refinement in the balconies. Finally, a total of 5,406,209 elements was obtained and, after running the simulation, the time to reach convergence was 55 h.

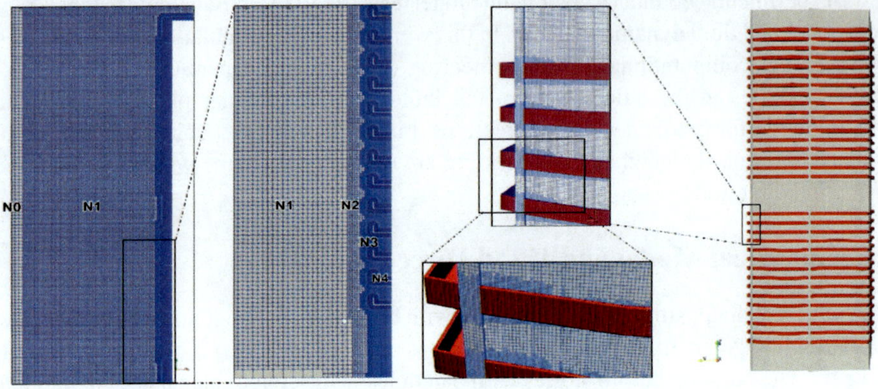

Fig. 2. Mesh: a) refinement levels; b) details the building and balconies.

The geometric characteristics of the case require a high refinement in the mesh, for example, in railings with a thickness of 15 cm, there are cells of approximately 10 cm, for which, for this study, the grid independence analysis is omitted, other studies have also omitted this analysis (Montazeri et al. 2013) due to the high level of refinement.

Wind meteorological information from 48 years (1973–2021) was provided by the National Oceanic and Atmospheric Administration (NOAA) to identified the most critical incident directions for the tall building and the wind frequency at each direction. As can be seen in the compass rose in Fig. 3, the prevailing wind directions are North, East-Northeast, and East. Based on the meteorological information, the wind directions considered were East ($\theta = 90°$) and North ($\theta = 0°$).

3 Numerical Analyses and Results

For all the numerical analyses with OpenFOAM, the 3D Steady RANS formulation and the k–ω SST turbulence model were used. The numerical simulations were performed following the best CFD practice guidelines (Franke 2004; Franke et al. 2007; Blocken et al. 2007; Tominaga et al. 2008; Franke et al. 2011). The wind profiles (wind speed

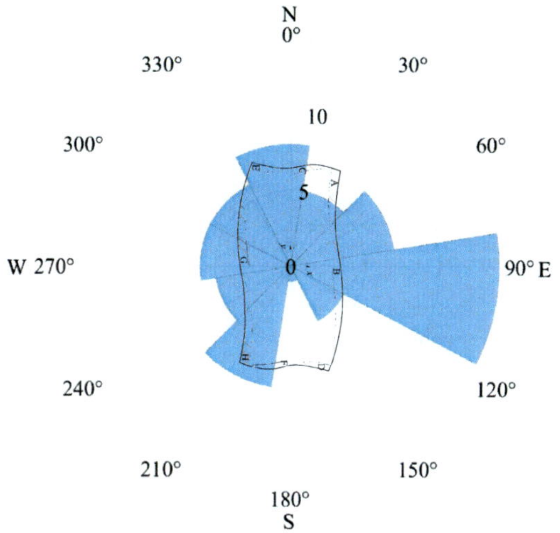

Fig. 3. Wind directionality.

(U), turbulence intensity (I_U), turbulence kinetic energy (k) and turbulence dissipation rate (ε)) were simulated by using the following equations:

$$U(y) = u^*{}_{ABL}/\kappa \ln\left[\left(y - y_g + y_0\right)/y_0\right] \tag{1}$$

$$k(y) = u^{*2}{}_{ABL}/{c_\mu}^{1/2} \tag{2}$$

$$\varepsilon(y) = u^{*3}{}_{ABL}/\left[\kappa\left(y - y_g + y_0\right)\right] \tag{3}$$

$$\omega(y) = \varepsilon(y)/(c_\mu k(y)) \tag{4}$$

where $U(y)$ is the velocity profile (m/s), $k(y)$ is the turbulence kinetic energy profile (m^2/s^2), $\varepsilon(y)$ is the turbulence dissipation rate (m^2/s^3), and $\omega(y)$ is the specific dissipation rate (s^{-1}); is the atmospheric boundary layer (ABL) friction velocity (m/s), y_g is the minimum height from the ground (m) which for practical purposes is considered equal to zero, y_0 is the aerodynamic roughness length (m) and c_μ is a turbulence model empirical constant (≈0.09).

In the horizontal homogeneity test, the refinement level of finite volumes is increased to correctly characterize the elements that make up the structure. Therefore, when studying horizontal homogeneity in the empty domain, 12,000 iterations are established, from which results are extracted every 1500 iterations; This is intended to ensure that the "simpleFoam" solver does not finish the analysis process before reaching convergence. The horizontal homogeneity test was performed in the empty computational domain, where the internal mesh of the real building is preserved. In Fig. 4 the inlet profile, incident profile and outlet profiles are compared for U, k, ε and I_U. It is observed in Fig. 4 that there is an absolute deviation less than 1% between the three profiles (inlet, incident and outlet) compared throughout the domain.

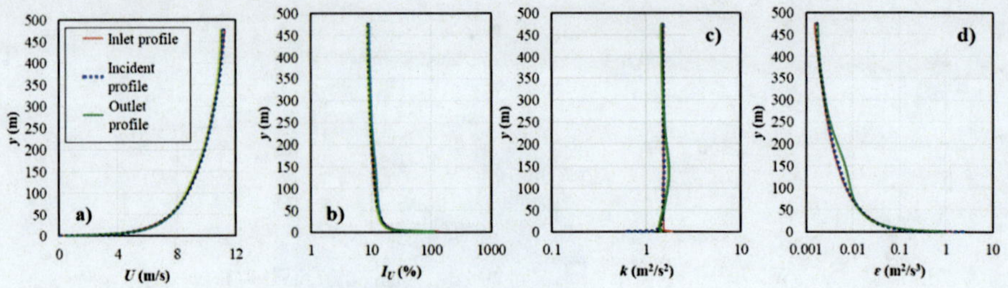

Fig. 4. ABL wind profiles: (a) mean velocity; (b) turbulence intensity; (c) turbulence kinetic energy and (d) turbulence dissipation rate

In the following, the wind behavior on the balconies for three different wind velocities (5, 10 and 15 m/s) and two wind directions ($\theta = 0°$ and 90°) is described. In addition, to study comfort conditions the building height is divided into four zones (Fig. 5). The first quarter of height is measured from the base to 40 m in height (ground floor to 10th floor), the second zone from 40 m to 79 m (11th floor to 20th floor), the third zone from 79 m to 118 m (21st floor to the 31st floor) and the fourth zone from 119 m to 158 m (32nd floor to 40th floor). Velocity contours for $U = 5$ m/s (U_5) from these analyses are presented in Figs. 6 and 7 for the four different building zones and $\theta = 0°$ and $\theta = 90°$ wind directions, respectively. Velocity contours for $U = 10$ (U_{10}) and 15 m/s (U_{15}) presented similar distribution as those for $U = 5$ m/s, except that the magnitude of local wind speed is different.

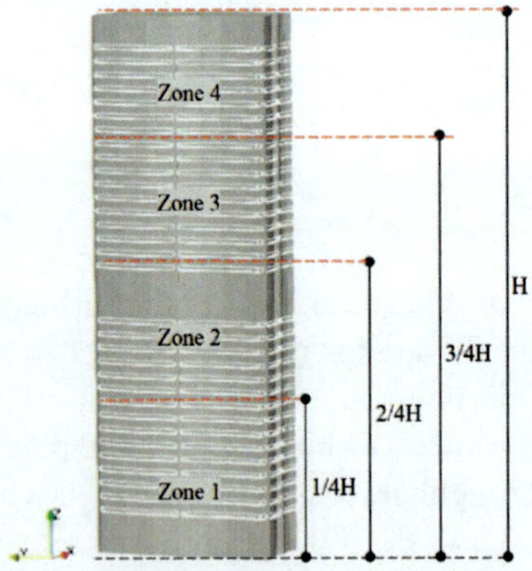

Fig. 5. Study areas in relation to H.

Tables 1, 2 and 3 present the results of a detail evaluation of comfort on balconies for people standing condition (B.L. + 1.70 m) for U_5, U_{10} and U_{15} at $\theta = 0°$ and $\theta = 90°$. The evaluation takes into account the effect and comfort level based on the parameter K_G, which is defined as the ratio of the undisturbed wind speed to the local wind speed.

From Tables 1, 2 and 3 the results indicate that for $K_G < 1$, the comfort condition is good; for values in the range between $1 \leq K_G \leq 2$, the condition is moderate; for $2 \leq K_G \leq 3$ the comfort condition is poor and for $K_G > 3$ the situation is dangerous while

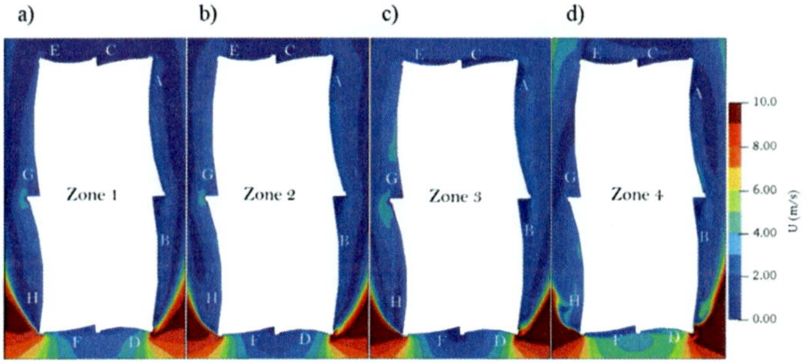

Fig. 6. Plan view of local wind velocity contours at balcony level + 1.70 m for $\theta = 0°$ for the different building zones: a) zone 1, b) zone 2, c) zone 3 and d) zone 4.

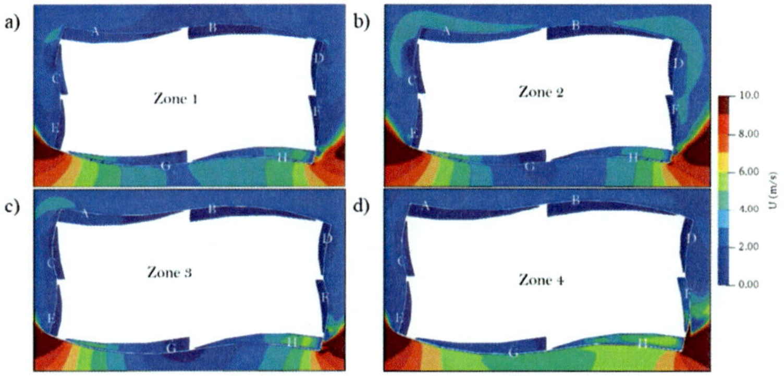

Fig. 7. Plan view of local wind velocity contours at balcony level + 1.70 m for $\theta = 90°$ for the different building zones: a) zone 1, b) zone 2, c) zone 3 and d) zone 4.

standing or walking. This exhibits that the most uncomfortable balconies are "E" and "H", being mainly those located on the windward surface for the $\theta = 90°$ wind direction. On the other hand, it is observed that for $\theta = 90°$ balcony "G" is the safest along the building height. Similarly, for $\theta = 0°$, it was observed that the most uncomfortable balconies are "D" and "H" while the safest is "F". It is observed that, in general, for both wind directions, $\theta = 0°$ and 90°, the most comfortable situation is for wind velocities $U \leq 5$ m/s while the least comfortable conditions are for $U > 5$ m/s.

To further evaluate wind comfort at balcony levels, the methodology from the NEN 8100 (2006) is adopted. This standard fixed the mean wind speed to 5 m/s for any activity. Five grades of wind comfort are stablished in NEN 8100 (2006), which depend on the type of activity and a probability of exceedance. In this study, the steps to evaluate the wind comfort at balconies suggested by Montazeri et al., (2013) are employed for two wind directions. The steps to determine the exceedance probability are as follows:

1) Obtain the wind speed ratios ($\gamma = U_{CFD}/U_{ref,H}$) from the CFD simulations.
2) Estimate the threshold wind speed at the top of the building ($U_{thrH} = 5/\gamma$).
3) Determine the percentage of time that the threshold value calculated in 2) is exceeded according to the wind statistics.

Table 1. Evaluation of comfort on balconies for people standing condition (B.L. + 1.70 m) at $\theta = 0°$ and $\theta = 90°$ for $U5$

Zone	Balcony	U_{BZ} (m/s)	U_L (m/s)	$K_L = U_L/U_{BZ}$	$K_G = U_5/U_L$	Effect	Comfort level
1	A	7	4.06	0.58	0.81	Gentle breeze: difficulty reading	Good
	B		4.06	0.58	0.81	Gentle breeze: difficulty reading	Good
	C		6.02	0.86	1.20	Light air: no noticeable wind	Moderate
	D		8.75	1.25	1.75	Strong breeze - unpleasant noise	Moderate
	E		5.04	0.72	1.01	Light breeze: wind perceived in the face	Moderate
	F		2.10	0.30	0.42	Moderate breeze: raises dust and loose papers	Good
	G		4.06	0.58	0.81	Gentle breeze: difficulty reading	Good
	H		8.75	1.25	1.75	Strong breeze - unpleasant noise in the ears	Moderate
2	A	8.23	5.18	0.63	1.04	Gentle breeze: difficulty reading	Moderate
	B		5.18	0.63	1.04	Gentle breeze: difficulty reading	Moderate
	C		7.24	0.88	1.45	Light air: no noticeable wind	Moderate
	D		10.21	1.24	2.04	Light gale: discomfort when walking or standing	Poor
	E		6.17	0.75	1.23	Light breeze: wind perceived in the face	Moderate
	F		3.13	0.38	0.63	Moderate breeze: raises dust and loose papers	Good
	G		4.12	0.50	0.82	Moderate breeze: raises dust and loose papers	Good
	H		9.22	1.12	1.84	Strong breeze - unpleasant noise	Moderate

(continued)

Table 1. *(continued)*

Zone	Balcony	U_{BZ} (m/s)	U_L (m/s)	$K_L = U_L/U_{BZ}$	$K_G = U_5/U_L$	Effect	Comfort level
3	A	9.07	5.99	0.66	1.20	Gentle breeze: difficulty reading	Moderate
	B		5.99	0.66	1.20	Gentle breeze: difficulty reading	Moderate
	C		8.07	0.89	1.61	Light air: no noticeable wind	Moderate
	D		10.16	1.12	2.03	Light gale: discomfort when walking or standing	Poor
	E		7.07	0.78	1.41	Light breeze: wind perceived in the face	Moderate
	F		3.99	0.44	0.80	Moderate breeze: raises dust and loose papers	Good
	G		4.99	0.55	1.00	Moderate breeze: raises dust and loose papers	Good
	H		10.16	1.12	2.03	Light gale: discomfort when walking or standing	Poor
4	A	9.72	5.54	0.57	1.11	Moderate breeze: raises dust and loose papers	Moderate
	B		6.61	0.68	1.32	Gentle breeze: difficulty reading the newspaper	Moderate
	C		8.65	0.89	1.73	Light air: no noticeable wind	Moderate
	D		10.40	1.07	2.08	Light gale: discomfort when walking or standing	Poor
	E		8.65	0.89	1.73	Light air: no noticeable wind	Moderate
	F		6.61	0.68	1.32	Gentle breeze: difficulty reading	Moderate
	G		6.61	0.68	1.32	Gentle breeze: difficulty reading	Moderate
	H		10.40	1.07	2.08	Light gale: discomfort when walking or standing	Poor

(continued)

Table 1. *(continued)*

Zone	Balcony	U_{BZ} (m/s)	U_L (m/s)	$K_L = U_L/U_{BZ}$	$K_G = U_5/U_L$	Effect	Comfort level
$\theta = 90°$							
1	A	7	3.00	0.42	0.60	Gentle breeze: difficulty reading the newspaper	Good
	B		2.00	0.28	0.40	Light breeze: wind perceived in the face	Good
	C		2.00	0.28	0.40	Light breeze: wind perceived in the face	Good
	D		1.00	0.14	0.20	Light air: no noticeable wind	Good
	E		8.94	1.25	1.79	Strong breeze: difficulty walking, unpleasant noise	Moderate
	F		1.00	0.14	0.20	Light air: no noticeable wind	Good
	G		6.94	0.97	1.39	Cool breeze: tripping hazard when entering a windy area	Moderate
	H		8.94	1.25	1.79	Strong breeze: difficulty walking, unpleasant noise	Moderate
2	A	8.23	6.17	0.75	1.23	Light breeze: wind perceived in the face	Moderate
	B		7.24	0.88	1.45	Light air: no noticeable wind	Moderate
	C		7.24	0.88	1.45	Light air: no noticeable wind	Moderate
	D		7.24	0.88	1.45	Light air: no noticeable wind	Moderate
	E		10.21	1.24	2.04	Light gale: discomfort when walking or standing	Poor
	F		7.24	0.88	1.45	Light air: no noticeable wind	Moderate
	G		1.07	0.13	0.21	Cool Breeze	Good
	H		10.21	1.24	2.04	Light gale: discomfort when walking or standing	Poor

(continued)

Table 1. *(continued)*

Zone	Balcony	U_{BZ} (m/s)	U_L (m/s)	$K_L = U_L/U_{BZ}$	$K_G = U_5/U_L$	Effect	Comfort level
3	A	9.07	5.99	0.66	1.20	Gentle breeze: difficulty reading the newspaper	Moderate
	B		7.07	0.78	1.41	Light breeze: wind perceived in the face	Moderate
	C		8.07	0.89	1.61	Light air: no noticeable wind	Moderate
	D		8.07	0.89	1.61	Light air: no noticeable wind	Moderate
	E		10.16	1.12	2.03	Light gale: discomfort when walking	Poor
	F		8.07	0.89	1.61	Light air: no noticeable wind	Moderate
	G		2	0.22	0.40	Cool Breeze	Good
	H		10.16	1.12	2.03	Light gale: discomfort when walking or standing	Poor
4	A	9.72	8.65	0.89	1.73	Light air: no noticeable wind	Moderate
	B		8.65	0.89	1.73	Light air: no noticeable wind	Moderate
	C		8.65	0.89	1.73	Light air: no noticeable wind	Moderate
	D		8.65	0.89	1.73	Light air: no noticeable wind	Moderate
	E		10.4	1.07	2.08	Light gale: discomfort when walking	Poor
	F		4.47	0.46	0.89	Moderate breeze: raises dust and loose papers	Good
	G		5.54	0.57	1.11	Moderate breeze: raises dust and loose papers	Moderate
	H		10.40	1.07	2.08	Light gale: discomfort when walking or standing	Poor

B.L. = Balcony level; U_{BZ} = Undisturbed velocity at corresponding building zone; U_L = Balcony local velocity; K_L = Local amplification factor

Table 2. Evaluation of comfort on balconies for people standing condition (B.L. + 1.70 m) at $\theta = 0°$ and $\theta = 90°$ for $U10$

Zone	Balcony	U_{BZ} (m/s)	U_L (m/s)	$K_L = U_L/U_{BZ}$	$K_G = U_5/U_L$	Effect	Comfort level
$\theta = 0°$							
1	A	14	8.12	0.58	1.62	Cool Breeze: Tripping hazard when entering a windy area	Moderate
	B		8.12	0.58	1.62	Cool Breeze: Tripping hazard when entering a windy area	Moderate
	C		12.04	0.86	2.41	Light breeze: wind perceived in the face	Poor
	D		17.50	1.25	3.50	Strong gale: people tend to rise	Dangerous
	E		10.08	0.72	2.02	Moderate breeze: raises dust and loose papers	Poor
	F		4.34	0.31	0.87	Light gale: discomfort when walking or standing	Good
	G		8.12	0.58	1.62	Cool Breeze: Tripping hazard when entering a windy area	Moderate
	H		17.50	1.25	3.50	Strong gale: people tend to rise	Dangerous
2	A	16.46	10.37	0.63	2.07	Cool Breeze: Tripping hazard when entering a windy area	Poor
	B		10.37	0.63	2.07	Cool Breeze: Tripping hazard when entering a windy area	Poor
	C		14.48	0.88	2.90	Light breeze: wind perceived in the face	Poor
	D		20.58	1.25	4.12	Strong gale: people tend to rise	Dangerous
	E		12.35	0.75	2.47	Moderate breeze: raises dust and loose papers	Poor
	F		6.25	0.38	1.25	Light gale: discomfort when walking or standing	Moderate

(continued)

Table 2. (*continued*)

Zone	Balcony	U_{BZ} (m/s)	U_L (m/s)	$K_L = U_L/U_{BZ}$	$K_G = U_5/U_L$	Effect	Comfort level
	G		10.37	0.63	2.07	Cool Breeze: Tripping hazard when entering a windy area	Poor
	H		18.44	1.12	3.69	Strong gale: people tend to rise	Dangerous
3	A	18.15	11.98	0.66	2.40	Cool Breeze: Tripping hazard when entering a windy area	Poor
	B		11.98	0.66	2.40	Cool Breeze: Tripping hazard when entering a windy area	Poor
	C		16.15	0.89	3.23	Light breeze: wind perceived in the face	Dangerous
	D		20.87	1.15	4.17	Strong gale: people tend to rise	Dangerous
	E		13.98	0.77	2.80	Moderate breeze: raises dust and loose papers	Poor
	F		7.80	0.43	1.56	Light gale: discomfort when walking or standing	Moderate
	G		9.80	0.54	1.96	Strong breeze - unpleasant noise in the ears	Moderate
	H		20.87	1.15	4.17	Strong gale: people tend to rise	Dangerous
4	A	19.45	10.89	0.56	2.18	Strong breeze - unpleasant noise in the ears	Poor
	B		13.03	0.67	2.61	Cool Breeze: Tripping hazard when entering a windy area	Poor
	C		17.31	0.89	3.46	Light breeze: wind perceived in the face	Dangerous
	D		21.59	1.11	4.32	Strong gale: people tend to rise	Dangerous

(*continued*)

Table 2. (*continued*)

Zone	Balcony	U_{BZ} (m/s)	U_L (m/s)	$K_L = U_L/U_{BZ}$	$K_G = U_5/U_L$	Effect	Comfort level
	E		17.31	0.89	3.46	Light breeze: wind perceived in the face	Dangerous
	F		13.03	0.67	2.61	Cool Breeze: Tripping hazard when entering a windy area	Poor
	G		13.03	0.67	2.61	Cool Breeze: Tripping hazard when entering a windy area	Poor
	H		21.59	1.11	4.32	Strong gale: people tend to rise	Dangerous
$\theta = 90°$							
1	A	14	8.12	0.58	1.62	Cool breeze: tripping hazard when entering a windy area	Moderate
	B		10.08	0.72	2.02	Moderate breeze: raises dust and loose papers	Poor
	C		10.08	0.72	2.02	Moderate breeze: raises dust and loose papers	Poor
	D		12.04	0.86	2.41	Light breeze: wind perceived in the face	Poor
	E		17.5	1.25	3.50	Strong gale: people tend to rise	Dangerous
	F		12.04	0.86	2.41	Light breeze: wind perceived in the face	Poor
	G		0.42	0.03	0.08	Gale: generally, prevents progress, great difficulty with balance due to gusts	Good
	H		17.5	1.25	3.50	Strong gale: people tend to rise	Dangerous
2	A	16.46	10.37	0.63	2.07	Cool Breeze: Tripping hazard when entering a windy area	Poor
	B		12.35	0.75	2.47	Moderate breeze: raises dust and loose papers	Poor
	C		12.35	0.75	2.47	Moderate breeze: raises dust and loose papers	Poor

(*continued*)

Table 2. *(continued)*

Zone	Balcony	U_{BZ} (m/s)	U_L (m/s)	$K_L = U_L/U_{BZ}$	$K_G = U_5/U_L$	Effect	Comfort level
	D		14.48	0.88	2.90	Light breeze: wind perceived in the face	Poor
	E		20.58	1.25	4.12	Strong gale: people tend to rise	Dangerous
	F		12.35	0.75	2.47	Moderate breeze: raises dust and loose papers	Poor
	G		0.07	0.00	0.01	Strong gale: people tend to rise	Good
	H		20.58	1.25	4.12	Strong gale: people tend to rise	Dangerous
3	A	18.15	11.98	0.66	2.40	Cool Breeze: Tripping hazard when entering a windy area	Poor
	B		13.98	0.77	2.80	Moderate breeze: raises dust and loose papers	Poor
	C		16.15	0.89	3.23	Light breeze: wind perceived in the face	Dangerous
	D		16.15	0.89	3.23	Light breeze: wind perceived in the face	Dangerous
	E		20.87	1.15	4.17	Strong gale: people tend to rise	Dangerous
	F		13.98	0.77	2.80	Moderate breeze: raises dust and loose papers	Poor
	G		1.45	0.08	0.29	Strong gale: people tend to rise	Good
	H		20.87	1.15	4.17	Strong gale: people tend to rise	Dangerous
4	A	19.45	17.31	0.89	3.46	Light breeze: wind perceived in the face	Dangerous
	B		17.31	0.89	3.46	Light breeze: wind perceived in the face	Dangerous
	C		15.17	0.78	3.03	Moderate breeze: raises dust and loose papers	Dangerous

(continued)

Table 2. (*continued*)

Zone	Balcony	U_{BZ} (m/s)	U_L (m/s)	$K_L = U_L/U_{BZ}$	$K_G = U_5/U_L$	Effect	Comfort level
	D		17.31	0.89	3.46	Light breeze: wind perceived in the face	Dangerous
	E		21.59	1.11	4.32	Strong gale: people tend to rise	Dangerous
	F		8.56	0.44	1.71	Light gale: discomfort when walking or standing	Moderate
	G		10.89	0.56	2.18	Strong breeze - unpleasant noise in the ears	Poor
	H		21.59	1.11	4.32	Strong gale: people tend to rise	Dangerous

Table 3. Evaluation of comfort on balconies for people standing condition (B.L. + 1.70 m) at $\theta = 0°$ and $\theta = 90°$ for $U15$

Zone	Balcony	U_{BZ} (m/s)	U_L (m/s)	$K_L = U_L/U_{BZ}$	$K_G = U_5/U_L$	Effect	Comfort level
$\theta = 0°$							
1	A	21	12.39	0.59	2.48	Strong breeze - difficulty walking, unpleasant noise in the ears	Poor
	B		12.39	0.59	2.48	Strong breeze - difficulty walking, unpleasant noise in the ears	Poor
	C		18.06	0.86	3.61	Difficulty reading the newspaper	Dangerous
	D		26.04	1.24	5.21	Strong gale: people tend to rise	Dangerous
	E		15.33	0.73	3.07	Cool Breeze: Tripping hazard when entering a windy area	Dangerous
	F		6.51	0.31	1.30	Strong gale: people tend to rise	Moderate
	G		12.39	0.59	2.48	Strong breeze - difficulty walking, unpleasant noise in the ears	Poor
	H		26.04	1.24	5.21	Strong gale: people tend to rise	Dangerous
2	A	24.68	15.55	0.63	3.11	Strong breeze	Dangerous
	B		15.55	0.63	3.11	Strong breeze	Dangerous
	C		21.72	0.88	4.34	Strong breeze	Dangerous
	D		30.36	1.23	6.07	Strong gale: people tend to rise	Dangerous
	E		15.55	0.63	3.11	Strong breeze	Dangerous
	F		9.38	0.38	1.88	Strong gale: people tend to rise	Moderate

(continued)

Table 3. (*continued*)

Zone	Balcony	U_{BZ} (m/s)	U_L (m/s)	$K_L = U_L/U_{BZ}$	$K_G = U_5/U_L$	Effect	Comfort level
	G		12.59	0.51	2.52	Light gale: discomfort when walking or standing	Poor
	H		30.36	1.23	6.07	Strong gale: people tend to rise	Dangerous
3	A	27.2	18.22	0.67	3.64	Strong breeze - difficulty walking, unpleasant noise in the ears	Dangerous
	B		18.22	0.67	3.64	Strong breeze - difficulty walking, unpleasant noise in the ears	Dangerous
	C		24.21	0.89	4.84	Gentle breeze: difficulty reading the newspaper	Dangerous
	D		30.19	1.11	6.04	Strong gale: people tend to rise	Dangerous
	E		18.22	0.67	3.64	Strong breeze - difficulty walking, unpleasant noise in the ears	Dangerous
	F		11.97	0.44	2.39	Strong gale: people tend to rise	Poor
	G		14.96	0.55	2.99	Light gale: discomfort when walking or standing	Poor
	H		30.19	1.11	6.04	Strong gale: people tend to rise	Dangerous
4	A	29.17	16.63	0.57	3.33	Light gale: discomfort when walking or standing	Dangerous
	B		19.84	0.68	3.97	Strong breeze - difficulty walking, unpleasant noise in the ears	Dangerous
	C		25.96	0.89	5.19	Gentle breeze: difficulty reading the newspaper	Dangerous
	D		31.21	1.07	6.24	Strong gale: people tend to rise	Dangerous

(*continued*)

Table 3. *(continued)*

Zone	Balcony	U_{BZ} (m/s)	U_L (m/s)	$K_L = U_L/U_{BZ}$	$K_G = U_5/U_L$	Effect	Comfort level
	E		25.96	0.89	5.19	Gentle breeze: difficulty reading the newspaper	Dangerous
	F		19.84	0.68	3.97	Strong breeze - difficulty walking, unpleasant noise in the ears	Dangerous
	G		19.84	0.68	3.97	Strong breeze - difficulty walking, unpleasant noise in the ears	Dangerous
	H		31.21	1.07	6.24	Strong gale: people tend to rise	Dangerous
$\theta = 90°$							
1	A	21	12.39	0.59	2.48	Strong breeze: unpleasant noise in the ears	Poor
	B		15.33	0.73	3.07	Cool breeze: Tripping hazard when entering a windy area	Dangerous
	C		15.33	0.73	3.07	Cool breeze: Tripping hazard when entering a windy area	Dangerous
	D		18.06	0.86	3.61	Gentle breeze: difficulty reading the newspaper	Dangerous
	E		26.04	1.24	5.21	Strong gale: people tend to rise	Dangerous
	F		18.06	0.86	3.61	Gentle breeze: difficulty reading the newspaper	Dangerous
	G		0.84	0.04	0.17	Strong gale: people tend to rise	Good
	H		26.04	1.24	5.21	Strong gale: people tend to rise	Dangerous
2	A	24.68	15.55	0.63	3.11	Strong breeze - unpleasant noise in the ears	Dangerous
	B		18.51	0.75	3.70	Cool Breeze: Tripping hazard when entering a windy area	Dangerous
	C		18.51	0.75	3.70	Cool Breeze: Tripping hazard when entering a windy area	Dangerous

(continued)

Table 3. (*continued*)

Zone	Balcony	U_{BZ} (m/s)	U_L (m/s)	$K_L = U_L/U_{BZ}$	$K_G = U_5/U_L$	Effect	Comfort level
	D		18.51	0.75	3.70	Cool Breeze: Tripping hazard when entering a windy area	Dangerous
	E		30.6	1.24	6.12	Strong gale: people tend to rise	Dangerous
	F		15.55	0.63	3.11	Strong breeze - unpleasant noise in the ears	Dangerous
	G		0.25	0.01	0.05	Strong gale: people tend to rise	Good
	H		30.36	1.23	6.07	Strong gale: people tend to rise	Dangerous
3	A	27.2	18.22	0.67	3.64	Strong breeze - unpleasant noise in the ears	Dangerous
	B		21.22	0.78	4.24	Cool Breeze: Tripping hazard when entering a windy area	Dangerous
	C		24.21	0.89	4.84	Gentle breeze: difficulty reading the newspaper	Dangerous
	D		24.21	0.89	4.84	Gentle breeze: difficulty reading the newspaper	Dangerous
	E		30.19	1.11	6.04	Strong gale: people tend to rise	Dangerous
	F		21.22	0.78	4.24	Cool Breeze: Tripping hazard when entering a windy area	Dangerous
	G		5.98	0.22	1.20	Strong gale: people tend to rise	Moderate
	H		30.19	1.11	6.04	Strong gale: people tend to rise	Dangerous
4	A	29.17	25.96	0.89	5.19	Gentle breeze: difficulty reading the newspaper	Dangerous
	B		25.96	0.89	5.19	Gentle breeze: difficulty reading the newspaper	Dangerous
	C		23.04	0.79	4.61	Cool Breeze: Tripping hazard when entering a windy area	Dangerous

(*continued*)

Table 3. *(continued)*

Zone	Balcony	U_{BZ} (m/s)	U_L (m/s)	$K_L = U_L/U_{BZ}$	$K_G = U_5/U_L$	Effect	Comfort level
	D		25.96	0.89	5.19	Gentle breeze: difficulty reading the newspaper	Dangerous
	E		31.21	1.07	6.24	Strong gale: people tend to rise	Dangerous
	F		10.5	0.36	2.10	Strong gale: people tend to rise	Poor
	G		16.63	0.57	3.33	Light gale: discomfort when walking	Dangerous
	H		31.21	1.07	6.24	Strong gale: people tend to rise	Dangerous

Table 4 presents a summary of the exceedance probability for wind nuisance in a horizontal plane at 1.7 m height above balcony levels for zones 1 to 4.

Table 4. Exceedance probability for wind nuisance in a horizontal plane at 1.7 m height above balcony levels for zones 1 to 4.

Zone	zb + 1.7	Balcony location	$KL = U_L/U_{BZ}$	U_{thr_H} (m/s)	*Pexc (0° + 90°)**	*Class***
1	41.2	A	0.406	2.03	15.30%	D
		B	0.406	2.03	21.74%	E
		C	0.602	3.01	19.61%	D
		D	0.875	4.375	18.76%	D
		E	0.504	2.52	2.65%	B
		F	0.21	1.05	25.86%	E
		G	0.406	2.03	6.14%	C
		H	0.875	4.375	1.65%	A
2	80.7	A	0.518	2.59	7.04%	C
		B	0.518	2.59	4.03%	B
		C	0.724	3.62	3.23%	B
		D	1.021	5.105	2.82%	B
		E	0.617	3.085	2.57%	B
		F	0.313	1.565	6.20%	C
		G	0.412	2.06	21.86%	E
		H	0.922	4.61	1.40%	A
3	120.2	A	0.599	2.995	7.0%	C
		B	0.599	2.995	4.0%	B
		C	0.807	4.035	3%	B
		D	1.016	5.08	2.8%	B
		E	0.707	3.535	2%	A
		F	0.399	1.995	6%	C

(continued)

It is observed from Table 4 that a variety of quality classes are assigned. The class with the highest frequency is B, indicating that Good conditions for traversing and strolling and Moderate for sitting. It must be emphasized that the classes and exceedance probabilities presented in Table 4 were calculated just for two wind directions. It is noted that a comprehensive analysis of the wind effects on the balconies include a broad range of wind directions that allow the application of the complete probabilistic approach as indicated in NEN 8100 (2006). For this study, only the results for two wind directions were considered.

Table 4. (*continued*)

Zone	zb + 1.7	Balcony location	$KL = U_L/U_{BZ}$	U_{thr_H} (m/s)	*Pexc* ($0° + 90°$)*	*Class***
		G	0.499	2.495	20%	D
		H	1.016	5.08	1.4%	A
4	147.85	A	0.554	2.77	4.0%	B
		B	0.661	3.305	3.6%	B
		C	0.865	4.325	3%	B
		D	1.04	5.2	2.8%	B
		E	0.865	4.325	2%	A
		F	0.661	3.305	7%	C
		G	0.661	3.305	7%	C
		H	1.04	5.2	1.4%	A

* The exceedance probability is calculated just for two directions. ** $P(U_{thr} > 5\ \text{m/s})$: < 2.5% quality class A (traversing = Good, Strolling = Good, Sitting = Good); 2.5–5% quality class B (traversing = Good, Strolling = Good, Sitting = Moderate); 5–10% quality class C (traversing = Good, Strolling = Moderate, Sitting = Poor; 10–20% quality class D (traversing = Moderate, Strolling = Poor, Sitting = Poor; > 20% quality class E (traversing = Poor, Strolling = Poor, Sitting = Poor) (NEN 8100, 2006)

4 Conclusions

The comfort of an irregular in plan tall building with balconies was studied by means of CFD simulations. To evaluate the comfort in this building, the criterion proposed by Lawson and Penwarden (1975) was adopted, based on the measurements of Penwarden (1974) where observations were made on people while they were standing or walking in the test section of a tunnel, as well as outside to evaluate the effects on them. Based on this, the model was studied for three mean wind speeds, 5, 10 and 15 m/s, in two wind directions ($\theta = 0°$ and $\theta = 90°$). From the simulations carried out for U_5, U_{10} and U_{15}, and with the comfort evaluation corresponding to each speed, the following conclusions are drawn:

- It is noted that at a qualitative level, bluff-body aerodynamics indicates that the results are certainly influenced by the specific building and balcony shapes and that the worst conditions are expected at predefined balconies, especially close to the building corners (wind flow separation); however, the latter requires to be numerically validated with CWE.
- The global amplification factor exhibits the most uncomfortable balconies for two wind incident directions, being mainly those located on half of the windward surface and one in the left side surface, i.e., for $\theta = 90°$ the critical balconies are "E" and "H", while for $\theta = 0°$ the critical balconies are "D" and "H".

- In the intermediate zone of the building where there are no balconies, there is no adverse effect that could intensify the wind speeds in the immediate upper and lower balcony.
- The level of comfort can be improved by applying the second skin concept (Montazeri et al. 2013), with the intention of evenly balancing the wind speeds inside the balconies.

Acknowledgements. The financial support received from the Institute of Engineering of the National Autonomous University of Mexico (UNAM) and the National Council on Science and Technology of Mexico (CONACYT) are gratefully acknowledged.

References

Blocken, B., Stathopoulos, T., Carmeliet, J.: CFD simulation of the atmospheric boundary layer: wall function problems. Atmos. Environ. **41**(2), 238–252 (2007)

Franke, J.: Introduction to the prediction of wind loads on buildings by computational wind engineering (CWE). In: Stathopoulos, T., Baniotopoulos, C.C. (eds.) Wind effects on buildings and design of wind-sensitive structures. Springer, Vienna, Austria, 67–103 (2004). https://doi.org/10.1007/978-3-211-73076-8_3

Franke, J., Hellsten, A., Schlünzen, H., Carissimo, B.: Best practice guideline for the CFD simulation of flows in the urban environment. In: COST Action 732, Quality Assurance and Improvement of Microscale Meteorological Models, The European Cooperation in Science and Technology, Brussels, Belgium (2007)

Franke, J., Hellsten, A., Schlunzen, K.H., Carissimo, B.: The COST 732 best practice guideline for CFD simulation of flows in the urban environment: a summary. Int. J. Environ. Pollut. **44**(1–4), 419 (2011)

Lawson, T.V., Penwarden, A.D., The effects of wind on people in the vicinity of buildings. 4th International Conference on Wind Effects on Buildings and Structures. Cambridge University Press. Heathrow, pp. 605–622 (1975)

Montazeri, H., Blocken, B., Jassen, W.D., van Hooff, T.: CFD evaluation of new second-skin facade concept for wind comfort on building balconies: case study for the Park Tower in Antwerp. Build. Environ. **68**, 179–192 (2013)

NEN. Wind comfort and wind danger in the built environment, NEN 8100 Dutch Standard (2006)

Omrani, S., Garcia-hansen, V., Capra, B., Drogemuller, R.: On the effect of provision of balconies on natural ventilation and thermal comfort in high-rise residential buildings. Build. Environ. **123**, 504–516 (2017)

Pozos-Estrada, A., Hong, H., Galsworthy, J.: Serviceability design factors for wind-sensitive structures. Department of Civil and Engineering, University of Western Ontario, London, ON N6A 5B9, Canada (2010)

Tominaga, Y., et al.: AIJ guidelines for practical applications of CFD to pedestrian wind environment around buildings. J. Wind Eng. Ind. Aerodyn. **96**(10–11), 1749–1761 (2008)

Zheng, X., Montazeri, H., Blocken, B.: CFD simulations of wind flow and mean surface pressure for buildings with balconies: comparison of RANS and LES. Build. Environ. **182**, 107017 (2020)

Effect of the Lateral Mean Recirculation Characteristics on Near-Wake and Bulk Quantities of a 5:1 Rectangular Cylinder

G. Lunghi(✉), E. Pasqualetto, A. Mariotti, and M. V. Salvetti

Dipartimento Di Ingegneria Civile E Industriale, Università Di Pisa, Pisa, Italy
gianmarco.lunghi@phd.unipi.it

Abstract. A database of several LES carried out for the high-Reynolds number flow around a 5:1 rectangular cylinder is analyzed to highlight the effects of the characteristics of the mean flow over the cylinder side on the near-wake features and on some bulk quantities. Different simulation set-ups are considered by varying the grid refinement in the spanwise direction and the amount of SGS dissipation for sharp upstream edges and with different values of the upstream-edge rounding. These simulations may be considered as possible instances of this kind of flow with different mean flow features on the cylinder side. Among the different characteristic lengths defining the mean recirculation, the streamwise length is the most important parameter. The wake width increases linearly with this parameter, while the vortex-shedding non-dimensional frequency shows a linear decrease. The drag coefficient also linearly decreases with increasing the recirculation length and this is due to a reduction of the suctions on the cylinder base. However, the overall variation of C_D is small. Finally, a significant, and once again linear, increase of the fluctuations of the lift coefficient is found for increasing the mean recirculation streamwise length.

Keywords: 5:1 rectangular cylinder · Large-Eddy Simulations · lateral and near-wake flow features

1 Introduction

The high-Reynolds number flow around a rectangular cylinder having the ratio between the streamwise length, B, and the crossflow one, D, equal to 5 is paradigmatic of many wind-engineering applications, such as tall buildings and bridge decks. The simplicity of this geometry does not translate into a simple flow, since, on the contrary, the flow shows complex, three-dimensional and unsteady features typical of massive-separated bodies. Indeed, the flow presents a boundary layer that develops on the front face of the cylinder and a shear-layer separation from the upstream edges. Vortical structures of different size form from the roll-up of these shear-layers, move downstream and interact each other. The corresponding time-averaged flow is characterized by the mean flow reattachment on cylinder side, followed by a second separation at the rear corners and by vortex shedding in the wake. This flow is the focus of the BARC international benchmark (Benchmark

P. Schito and A. Zasso (Eds.): IN VENTO 2022, LNCE 461, pp. 137–146, 2024.
https://doi.org/10.1007/978-3-031-53059-3_12

on the Aerodynamic of a Rectangular 5:1 Cylinder), launched in 2008 (Bartoli et al. 2008). This is a blind benchmark, since no reference data are available. A review of the experimental and numerical investigations collected in the first years is presented in Bruno et al. (2014), where a significant dispersion was observed on some quantities of interest. More recent data are available from the experiments in Schewe (2013), Mannini et al. (2017), Nguyen et al. (2018), Moore et al. (2019), Moore and Amitay (2021), Wu et al. (2020), Cardenas-Rondon et al. (2022), Guissart et al. (2022) and Yang et al. (2022), from the unsteady Reynolds-Averaged Navier-Stokes (RANS) simulations in Patruno et al. (2016) and Mariotti et al. (2016), from the Large-Eddy Simulations (LES) in Patruno et al. (2016), Mariotti et al. (2017), Ricci et al. (2017), Rocchio et al. (2020), Zhang and Xu (2020), Wu et al. (2020) and Sakuma et al. (2022), and from the low-Reynolds Direct Numerical Simulations (DNS) in Cimarelli et al. (2018), Cimarelli et al. (2018b), Chiarini and Quadrio (2021, 2022). From an experimental viewpoint, a high sensitivity to the incoming turbulence and small misalignment of the flow was found in Mannini et al. (2017). As regard simulations, Bruno et al. (2012) found out that the flow is highly sensitive to the spanwise grid refinement and, in particular, the increase of the spanwise grid resolution leads a shortening of the mean recirculation region on the cylinder side. Similar results were provided by the stochastic analysis carried out on the LES simulations in Mariotti et al. (2017), where it was also shown how the increase of the subgrid-scale (SGS) dissipation causes the mean recirculation region to shorten. Another source of flow variability is the degree of roundness of the upstream edges, as presented in Rocchio et al. (2020), where it was found that small values of these radii lead to different results compared with perfectly sharp edges. The different mean recirculation regions are due to a different location of the instability onset in the shear layers from the upstream corners. The flow on the cylinder side, in turn, affects the near-wake flow features and the bulk quantities, e.g. the drag coefficient and the vortex-shedding Strouhal number.

We consider herein the LES simulations in Mariotti et al. (2017), Rocchio et al. (2020) and Lunghi et al. (2022), in which the mean recirculation length is highly sensitive to the grid refinement, the SGS modeling and the upstream-edge treatment. Furthermore, we carry out additional LES simulations carried out for higher values of the upstream-corner rounding than in Rocchio et al. (2020) and Lunghi et al. (2022). All the simulations have been carried out at $Re = Du_\infty/\nu = 40000$, being u_∞ the free-stream velocity and ν the kinematic viscosity of air. These simulations give significantly different mean-flow features on the cylinder side and, thus, represent a useful database for our analysis. We investigate possible quantitative links between the features of the mean recirculation region and some quantities of practical interest, such as force coefficients and pressure distributions, to quantify the effect of the lateral mean recirculation characteristics on the near-wake and the bulk quantities and to possibly derive flow-control strategies. Some characteristic lengths of the mean recirculation region are identified and the relevant effects on the wake width, the vortex-shedding Strouhal number, the base pressure and the lift-coefficient fluctuations are addressed.

2 Problem Definition and Investigation Methodology

Large-Eddy Simulations of the incompressible flow around the 5:1 rectangular cylinder are carried out through the open-source code Nek5000, based on a high-order spectral-element method (Fischer et al. 2008). The numerical methodology and the simulation set-up described in Mariotti et al. (2017) and Rocchio et al. (2020) are used, and they are briefly recalled herein. Each spectral element in Nek5000 is rectangular or a suitable coordinate mapping of a rectangle. This basis functions inside each element consist of Legendre polynomials of order N for velocity and N - 2 for pressure in each direction, where the polynomial order is kept constant to $N = 6$. The method has a spectral convergence in N (*p*-refinement) and the convergence of a high-order finite-element method in the number of elements (*h*-refinement). The time resolution is based on a high-order splitting method (Maday et al. 1990), and a third-order backward differentiation scheme is used for the time derivatives. The viscous terms are treated implicitly while an explicit scheme is considered for the non-linear convective term, with a third-order forward extrapolation in time. The time step, made non dimensional by the freestream velocity and the crossflow dimension, is equal to 0.003 and it corresponds to approximately 3000 time steps for each vortex-shedding cycle.

The Reynolds number, based on the freestream velocity and on the cylinder depth, D, is equal to 40000. The computational domain is rectangular with $-15.5 \leq x/B \leq 25.5$, $-15.1 \leq y/B \leq 15.1$, where x and y are the streamwise and crossflow directions, respectively, and $B = 5D$. The domain size in spanwise direction is B. The cylinder center is located at $x = y = 0$. A uniform velocity profile with no turbulence is imposed at the inflow, no-slip conditions are applied at the solid walls and traction-free boundary conditions are used for the outflow and for the upper and lower boundaries of the computational domain. Finally, periodic boundary conditions are imposed in the spanwise direction.

The element size in the streamwise and crossflow directions are $\Delta x/D = \Delta y/D =$ 0.125 near the cylinder; in the spanwise direction, four different values are considered in the simulations with the sharp edges, viz. $\Delta z_i/D = (0.649, 0.558, 0.437, 0.346)$, where $i = 1, 2, 3, 4$, as in Mariotti et al. (2017), while the simulations with rounded corners has the spanwise grid resolution fixed at $\Delta z_2/D = 0.558$ (Rocchio et al. 2020). A low-pass explicit filter is applied to velocity field in the modal space at the end of each time-integration step. The adopted filter is a sharp cut-off for the modes up to k_c (unfiltered) and it has a quadratic transfer function for the modes inside $k_c \leq k \leq N$. Thus, the transfer function of the filter is written as:

$$\begin{cases} \sigma_k = 1 k < k_c \\ \sigma_k = 1 - w\left(\frac{k-k_c}{N-k_c}\right)^2 k_c \leq k \leq N \end{cases} \tag{1}$$

where w is a weighting parameter. This modal filter introduces dissipation of the highest resolved modes only, therefore it can be considered as a SGS dissipation (Domarazdki, 2010), in which the amount of dissipation introduced is controlled by the filter weight, w. As in Mariotti et al. (2017), four values of the filter weight are considered in the sharp-edge configurations, viz. $w_i = (0.018, 0.05, 0.0912, 0.1218)$, where i = 1, 2, 3, 4, whereas the rounded-edge configurations have a fixed value of the filter weight equal to $w_2 = 0.05$ (Rocchio et al. 2020).

Sharp upstream edges ($r_0/D = 0$) are considered in the simulations in Mariotti et al. (2017), whereas the different upstream-edge rounding in Rocchio et al. (2020) are $r_i/D =$ (0.0037, 0.0177, 0.0360, 0.0500), where $i = 2, 3, 4, 5$. An additional simulation is carried out to match the value of the roundness of the experiments in Pasqualetto et al. (2022), i.e. $r_1/D = 0.0005$. Finally, two simulations consider higher values of the upstream roundings, i.e. $r_6/D = 0.0781$ and $r_7/D = 0.1104$. All the cases are is summarized in Table 1.

Table 1. Summary of the values of the parameters for the 23 Large-Eddy Simulations.

Upstream-edge rounding	Filter weight	Spanwise grid resolution
$r/D = 0$	$w =$ [0.018, 0.05, 0.0912, 0.1218]	$\Delta z/D =$ [0.649, 0.558, 0.437, 0.346]
$r/D =$ [0.0005, 0.0037, 0.0177, 0.0360, 0.0500, 0.0781, 0.1104]	$w = 0.05$	$\Delta z/D = 0.558$

3 Results and Discussion

We compare the results of 23 Large-Eddy Simulations to highlight the links between the lateral mean recirculation features and the near-wake and bulk quantities on the flow around a 5:1 rectangular cylinder. Two different flow fields are found among the results of the simulations: a long lateral mean-recirculation (see, e.g., the results for the simulation with w_2 and Δz_1 in Mariotti et al., 2017, shown in Fig. 1a) and a short lateral mean-recirculation (see, e.g., the results for w_2 and Δz_2 in Mariotti et al., 2017, reported in Fig. 1b). It was indeed found that the increase of the spanwise grid resolution (decrease of the value of Δz) leads to a shortening of the mean reattachment length, l_r.

As discussed in Rocchio et al. (2020), a governing quantity of the mean recirculation length is the position of the peak of the turbulent kinetic energy evaluated on the outer border of the separated shear layers, shown in Fig. 2a. The distance of this position from the upstream edge, l_t, plays a fundamental role in predicting the mean recirculation length on the cylinder side. Indeed, all the data in Fig. 2b collapse on a single straight line, proving the linear correlation between the location of the shear-layer instability and the length of the mean recirculation. In particular, the delay in the onset of the instability on the shear layers causes the maximum fluctuations to occur further downstream and, thus, a lengthening of the mean recirculation.

Now we consider the distribution on the cylinder walls of the pressure coefficient, $Cp = (p - p_\infty)/(1/2\rho u_\infty{}^2)$, being p the local pressure, ρ the density of air and p_∞ and u_∞ the freestream pressure and velocity, respectively. The distribution of the side-, spanwise- and time-averaged pressure coefficient, $\langle C_p \rangle$, and its standard deviation, $\sigma(C_p)$, are reported in Fig. 3a and 3b, respectively. The results for the long-recirculation case are shown in the top panel, whereas the ones for the short recirculation in the

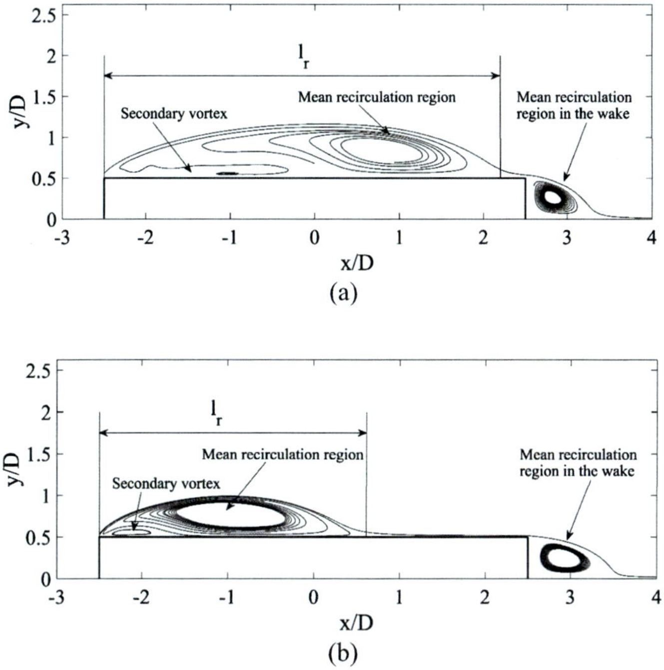

Fig. 1. Mean-flow streamlines: (a) long mean-recirculation of the case (w_2, Δz_1) from Mariotti et al. (2017) and (b) short mean-recirculation of the case (w_2, Δz_2) from Mariotti et al. (2017). The quantity l_r indicates the mean reattachment length.

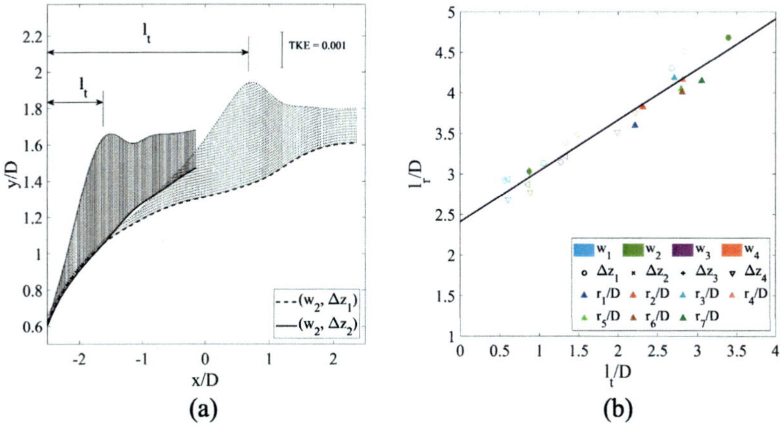

Fig. 2. (a) Turbulent Kinetic Energy (TKE) evaluated on the external border of the mean separated shear-layer and (b) correlation between the streamwise distance of the maximum TKE along the shear layer borders, l_t, and the mean reattachment length on the cylinder side, l_r.

bottom panel. For both recirculation lengths, a clear peak of the pressure coefficient fluctuations along the cylinder side, l_p, measured from the upstream edge, is found and

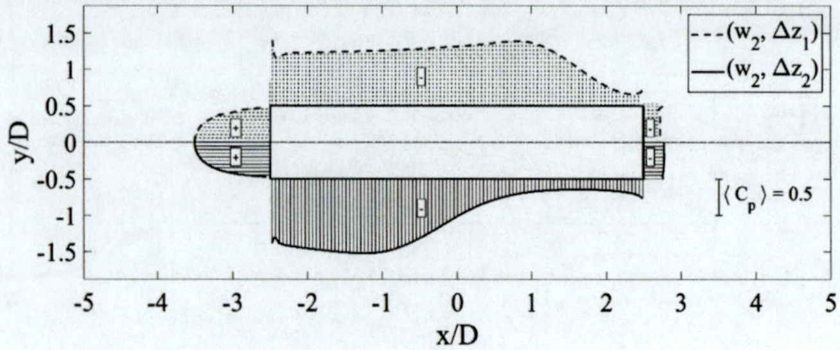

(a)

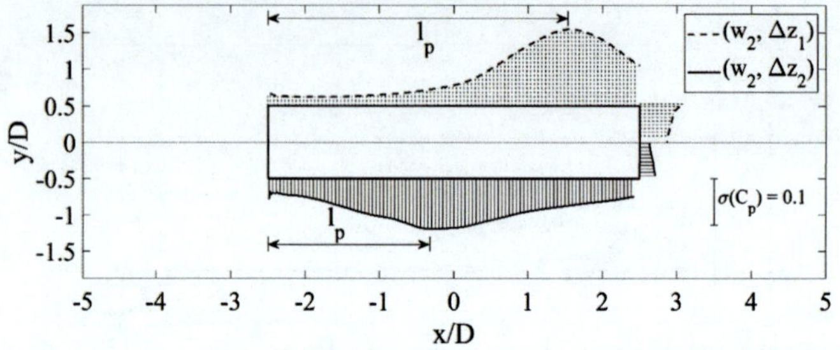

(b)

Fig. 3. Distribution (a) of the side-, spanwise- and time-averaged pressure coefficient, $\langle C_P \rangle$, and (b) of its standard deviation, $\sigma(C_P)$, on the cylinder surface for the long-recirculation case (top panel) and for the short-recirculation case (bottom panel). The quantity l_p corresponds to the distance of the peak of the pressure coefficient fluctuations from the upstream edge.

a linear relation between l_r and l_p was reported in Lunghi et al. (2022) as $l_r = l_p + 0.614$. Both lengths l_r and l_p can be, thus, considered as measures of the lateral mean recirculation. The intensity of the maximum pressure coefficient fluctuations, $\sigma(C_P)_{max}$ is linearly correlated with the length l_p (Fig. 4) and the longer the mean recirculation is, the higher the intensity of the peak is. The additional LES simulations carried out in this work confirm this trend and the experimental data from Pasqualetto et al. (2022) and Mannini et al. (2017) also collapse on the same linear correlation.

In Fig. 5, we shown the effect of the mean reattachment length, l_p, on the pressure coefficient acting on the rear face of the cylinder averaged in time and on the base itself, $\langle C_P \rangle_{base}$. A linear decrease of the suction acting on the base is found by decreasing the mean reattachment length. As shown in Mariotti et al. (2017) and Rocchio et al. (2020), the reduction of the base suction is related to the reduction of suction in the rear part of the cylinder lateral side; the same trend can be found in the experiments in Pasqualetto et al. (2022).

The length of the lateral mean recirculation influences also the wake borders. In Fig. 6a the external borders of the lateral separated flow (evaluated as in Mariotti et al., 2013) and the wake borders are shown for the last portions of the cylinder and for the near-wake up to 2.5D downstream of the cylinder base. The wake width, t_w, is evaluated at 2D downstream of the cylinder base, and it is linearly linked with the mean recirculation length l_p, as can be seen in Fig. 6b, and a shorter recirculation leads to a

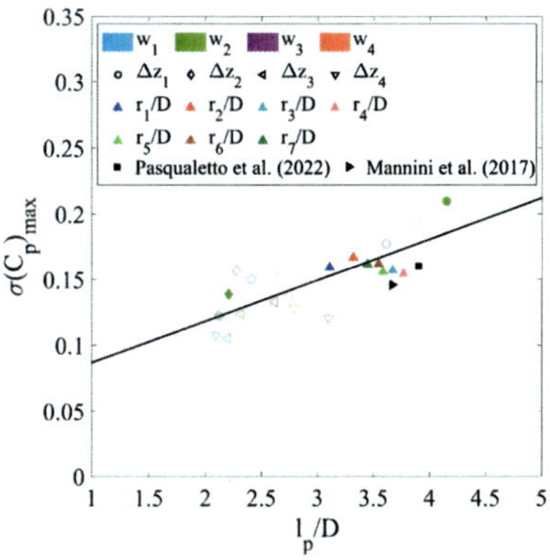

Fig. 4. Correlation between the position of the maximum pressure coefficient fluctuations on the cylinder side, *lp*, and the intensity of the maximum pressure coefficient fluctuations, $\sigma(Cp)max$.

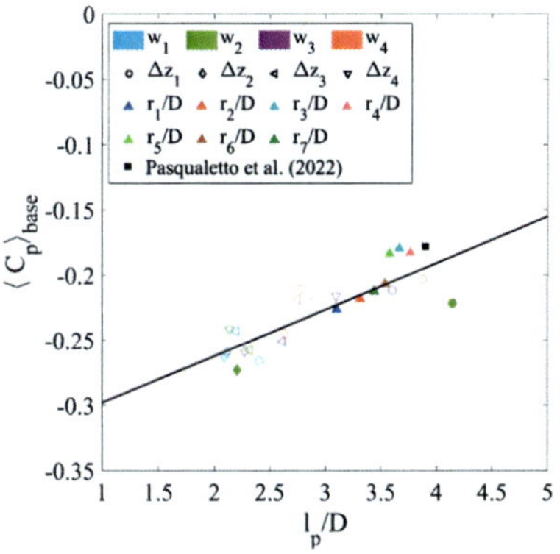

Fig. 5. Effect of the mean recirculation length, l_p, on the time- and base-averaged pressure coefficient, $\langle C_p \rangle_{base}$.

narrower wake. The increase and the upstream movement of the inward curvature of the mean flow streamlines bounding the recirculation region seems to be the main cause of the narrowing of wake.

The length l_p has a significant impact on the Strouhal number of the vortex-shedding frequency, *St* (Fig. 7a), computed following the procedure in, e.g., Mariotti (2018) and Perna et al. (2020). A clear trend is found: a shorter recirculation leads to a to a higher vortex-shedding Strouhal number. Finally, the lengthening of the lateral mean recirculation region leads the lift coefficient to highly fluctuate, as shown in Fig. 7b. As explained in Bruno et al. (2010), the lengthening of the lateral mean recirculation favors the interaction between the vortices shed from the shear layers and from the trailing edges, and this may be reasonably ascribed as cause of the increased lift coefficient fluctuations, C_L'. Therefore, it is reasonable to assume the mean recirculation length as the controlling parameter also for these quantities.

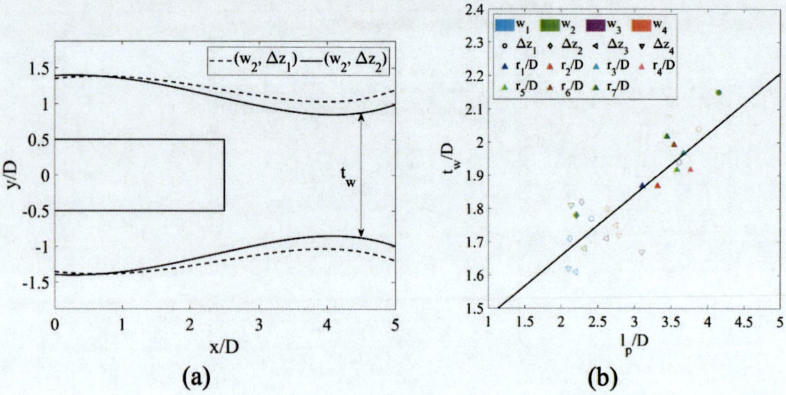

Fig. 6. (a) Sketch of the external border of the mean separated shear-layer for $x/D > 0$ and of the wake border. The quantity t_w represents the wake width evaluated 2D downstream of the cylinder base (at $x/D = 4.5$). (b) Effect of the mean recirculation length, l_p, on the wake width, t_w.

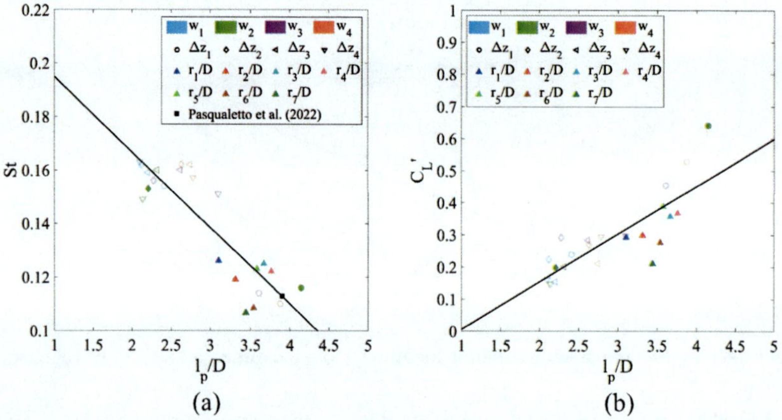

Fig. 7. Effect of the mean recirculation length, l_p, (a) on the vortex-shedding Strouhal number based on the cylinder depth, St, and (b) on the fluctuations of the lift coefficient, $C_L{}'$.

4 Conclusions

We analyze a database of 23 Large-Eddy Simulations, representative of different lateral mean-flow features, to quantitatively describe the effect of the lateral mean-flow recirculation on the near-wake and bulk quantities. The mean streamwise length of the recirculation region is directly related to the onset of the roll-up in the detaching shear layers, and it can be measured both in terms of the streamwise distance between the upstream edge and the mean reattachment point, l_r, or of the streamwise distance between the upstream edge and the position of the peak of the pressure coefficient fluctuations on the cylinder side, l_p. These two quantities differs only by a constant and, in this work, l_p is chosen because it is easier to be measured in experiments. The length of the lateral mean recirculation, l_p, is a controlling parameter for many near-wake flow features and bulk quantities: its increase causes a thickening of the wake width, a linear decrease of the vortex-shedding Strouhal number, a linear decrease of the base suction, and an increase of the fluctuations of the lift coefficient. These effects are caused by the reduction of the inward curvature of the streamlines in the configurations with a longer lateral mean recirculation and to the interactions between the vortices separated from the shear layers detaching from the upstream edges and those shed at the trailing edges.

References

1. Bartoli, G., Bruno, L., Buresti, G., Ricciardelli, F., Salvetti, M.V., Zasso, A.: BARC overview document (2008). http://www.aniv-iawe.org/barc
2. Bruno, L., Coste, N., Fransos, D.: Simulated flow around a rectangular 5:1 cylinder: spanwise discretization effects and emerging flow features. J. Wind Eng. Ind. Aerod. **104–106**, 203–215 (2012)
3. Bruno, L., Fransos, D., Coste, N., Bosco, A.: 3D flow around a rectangular cylinder: a computational study. J. Wind Eng. Ind. Aerod. **98**(6–7), 263–276 (2010)
4. Bruno, L., Salvetti, M.V., Ricciardelli, F.: Benchmark on the aerodynamics of a rectangular 5:1 cylinder: an overview after the first four years of activity. J. Wind Eng. Ind. Aerod. **126**, 87–106 (2014)
5. Cardenas-Rondon, J., Ogueta-Gutierrez, M., Franchini, S., Gomez-Ortega, O.: Experimental analysis of the aerodynamic characteristics of a rectangular 5:1 cylinder using pod. Wind Struct. **34**(1), 29–42 (2022)
6. Chiarini, A., Quadrio, M.: The turbulent flow over the BARC rectangular cylinder: a DNS study. Flow Turbul. Combust. **107**, 875–899 (2021)
7. Chiarini, A., Quadrio, M.: The importance of corner sharpness in the BARC test case: a numerical study. Wind Struct. **34**(1), 43–58 (2022)
8. Cimarelli, A., Leonforte, A., de Angeli, D.: Direct numerical simulation of the flow around a rectangular cylinder at a moderately high Reynolds number. J. Wind Eng. Ind. Aerod. **174**, 39–49 (2018)
9. Cimarelli, A., Leonforte, A., de Angeli, D.: On the structure of the self-sustaining cycle in separating and reattaching flows. J. Fluid Mech. **857**, 907–936 (2018)
10. Domarazdki, J.A.: Large eddy simulations without explicit eddy viscosity models. International J. Comput. Fluid Dynamic **24**(10), 435–447 (2010)
11. Fischer, P.F., Lottes, J.W., Kerkemeier, S.G.: Nek5000 Web page (2008). http://nek5000.mcs.anl.gov
12. Guissart, A., Elbæk, E., Hussong, J.: PIV study of the flow around a 5:1 rectangular cylinder at moderate Reynolds number and small incidence angle. Wind Struct. **34**(1), 15–27 (2022)
13. Lunghi, G., Pasqualetto, E., Rocchio, B., Mariotti, A., Salvetti, M.V.: Impact of the lateral mean recirculation characteristics on the near-wake and bulk quantities of the BARC configuration. Wind & Struct. **34**(1), 115–125 (2022)
14. Maday, Y., Patera, A.T., Rønquist, E.M.: An operator-integration-factor splitting method for time-dependent problems: application to incompressible fluid flow. J. Sci. Comput. **5**(4), 263–292 (1990)
15. Mannini, C., Marra, A.M., Pigolotti, L., Bartoli, G.: The effects of free-stream turbulence and angle of attack on the aerodynamics of a cylinder with rectangular 5:1 cross section. J. Wind Eng. Ind. Aerod. **161**, 42–58 (2017)
16. Mariotti, A., Grozescu, A.N., Buresti, G., Salvetti, M.V.: Separation control and efficiency improvement in a 2D diffuser by means of contoured cavities. Eur. J. Mech B-Fluids **41**, 138–149 (2013)

17. Mariotti, A.: Axisymmetric bodies with fixed and free separation: base-pressure and near-wake fluctuations. J. Wind Eng. Ind. Aerod. **176**, 21–31 (2018)
18. Mariotti, A., Salvetti, M.V., Shoebi-Omrani, P., Witteveen, J.A.S: Stochastic analysis of the impact of freestream conditions on the aerodynamics of a rectangular 5:1 cylinder. Comput. Fluids **136**, 170–192 (2016)
19. Mariotti, A., Siconolfi, L., Salvetti, M.V.: Stochastic sensitivity analysis of large-eddy simulation predictions of the flow around a 5:1 rectangular cylinder. Eur. J. Mech./B Fluids **62**, 149–165 (2017)
20. Moore, D.M., Amitay, M.: Production and migration of turbulent kinetic energy in bluff body shear layers. Int. J. Heat Fluid Flow **88**, 108716 (2021)
21. Moore, D.M., Letchford, C.W., Amitay, M.: Energetic scales in a bluff body shear layer. J. Fluid Mech. **875**, 543–575 (2019)
22. Nguyen, D.T., Hargreaves, D.M., Owen, J.S.: Vortex-induced vibration of a 5:1 rectangular cylinder: a comparison of wind tunnel sectional model tests and computational simulations. J. Wind Eng. Ind. Aerod. **175**, 1–16 (2018)
23. Pasqualetto, E., Lunghi, G., Rocchio, B., Mariotti, A., Salvetti, M.V.: Experimental characterization of the lateral and near-wake flow for the BARC configuration. Wind Struct. **34**(1), 101–113 (2022)
24. Patruno, L., Ricci, M., de Miranda, S., Ubertini, F.: Numerical simulation of a 5:1 rectangular cylinder at non-null angle of attack. J. Wind Eng. Ind. Aerod. **151**, 146–157 (2016)
25. Perna, R., et al.: Flow characterization of pulsating heat pipe through the wavelet analysis of pressure signals. App. Ther. Eng. **171**, 115128 (2020)
26. Ricci, M., Patruno, L., de Miranda, S., Ubertini, F.: Flow field around a 5:1 rectangular cylinder using LES: Influence of inflow turbulence conditions, spanwise domain size and their interaction. Comput. Fluids **149**, 181–193 (2017)
27. Rocchio, B., Mariotti, A., Salvetti, M.V.: Flow around a 5:1 rectangular cylinder: effects of upstream-edge rounding. J. Wind Eng. Ind. Aerod. **204**, 104237 (2020)
28. Sakuma, M., Popper, N., Warnakulasuriya, S., Montomoli, F., Wüch-ner, R., Bletzinger, K.: Multi-fidelity uncertainty quantification of high Reynolds number turbulent flow around a rectangular 5:1 cylinder. Wind Struct. **34**(1), 127–136 (2022)
29. Schewe, G.: Reynolds-number-effects in flow around a rectangular cylinder with aspect ratio 1:5. J. Fluids Struct. **39**, 15–26 (2013)
30. Wu, B., Li, S., Li, K., Zhang, L.: Numerical and experimental studies on the aerodynamics of a 5:1 rectangular cylinder at angles of attack. J. Wind Eng. Ind. Aerod. **119**, 104097 (2020)
31. Yang, Y., Li, M., Yang, X.: Fluctuating lift and drag acting on a 5:1 rectangular cylinder in various turbulent flow. Wind & Struct. **34**(1), 137–149 (2022)
32. Zhang, Z., Xu, F.: Spanwise length and mesh resolution effects on simulated flow around a 5:1 rectangular cylinder. J. Wind Eng. Ind. Aerod. **202**, 104186 (2020)

Wind Field Simulation and Wind Damage Assessment of Typhoon Rai in the Philippines

J. C. Agar(✉), T. S. Acosta, A. A. Barros, and J. Y. Hernandez Jr.

University of the Philippines, Quezon City, Philippines
jcagar@up.edu.ph

Summary. Typhoon Rai (Local Name: Odette) was regarded as the strongest typhoon to have struck the Southern Philippines in the last 31 years. It is also the first violent typhoon classified by the new classifications by the Japan Meteorological Agency. A forensic meteorological analysis was performed to extrapolate the surface winds and 10-min sustained winds experienced by the affected areas using the gradient wind equation. Gradient winds are determined to be closer to the surface winds during land interaction because of dissipation in the pressure field. Wind exposure plays a major factor in how the winds amplify or decrease over different exposure categories. A subsequent wind damage assessment is performed in the Cebu Province to relate the documented building damages to the winds determined by the wind field simulation, which identified 73% of building damage outliers for concrete structures and 87% of building damage outliers for steel structures.

Keywords: Typhoon modeling · Typhoon Rai · severe wind building vulnerability

1 Introduction

Lying on the pathway of Pacific typhoons, the Philippines monitors about 20 typhoons a year (Cinco et al. 2010) within the Philippine Area of Responsibility, where the Philippine Atmospheric, Geophysical, and Astronomical Services (PAGASA) assigns local names to the typhoons within, apart from the international designation by the Japan Meteorological Agency (JMA). Among the hazards that typhoons bring are severe winds, which are considered in the design of buildings in the Philippines.

The National Structural Code of the Philippines (NSCP) has wind design provisions implemented across engineered structures in the Philippines (ASEP, 2015). The design provisions were then put to the test when Typhoon Rai (Local Name: Odette) caused extensive infrastructural damages across Southern Philippines in December 2021, where 10-min sustained winds were estimated to be 185 kph at landfall. The areas affected by Typhoon Rai were previously affected by Typhoon Mike in 1990. The 31-year gap may have explained the social factors that resulted in the lack of wind resistance of non-engineered structures. But that doesn't explain some engineered buildings that suffered catastrophic failure, most notably the recently built Siargao Sports Complex, shown in

P. Schito and A. Zasso (Eds.): IN VENTO 2022, LNCE 461, pp. 147–159, 2024.
https://doi.org/10.1007/978-3-031-53059-3_13

Fig. 1. The Association of Structural Engineers of the Philippines (ASEP), when making the latest design provisions in the NSCP, imposed a design wind speed of 310 kph. The synoptic data from PAGASA Surigao Station, 60 km from the sports complex, recorded peak gusts of 208.8 kph. Since the sports complex was closer to Typhoon Rai than the station, the peak winds may have been higher.

Fig. 1. Siargao Sports Complex (Courtesy of ASEP)

The building failures documented in the aftermath of Typhoon Rai require detailed simulation, reanalysis, and investigation. To assess severe wind damage and to determine the building vulnerability of the affected area, the magnitude of the peak gusts experienced must be determined. This study aims to simulate the typhoon wind field and validate it using ground surface data to determine the winds experienced by the buildings that are assessed in the wind damage assessment.

2 Methods

The study used ground surface data from weather stations listed in Table 1. The weather stations are managed by either PAGASA or the Advanced Sciences and Technology Institute (ASTI). The station data over time are related to the position of Typhoon Rai over time using the storm track from Japan Meteorological Agency shown in Fig. 2.

Relating the position of Typhoon Rai to the weather stations resulted in determining the distance for every barometric pressure reading. The pressure vs distance data comprised the pressure profile of Typhoon Rai shown in Fig. 3.

With the compiled pressure profile resembling a rectangular hyperbola, the study fitted pressure profiles using Schloemer (1954) by determining the shape parameters A and B:

$$p(r) = p_0 + (p_n - p_0)e^{-\frac{A}{r^B}} \quad (1)$$

where p_0 is the minimum central pressure, p_n is the ambient pressure assumed to be equal to 1013 hPa, A is the shape parameter that determines the radius of maximum

Table 1. List of Weather Stations

Location	Agency	Type of Station	Latitude	Longitude	Peak Gust (m/s)
Guiuan	DOST-PAGASA	Synoptic/Radar	11.045199	125.755482	57
Maasin 1	DOST-PAGASA	Synoptic	10.139008	124.8604	55
Surigao	DOST-PAGASA	Synoptic	9.782717	125.489417	58
Mactan	DOST-PAGASA	Synoptic	10.32257	123.980488	46
Mainit	DOST-ASTI	Automatic Rain Station	9.540661	125.522986	
Malimono	DOST-ASTI	Automatic Rain Station	9.61856	125.40162	
Tubod	DOST-ASTI	Automatic Rain Station	9.553854	125.569493	
Madrid	DOST-ASTI	Automatic Rain Station	9.26147	125.96407	
Santiago	DOST-ASTI	Automatic Rain Station	9.2494262	125.549146	
Tago 1	DOST-ASTI	Automatic Rain Station	9.016448	126.233591	
Tago 2	DOST-ASTI	Automatic Rain Station	8.94671	126.038571	
Maasin 2	DOST-ASTI	Automatic Rain Station	10.13175	124.857183	
Malitbog	DOST-ASTI	Automatic Rain Station	10.16725	124.9866111	
Anahawan	DOST-ASTI	Automatic Rain Station	10.2741666	125.2581	
Giporlos	DOST-ASTI	Automated Rain Station	11.12049	125.45	
Palo	DOST-ASTI	Automatic Rain Station	11.178882	125.003715	
Leyte-Leyte	DOST-ASTI	Automatic Rain Station	11.371111	124.486444	
Dulag	DOST-ASTI	Automated Weather Station	10.91905556	125.0101111	19
Basilisa	DOST-ASTI	Automatic Rain Station	10.065417	125.596733	

(*continued*)

Table 1. (*continued*)

Location	Agency	Type of Station	Latitude	Longitude	Peak Gust (m/s)
Dinagat	DOST-ASTI	Automatic Rain Station	9.954167	125.604433	
Libjo	DOST-ASTI	Automatic Rain Station	10.195108	125.533495	
Loreto	DOST-ASTI	Automatic Rain Station	10.359748	125.57843	

Fig. 2. Location of the weather stations (red) and the track of Typhoon Rai (black)

winds (Holland, 1980), and B is the shape parameter that relates the pressure gradient to the magnitude of the maximum winds.

Mainly considering the primary circulation of Typhoon Rai, three main forces have been identified and balanced, namely the Coriolis force (fictitious acceleration due to Earth's rotation and differences in relative velocities), the Centrifugal force, and the pressure gradient force:

$$\frac{U^2}{r} + f|U| + \frac{1}{\rho_{air}}\frac{\partial p}{\partial r} = 0 \tag{2}$$

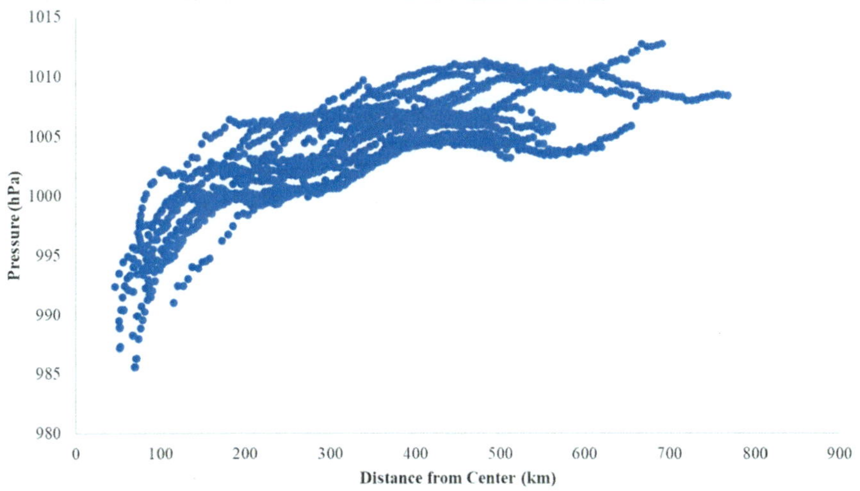

Fig. 3. Barometric Pressure vs Distance Measures

where U is the velocity field, r is the radius of curvature or the radial distance (given that the pressure field is assumed to be composed of concentric and circular pressure isobars), f is the Coriolis parameter, which is dependent on the latitude, ρ_{air} is the density of air, and $\frac{\partial p}{\partial r}$ is the pressure gradient. (Holmes, 2001)

Substituting Eq. (2) to Eq. (1) results in the gradient wind profile:

$$U(r) = -\frac{fr}{2} + \sqrt{\left(\frac{fr}{2}\right)^2 + \frac{AB}{\rho_{air} r^B} e^{-\frac{A}{r^B}}} \quad (3)$$

For every period of observation, a pressure profile was fitted over the station mean sea level pressure data, with an example shown in Fig. 4. The goodness of fit is used as an indicator of the reliability of the gradient wind profile that has been derived from it.

The 10-min sustained wind profiles, on the other hand, are fitted using Eq. (3) from the detailed wind information data of the Japan Meteorological Agency, which contained the maximum 10-min sustained wind speed, minimum central pressure, the storm wind radius, and the gale wind radius, shown in Fig. 5.

The study also considers that the gradient wind profile obtained using Eq. (3) is relative to the center of Typhoon Rai, which also has a translation velocity:

$$\overrightarrow{U_{surface}(r)} = \overrightarrow{U(r)} + U_{translation} \quad (4)$$

c Complex, the authors mainly focused on Cebu Province, where they conducted a building assessment survey before Typhoon Rai made landfall.

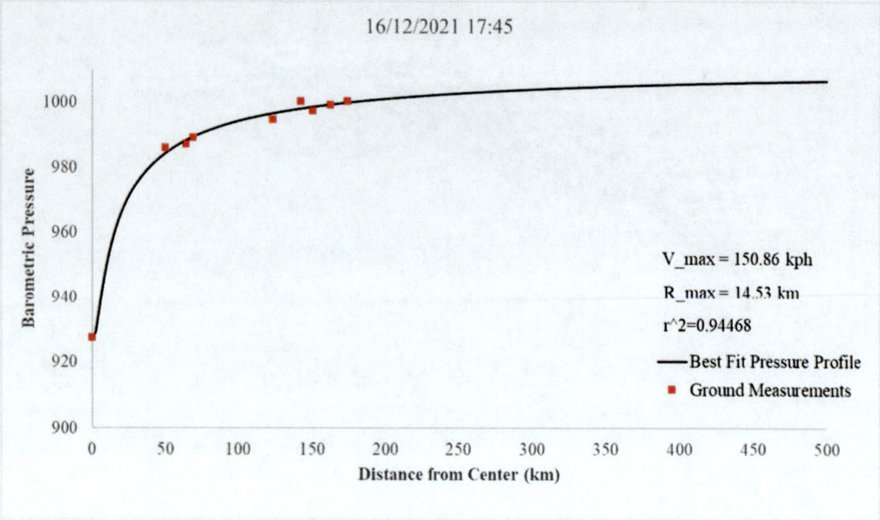

Fig. 4. Pressure

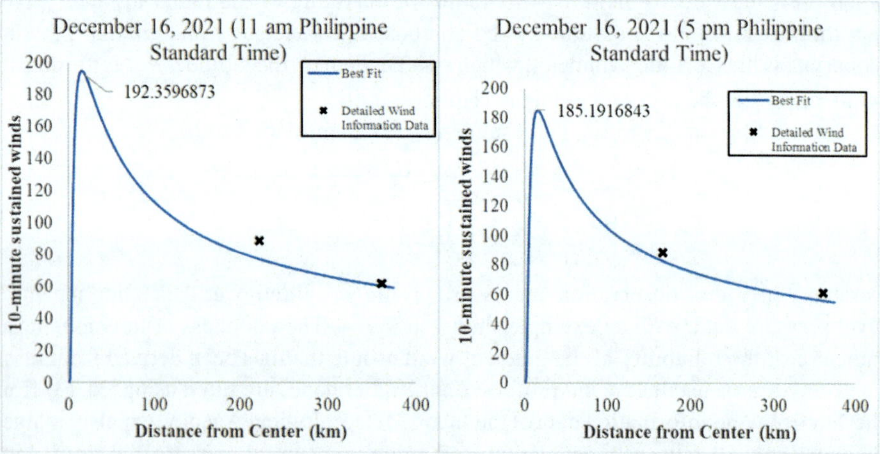

Fig. 5. Deriving velocity profile from the detailed wind information data. (Left – before landfall; Right – during landfall)

3 Results

The maximum wind speeds derived from the pressure field modeling using ground surface data are plotted in Fig. 6. The goodness of fit is also plotted in Fig. 6. The periods of observation with the best goodness of fit are determined to be at 5:45 pm and 9:00 pm (Philippine Standard Time) shown in Figs. 4 and 7, respectively.

The simulated wind field shown in Fig. 8 enabled the authors to determine survey areas within the province of Cebu. Building damage across 20 engineered buildings in

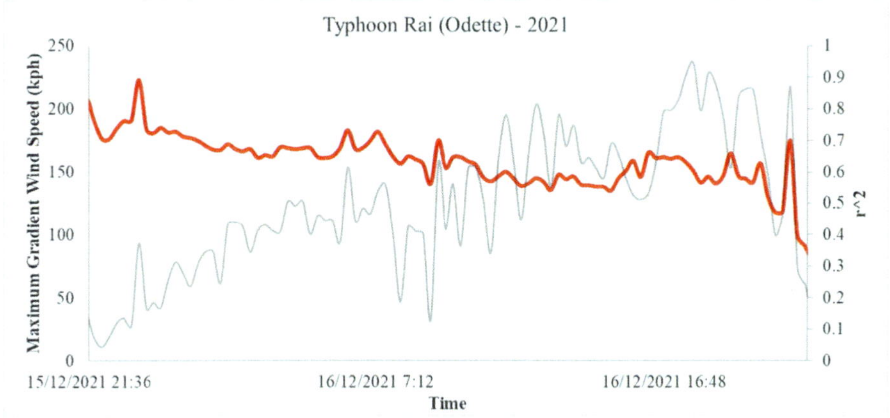

Fig. 6. Maximum wind speeds (red) determined from pressure profiles at different periods of observation and goodness of fit (grey)

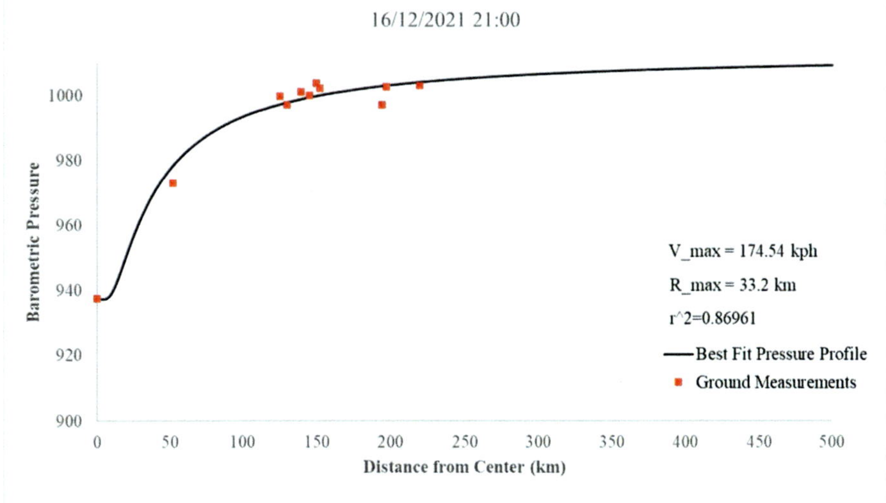

Fig. 7. Pressure Field fitting at the time when Typhoon Rai made landfall at Cebu.

Cebu City, Lapu-Lapu City, Mandaue City, and Argao are evaluated and listed in Table 2. The building typology and damage assessment were determined based on the UPD-ICE classifications (Gumaro et al. 2021).

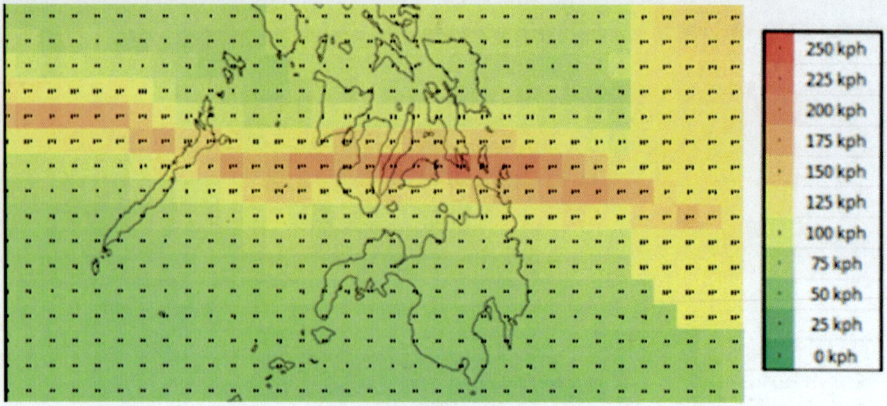

Fig. 8. Simulated 10-min sustained wind field derived from JMA's detailed wind information data.

4 Discussions

4.1 Wind Field Modeling

Even with a good fit and managing to capture the pinhole eye profile of Typhoon Rai before landfall in Fig. 4, the gradient winds derived from the pressure data are less in magnitude compared to the evaluated 10-min sustained winds, shown in Fig. 9. The existence of surface data reported by PAGASA enabled the comparison of the wind field values to the actual data, shown in Table 3.

The 10-min wind field values, which are compared to the actual surface data measured by the PAGASA synoptic stations in Table 3, are shown to have a relative difference of not less than 10 m/s. The difference is due to the wind exposure, which mainly affects the gustiness. As the Mactan and Surigao station of PAGASA is situated near the coast, it has exhibited lower gust factors, while the Maasin Station, which is situated inland, has higher gust factors. The gust factors determined in this study are consistent with the relationship of the gust factors to the wind exposure identified by Harper et al. (2010). Since it is a radar station situated on top of a hill, Guiuan Station may have experienced topographic amplification resulting in higher peak winds compared to the flat-surface conditions.

On the other hand, the gradient wind captures the larger scale of flows. When dissipation occurs when the typhoon begins to interact with land surfaces, the energy of the gradient wind cascades into smaller scales resulting in sporadic amplification of winds compared to the magnitude of the winds in the production scale, therefore affecting the wind spectra. The more dissipative the environment, the closer the production scale of the gradient wind to the scale of the surface wind.

This is why Typhoon Rai had lesser gradient winds than the surface winds when it first made landfall at Siargao island while having gradient winds closer to the surface winds when it was in the middle of making landfall across the Visayas region.

At the location of the Siargao Sports Complex, the peak gradient wind was determined to be 142.67 kph, shown in Fig. 10, while the peak 10-min sustained wind was

Table 2. Wind Damage Assessment Survey on Engineered Buildings in Cebu Province

Lat	Long	10-min Wind Speed (m/s)	Building Type	Roof Cover Damage (%)	Purlin Damage (%)	Column Damage (%))	Roof Truss Damage (%)	Window Damage (%)	Wall Damage (%)
9.880869	123.6084	52	Concrete	100	100	0	0	0	0
10.31122	123.8923	43	Concrete	10	0	0	0	10	0
10.10624	123.6391	49	Concrete	86	50	0	50	5	0
10.34341	123.9605	42	Concrete	0	0	0	0	0	0
10.34444	123.9598	42	Steel	3	1	0	0	0	0
10.34291	123.9604	43	Concrete	25	0	0	0	0	10
10.31205	123.8861	43	Concrete	0	0	0	0	0	0
10.28326	123.9553	44	Steel	3.33	1.5	0	0	0	0
10.28255	123.958	44	Concrete	50	0	0	0	0	0
10.28257	123.9579	44	Concrete	50	0	0	0	0	0
10.28264	123.9578	44	Concrete	50	0	0	0	0	0
10.28272	123.9577	44	Concrete	25	0	0	0	0	0
10.28515	123.96	44	Steel	80	75	0	75	90	4
10.28818	123.9616	44	Steel	2	2	0	0	0	0
10.28788	123.9624	44	Steel	2.27	2.27	0	0	0	0
10.29047	123.9628	44	Steel	35	0	0	0	0	0
10.29046	123.965	44	Steel	11	0	0	0	0	0
10.29126	123.9656	44	Concrete	30	0	0	0	0	0
10.29319	123.9658	44	Concrete	22	0	0	0	0	0
10.29355	123.9662	44	Concrete	100	95	0	5	13	10

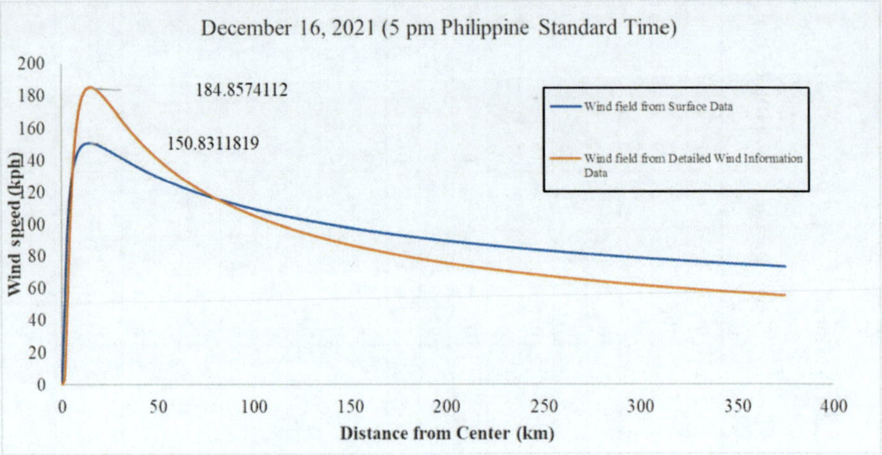

Fig. 9. Comparison of the 10-min sustained wind field from detailed wind information data (orange) and the gradient wind field from the pressure data.

Table 3. Comparison of Surface Data to Simulated Data

Location	Peak Gust (m/s)	10-min Wind (m/s)	Percent Difference	Gradient Wind(m/s)	Percent Difference
Maasin, Southern Leyte	55	45.29	17.65	40.31	26.71
Surigao City, Surigao del Norte	58	53.57	7.64	42.77	26.26
Mactan, Cebu	46.5	43.19	7.12	46.05	1.81
Guiuan, Eastern Samar	56	20.0	64	26.43	0.01

determined to be 184.86 kph. This could be a benchmark for forensic structural analysis when checking whether the as-built building resistance of the collapsed sports complex can sustain the wind loads that a 10-min sustained wind of 184.86 kph can induce upon it.

4.2 Wind Damage Assessment

Gumaro et al. (2021) have previously determined the computational vulnerability curves from the analysis of the buildings and building types present in the Cebu Province, identifying the minimum, average, and maximum vulnerability curves that reflect the expected performance of the buildings there against severe winds, given the typology and structural configurations. The higher confidence in gradient winds capturing the

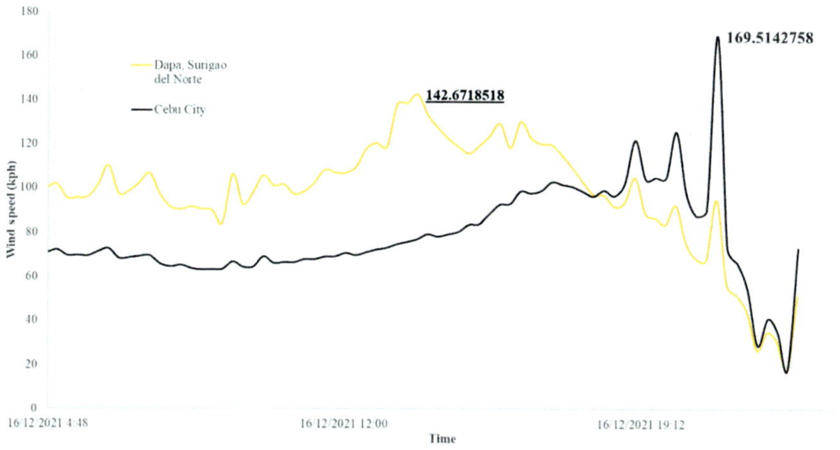

Fig. 10. Time history of gradient winds for the Siargao Sports Complex (Yellow) and Cebu City (Black)

Fig. 11. Documented building damage in Cebu Province

surface data at Typhoon Rai's landfall in the province has enabled the authors to project the peak winds the documented damaged buildings experienced.

The comparison of the documented building damages shown in Fig. 11 to the computational vulnerability curves and to the empirical data from a previous survey in the Bicol region (Region V), which was affected by a typhoon in 2019 but not Typhoon Rai, are shown in Figs. 12 and 13.

The study has identified 73% building damage outliers for concrete structures and 87% building damage outliers for steel structures, indicating that the building resistance against severe wind might be lesser than the wind demand that it is expected to resist. A careful evaluation and recalibration must be done to improve the performance

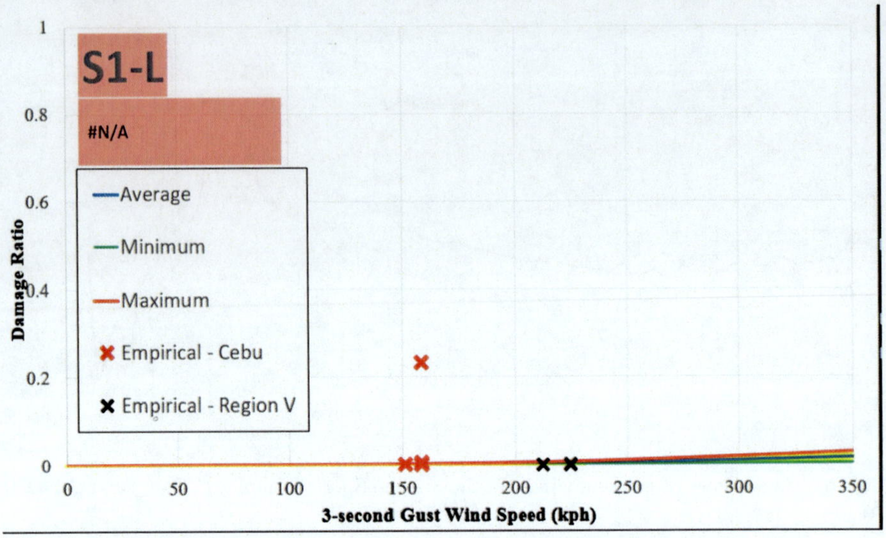

Fig. 12. Vulnerability Curves for Steel Structures and Empirical Data

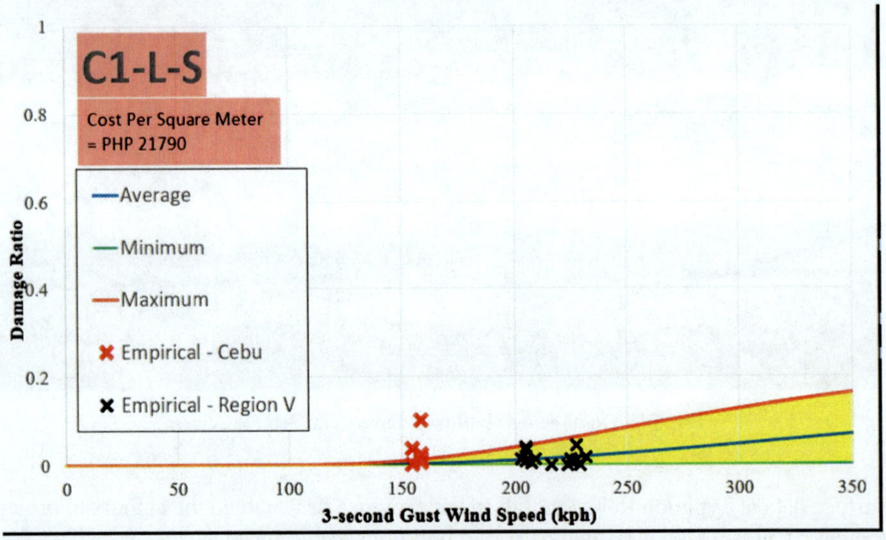

Fig. 13. Vulnerability Curves for Concrete Structures and Empirical Data

of buildings against severe winds in the future. Unlike the Cebu Province, which has not experienced severe winds since Typhoon Mike in 1990, Bicol Region has been more frequented by typhoons, with a design wind speed of 310 kph (compared to Cebu Province's 260 kph), hence people in the region have built their structures more resistant and resilient against severe winds.

5 Conclusions

The wind fields generated are determined to be consistent with identifying the gust positively related to the wind exposure (terrain and surface roughness). Wind exposure plays a part in amplifying the winds that the affected areas have experienced. Relating the wind field to the surveyed building damage shows a sudden increase of roof cover damage and purlin damage over the 40–50 m/s wind speed range. However, the documented damage indicated lower overall wind resistance of buildings in Cebu province compared to what was previously assessed in the last building vulnerability analysis. A careful reassessment of the capacity of structural components used in construction must be performed.

Acknowledgements. The authors would like to acknowledge the Association of Structural Engineers of the Philippines and the Philippine Institute of Civil Engineers for initiating and funding the severe wind damage assessment across the affected areas. Also, the authors acknowledge Engr. Jihan S. Pacer, Engr. Liezl Tan, and Engr. Dean Ashton Plamenco and Dr. Oscar Victor Antonio from the UPD Institute of Civil Engineering, who joined the authors in the field survey conducted in Cebu Province. The authors also acknowledge the DOST project titled "Enhanced Severe Wind Vulnerability Curves for Different Building Types in Cebu Province". The authors also acknowledge DOST-ASTI and DOST-PAGASA for the surface data.

References

Acosta, T., et al.: Enhanced Severe Wind Vulnerability Curves of Key Building Types in the Philippines (2021)

Association of Structural Engineers of the Philippines, 2015. National Structural Code of the Philippines (2015)

Cinco, T., et al.: Updating tropical cyclone climatology in the Philippines. In: Proceedings on Earthquake and Severe wind Exposure and Vulnerability Workshops. PHIVOLCS, PAGASA, and Geoscience Australia (2010)

Joshua, J.C., et al.: Identification of key components for developing building types for risk assessment against wind loadings: the case of Cebu Province, Philippines. Int. J. Dis. Risk Reduct. **67**, 102686 (2022). https://doi.org/10.1016/j.ijdrr.2021.102686

Harper, B., Kepert, J., Ginger, J.: Guidelines for converting between various wind averaging periods in tropical cyclone conditions. World Meteorological Or (2010)

Holland, G.: An analytic model of the wind and pressure profiles. Mon. Weather Rev. **108**(9), 1212–1218 (1980)

Holmes, J.: Wind Loadings on Structures (2001)

Schloemer, R.: Analysis and synthesis of hurricane wind patterns over Lake Okeechobee. FL. Hydromet Rep. **31**, 49 (1954)

Liezl, R.E., et al.: Investigation of the effects of the classification of building stock geometries determined using clustering techniques on the vulnerability of galvanized iron roof covers against severe wind loading. IOP Conf. Ser. Mater. Sci. Eng. **1150**(1), 012024 (2021). https://doi.org/10.1088/1757-899X/1150/1/012024

Gust Factors, Turbulence Intensities, Roughness Lengths, and Wind Direction Factors of Typhoon Winds in the Philippines

J. C. Agar(✉) and J. Y. Hernandez Jr.

University of the Philippines, Quezon City, Philippines
{jcagar,jyhernandez}@up.edu.ph

Summary. The exposure of a location against winds in the wind design provisions from the National Structural Code of the Philippines is mainly dependent on the topography and the degree of vegetation and urbanization. The wind exposure includes the vertical wind profile, the gust factors, and angles of approach. The current code provisions on the parameters for wind exposure are mainly dependent on surface roughness, which has been derived from steady wind conditions. However, winds from typhoons, the extreme winds being designed against, are, in fact, unsteady winds. The current code provisions also cover only area-based exposure, which assumes that wind exposure is constant. The provisions do not cover feature-based wind exposure, where winds can vary over short distances depending on the placement of the buildings and the angle of approach. To evaluate the exposure parameters at these conditions, two typhoon events are simulated using an unsteady-state CFD analysis. Wind exposure for typhoon winds reveals higher local gusts factors, which are related to the direction of the local wind and change in elevation. The upper limit of the surface roughness is determined by the building density and average roof height of the buildings, while its variation is dependent on the wind flow channeling, separation, and reattachment.

Keywords: Wind exposure · Typhoons

1 Introduction

Severe wind is one of the multiple destructive hazards that the Philippines is exposed to. In its 2015 edition, the National Structural Code of the Philippines (NSCP) updated its basic wind speed maps to consider the newer data that were compiled in the Generalized Extreme Value analysis (De Leoz et al. 2014). Being patterned after the provisions of ASCE-07, the NSCP 2015 basic wind speed maps have supposedly return periods of 1700 years (3% probability in 50 years), shown in Fig. 1.

However, upon checking the GEV functions from the peak gusts data of synoptic stations from PAGASA, inconsistencies between the extreme values and the basic wind speeds were observed, as stations in Mindanao returned less extreme wind speed values compared to that of the NSCP 2015 wind maps, while the Eastern seaboards of Luzon

P. Schito and A. Zasso (Eds.): IN VENTO 2022, LNCE 461, pp. 160–174, 2024.
https://doi.org/10.1007/978-3-031-53059-3_14

and Visayas returned higher wind speed values compared to that of the NSCP 2015, shown in Table 1.

One overlooked aspect in the inception of the current wind speed maps was the nature of the extreme winds at each of the weather stations. Evaluating the history of the recorded winds across the weather stations uncovered that the severe winds experienced by some weather stations, like Davao City, Cagayan de Oro City, and Butuan City, are not cyclonic in nature and are not caused by typhoons.

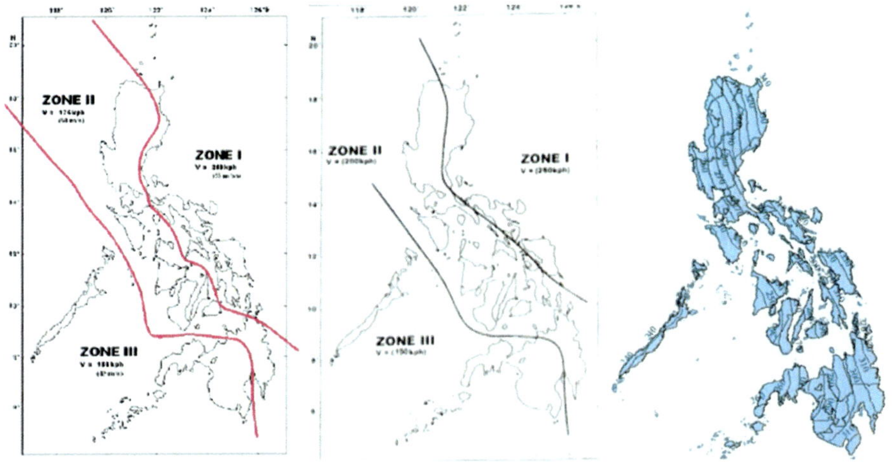

Fig. 1. Basic Wind Speed Maps of NSCP

Table 1. Comparison of GEV values to the Basic Wind Speed Map Values

Station	Extreme Winds (kph)	NSCP 2015 (kph)	Difference (kph)	Cyclonic/Non-Cyclonic
Virac Synoptic	402	330	−72	Cyclonic
Guiuan	394	320	−64	Cyclonic
Itbayat	369	340	−29	Cyclonic
Casiguran	366	315	−51	Cyclonic
Manila	259	265	+5	Cyclonic
Mactan	170	285	+115	Cyclonic
Davao City	117	305	+188	Non-cyclonic
Cagayan de Oro	97	280	+183	Non-cyclonic
Butuan City	128	300	+172	Non-cyclonic
Tacloban City	212	310	+98	Cyclonic

Another aspect overlooked, particularly in the extrapolation of the extreme values from the synoptic stations, was the factor brought by wind exposure. The weather stations experienced their respective peak winds, which was also a result of their distinct wind exposure, which was a factor of topography, vegetation, the direction of the wind, and other possible local effects. Local effects of wind exposure can be divided into two by evaluating the wind spectra: the mesoscale and the microscale. The mesoscale and the microscale are distinct from each other, for they are separated by what's called the "spectral gap", where monsoon and seasonal winds have longer time scales and larger spatial scales compared to cyclonic winds (Powell et al. 1996), which can occur over shorter periods of time and have smaller times scales. (Van der Hoven 1957). The nature of the dissipation will affect the cascade with the wind energy density being concentrated over a certain time scale, resulting in higher wind speeds over the averaging period that is equal to that time scale.

The distinction also identified two independent energy cascades, with the energy cascade of the mesoscale often associated with the wind exposure for the seasonal winds, which are steady or stationary in nature, and the energy cascade of the microscale often associated with the wind exposure of typhoon winds which are unsteady and non-stationary in nature.

Three key notions were observed for both energy cascades: (1) more dissipative effects like rougher exposures (hilly and heavily vegetated terrain) led to higher turbulent kinetic energy, (2) more turbulent kinetic energy meant a wider range of eddy sizes that would carry and transfer the kinetic energy, and (3) a wider range of eddy sizes meant higher peaks in instantaneous winds, resulting to higher gustiness. Indicative of the degree of turbulence, the gustiness is further simplified into what's called turbulence intensity (Harper et al. 2010).

There are parameters, other than the turbulence intensity, identified when concerning wind exposure. Roughness length represents the degree how the velocity changes along the vertical in atmospheric boundary layer winds. Wind direction factors, which were not present in the NSCP 2015, indicate in which direction an area is exposed to the strongest winds.

Due to the limitations during the time that they were conceived, the wind exposure was simplified to be related to the terrain classifications alone. Simplifications (Davenport 1961) were made previously made by in the horizontal gustiness (Charnock 1955) in evaluating the logarithmic wind profile for boundary layer winds and in relating turbulence intensity to roughness lengths in higher latitudes (Lumley and Panofsky 1964). In that manner, boundary layer winds were simulated in order to predict wind loadings on structures in a wind tunnel (Davenport and Isyumov 1967), which became the basis for the wind exposure conditions codified in the first wind code provisions in ASCE-1969. As adapted in the ASCE wind code provisions and in, extension, the NSCP 2015 wind code provisions, for the wind tunnel wind to develop the boundary layer profile, stationarity should be observed, in contrast to the nonstationary nature of typhoon winds that engineers are designing against.

This study aims to provide feature-based parameters for gust factors, roughness lengths, wind direction factors, and turbulence intensities during typhoon events, considering the urban effects and nonstationary effects. In defining the wind exposure parameters of typhoon winds, a wind field over a complex and urban terrain was simulated using Computational Fluid Dynamics. To capture the nonstationary effects, Large Eddy Simulation is used to model the eddies that carry the turbulent effects.

This study covered two cases: (1) winds brought by Typhoon Haiyan (Local Name: Yolanda) in Bantayan, Cebu, at 8:30 am on November 8, 2013, and (2) winds brought by Typhoon Nock-ten (Local Name: Nina) in Virac, Catanduanes at 10:30 am on December 25, 2016.

2 Methods

2.1 Typhoon Wind Field

The evaluation of the typhoon wind field marked the start of the hybrid procedure that will simulate the wind field over the subject areas. Focusing on the gradient wind balance of typhoons, the pressure gradient force, centrifugal force, and Coriolis force would be considered, along with the radius of curvature of the pressure isobars.

$$\frac{U^2}{r_{curv}} + f|U| + \frac{1}{\rho_{air}}\left|\frac{\partial p}{\partial r}\right| = 0 \tag{1}$$

where U is the gradient wind, assumed as both the friction-free surface wind and 10-min sustained winds, ρ_{air} is the air density, r_{curv} is the radius of curvature of the pressure isobar, f is the Coriolis parameter, $\partial p / \partial r$ is the pressure gradient fitted using a rectangular hyperbolic profile (Schloemer 1954) (Holland 1980) from the barometer pressures from synoptic weather stations by PAGASA and the Advanced Sciences and Technology Institute (ASTI) along the path of the typhoons subjected to this study listed in Table 2 and shown in Figs. 2 and 3.

Table 2. Weather Stations Included in the Analysis

Station	Agency	Type	Latitude	Longitude	Typhoon
Virac Synoptic	DOST-PAGASA	Synoptic	13.57914	124.2121	Nock-ten (2016)
Legazpi	DOST-PAGASA	Synoptic	13.15079	123.7284	Nock-ten (2016)
Calapan	DOST-PAGASA	Synoptic	13.40977	121.1896	Nock-ten (2016)
Romblon	DOST-PAGASA	Synoptic	12.59066	122.2753	Nock-ten (2016)
Ambulong	DOST-PAGASA	Synoptic	14.09008	121.0552	Nock-ten (2016)

(continued)

Table 2. (*continued*)

Station	Agency	Type	Latitude	Longitude	Typhoon
Guiuan	DOST-PAGASA	Synoptic	11.04541	125.7557	Haiyan (2013)
Tacloban	DOST-PAGASA	Synoptic	11.22561	125.0249	Haiyan (2013)
Bantayan, Cebu	DOST-ASTI	Automated Weather Station	11.166528	+ 123.716861	Haiyan (2013)
San Carlos, Negros Occidental	DOST-ASTI	Automated Weather Station	10.502866	123.433944	Haiyan (2013)
Ajuy, Iloilo	DOST-ASTI	Automated Weather Station	11.169037	123.018532	Haiyan (2013)

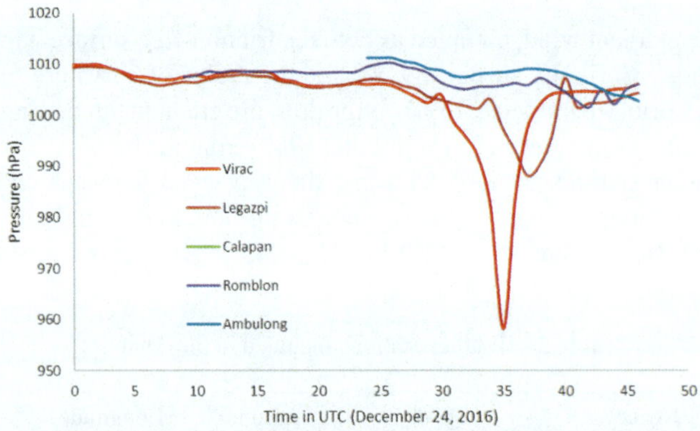

Fig. 2. Pressure Time History Across Weather Stations During Typhoon Nock-ten 2016

2.2 Complex-Terrain Wind Simulation

The gradient wind determined using Eq. (1) was then carried over as inlet velocities for the micro-scale analysis where the unsteady dissipation of the wind over the complex terrain - which were the subject areas - was simulated.

The complex terrain was divided into two parts: the topography and the buildings. The topographic model was obtained from the elevation data from OpenStreetMaps. OpenStreetMaps also provided the urban layout of buildings in the sample area. Field surveys provided further insight into the key dimensions of buildings found in the study areas.

To reduce the computational load for feasibility purposes, two measures were made. First, as the building edges indicated higher characteristic changes in the flow leading to higher resolution and higher computational load, the need to reduce the edges led into

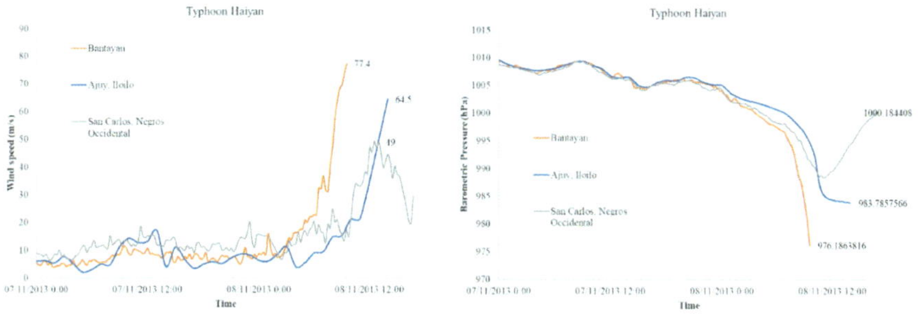

Fig. 3. Wind Speed and Pressure Time History Across Automated Weather Stations During Typhoon Haiyan 2013

modeling the buildings as rectangular blocks whilst retaining the key dimensions such as height, width, and length (Fig. 4).

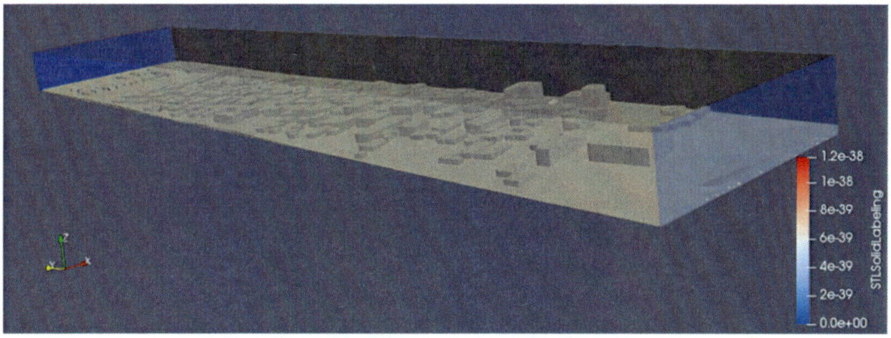

Fig. 4. Computational Model – Bantayan, Cebu

Second, the computational model was scaled down by 1:10 in both the length scale and velocity scale. Even though this scaling followed the similitude laws involving the Reynolds Number, the dynamic similarity involving the Navier-Stokes Equations would not be conserved as the viscosity of the air remained unchanged in the setup. The non-conservation of the dynamic similarities in the energy cascade results in the smaller eddies not being modeled. However, it was also assumed that these smaller eddies contributed little to gustiness, hence their effects can be neglected.

The simulation is done using Computational Fluid Dynamics (CFD). Finite volume method was used (Demirdzic 1983) as the numerical method in solving for the Navier-Stokes Equations in a manner of balancing the pressure field first in an algorithm called Pressure-Implicit with Splitting of Operators or PISO algorithm (Issa, 1986):

$$\frac{\partial p}{\partial x_i} = -\frac{\partial u_i}{\partial t} - u_j\frac{\partial u_i}{\partial x_j} + (\nu + \nu_t)\left(\frac{\partial^2 u_i}{\partial x_j \partial x_j}\right) \tag{2}$$

where p is the pressure, ui is the velocity field, ν is the kinematic viscosity of air, and νt is the eddy viscosity derived from modeling the eddies using the Filtered Turbulence Kinetic Energy Transport Large Eddy Simulation Model (You and Moin 2007).

2.3 Extraction of Gust Factors, Turbulence Intensities, Roughness Lengths, and Wind Direction Factors

To treat the feature-based wind exposure parameters as if they are area-based wind exposure parameters, the gust factors, turbulence intensities, and roughness lengths are extracted in a manner that assumes an exposed flat terrain where only the ground surface is causing shear stresses that resulted in a logarithmic profile. A filtering procedure during postprocessing was done using three main criteria: (1) The goodness of fit of the logarithmic profile was greater than 0.95 to ensure that a boundary layer profile was observed, (2) the vertical profile contained more than 10 points to ensure robustness and to eliminate the chance of overestimation of the reference velocity which was set at 10 m from the ground, and (3) the ground elevation computed in the curve fitting was consistent with the modeled profile which was expressed in a power law equation shown below:

$$U(z) = U_{10m}\left(\frac{z}{10}\right)^{\frac{1}{\ln\left(\frac{10}{z0}\right)}} \tag{3}$$

where U(z) is the velocity function with respect to a height z, U_{10m} is the reference velocity at 10 m high, and z0 is the roughness length being extracted. Consequently, U_{10m} was reflected on the computed gradient wind in Eq. (1a) to obtain the gust factors.

$$I_{u,prescribed}(z) = \frac{1}{\ln\left(\frac{z}{z0}\right)} \tag{4}$$

Once the roughness lengths were obtained, the turbulence intensities were computed using the Eq. (1d) from Lumley and Panofsky (1964) and were compared with measured turbulence intensities in the simulation using the equation below:

$$I_{u,\tau} = \frac{\sigma_{u,\tau}}{\overline{U_{T0,\tau}}} \tag{5}$$

where $\sigma_{u,\tau}$ is the standard deviation of the fluctuating wind and U_{T0}, τ is the average value of the fluctuating wind.

Wind direction factors are then obtained locally by determining the angle between the local wind from the general direction of the simulated wind.

The postprocessing of results involved a single section in the Virac case and 5 sections which included four alleyways in the Haiyan case, shown in Fig. 5.

This study presents the profile of the exposure values (values will be presented in the axis on the right) in the sections mentioned in a manner that the winds travel from right to left. Also presented in the profile was the elevation (values will be presented in the axis on the left) of the surfaces involved which will include escarpments, topographic changes, and buildings.

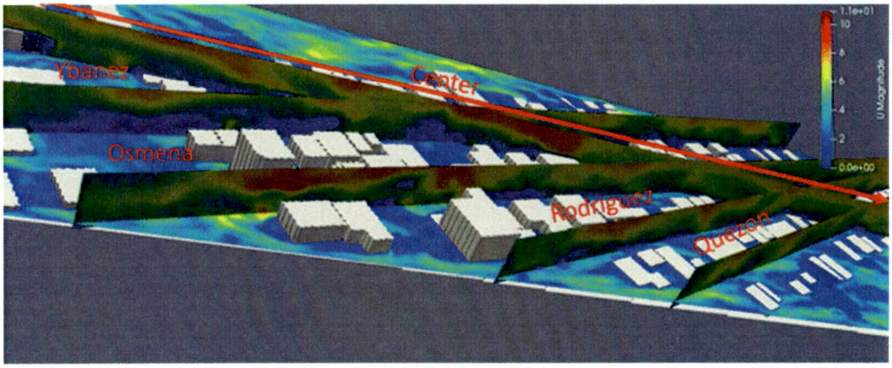

Fig. 5. Simulated Peak Gust at Bantayan Cebu

3 Results and Discussion

3.1 Validation

Within the simulated wind field of each of the simulated wind fields were validation points where an anemometer was installed. These anemometers recorded peak winds which were used to validate the simulated winds.

For the case of Typhoon Haiyan (Yolanda), at 8:30 am on November 8, 2013, the automated weather station, situated at the municipal hall of Bantayan, Cebu, recorded peak gusts of 64.1 m/s. The peak gust in the simulation was measured to be at 66.26 m/s, shown in Fig. 6.

In the case of Typhoon Nock-ten (Nina), at 10:30 am on December 25, 2016, the synoptic station near the airport in Virac, Catanduanes recorded a peak gust of 54 m/s. The peak gust in the simulation was measured to be at 57.83 m/s, shown in Fig. 7.

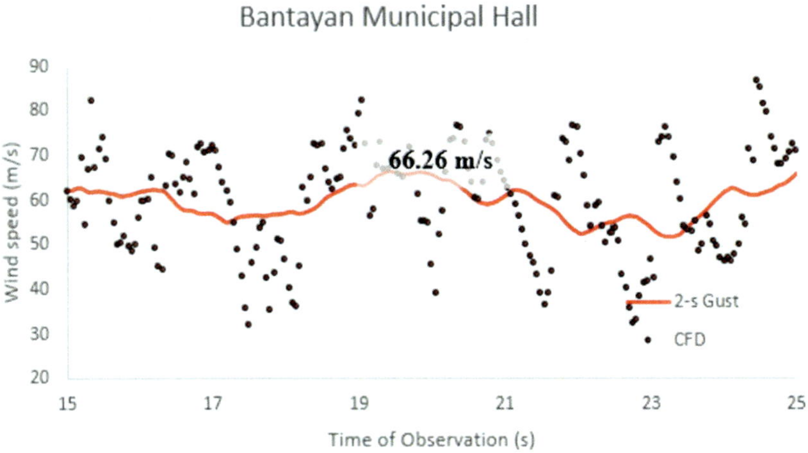

Fig. 6. Simulated Peak Gust at Bantayan Cebu

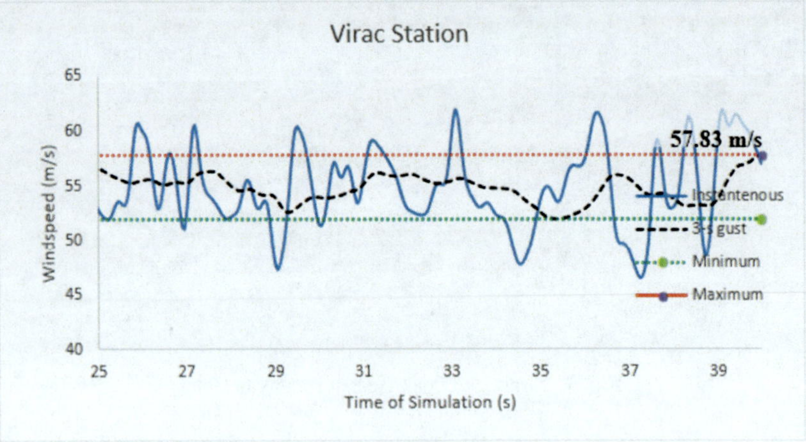

Fig. 7. Simulated Peak Gust at Virac, Catanduanes

3.2 Effects of Building Density

As the early settlement in Bantayan, Cebu is centered at the municipal hall, residential areas of the affluent families near the municipal hall exhibit lesser building densities than the residential areas farther from the municipal hall.

As shown in Figs. 8 and 9, the areas with higher building densities are shown to have higher peak roughness lengths while also having higher variations of roughness length when compared to the distribution of roughness lengths at the areas of lower building densities. Higher peak but more varied roughness lengths result in high variations of local winds at the mean roof height which was used in evaluating the wind loads in NSCP 2015.

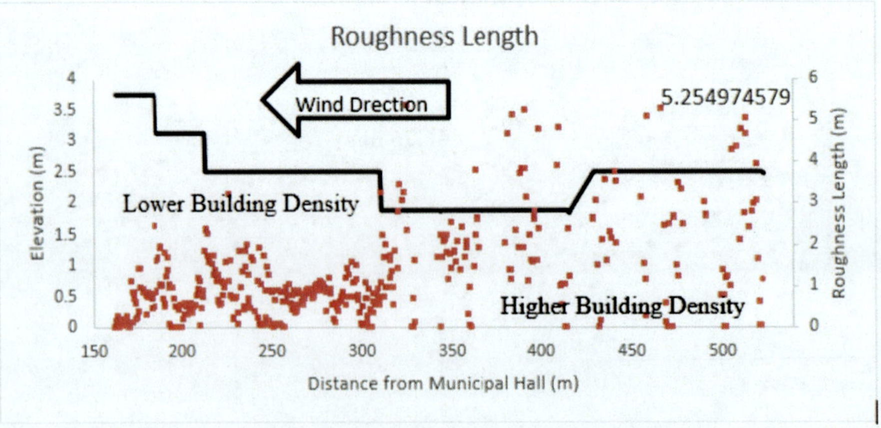

Fig. 8. Roughness Length over Rodriguez Street

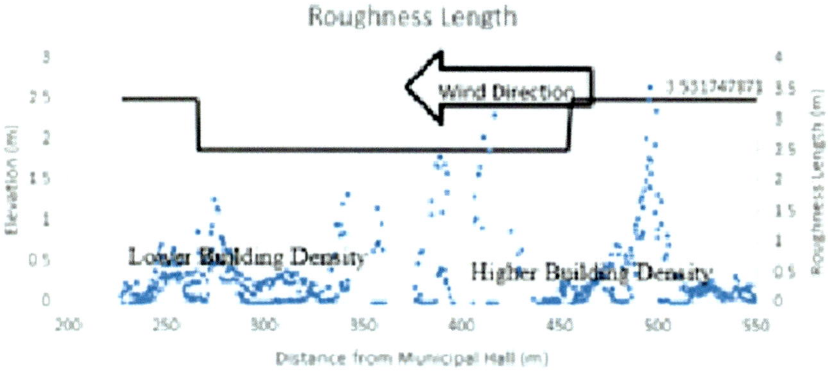

Fig. 9. Roughness Length over Quezon Street

3.3 Local Wind Direction

Due to the existence of buildings or obstacles along the path of the wind, some of the local winds exhibit a different direction compared to the general direction of winds. When plotting the directional deviations to the gust factors of the local winds, a fan-shaped envelope is observed, shown in Fig. 10. This indicates that when the local wind is near the same direction as the general wind, the more that the gustiness picks up, increasing in value.

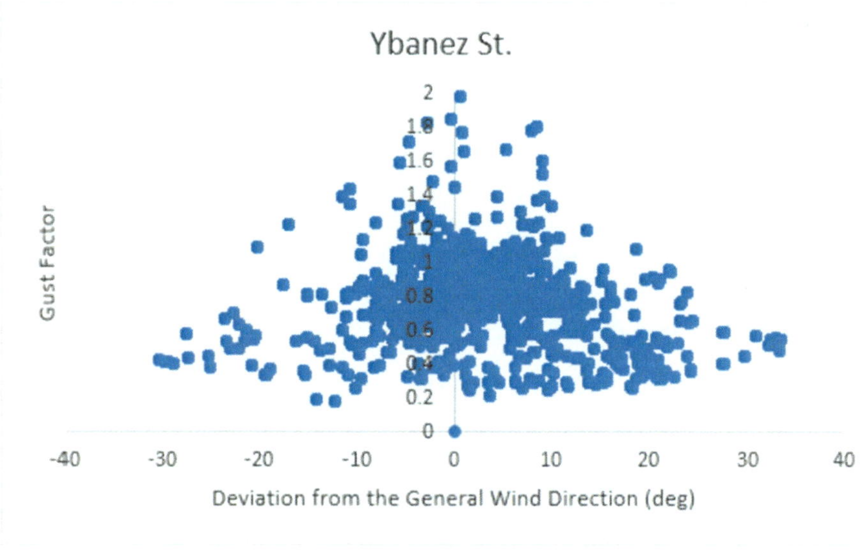

Fig. 10. Local Directional Derivation vs Gust Factors

3.4 Effects of Mid-Rise Buildings on the Boundary Layer

When the extraction was done on Osmeña St. in Bantayan, Cebu, where most of the mid-rise buildings were situated, out of 15775 points extracted, only 600 remained after the filtering process mentioned in Sect. 2.3. This indicates that the wind flows in the street, due to the existence of the mid-rise buildings. In that manner, Charnock's (1955) assumptions of the shear layer being only present near the ground which resulted in the logarithmic profile no longer applied. Additional shear layers caused by the mid-rise buildings also resulted in downdrafts which damaged the already deteriorating Escario house at the end of Osmeña St. Shown in Fig. 11.

Fig. 11. Escario House after Typhoon Haiyan

3.5 Topography

Apart from the directional deviations, along with the roughness length, the gust factor, when plotted as shown in Figs. 12 and 13 with respect to the elevation of the surfaces which included the topography and buildings, in contrast when at the sea, are shown to exhibit chaotic patterns when over land and the over the urban setting.

3.6 Turbulence Intensities

With the roughness length also being among the post-processed data, these enabled comparisons of the simulated turbulence intensity derived from the simulated time history and the turbulence intensity prescribed, shown in Fig. 14. The comparison revealed that the simplifying approximation was not applicable in estimating turbulence intensities, especially in clustered and urban conditions like in Bantayan. However, even though they were not equal in value, a correlation can be drawn between the turbulence intensities in the case of isolated terrain, like in Virac in Fig. 14. This implied that simplifying approximations like the one Lumley and Panosky (1964) did would still hold for isolated cases but with little or more modifications, especially since their assumption stemmed mainly

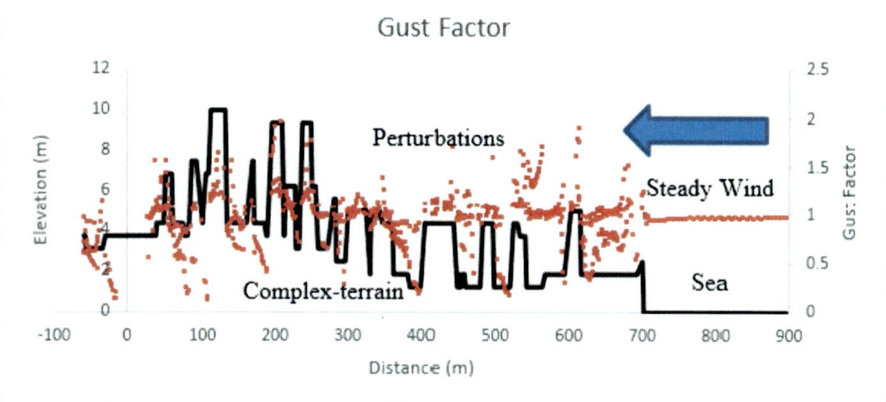

Fig. 12. Roughness Lengths over Bantayan, Cebu

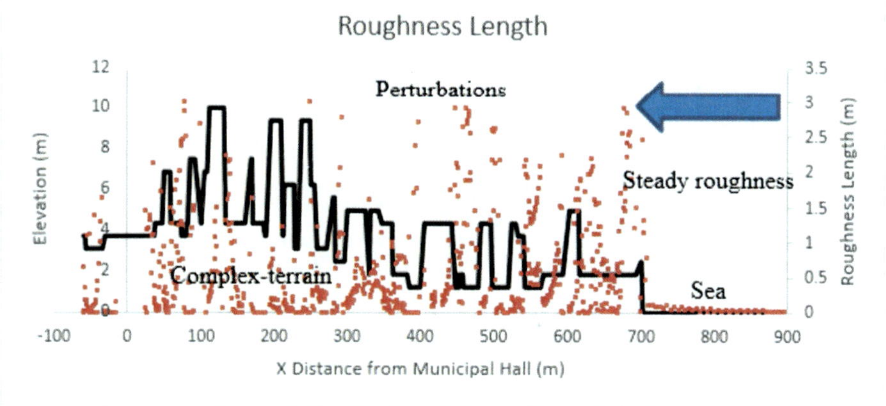

Fig. 13. Roughness Lengths over Bantayan, Cebu

from the determination of the friction velocity, which represented the shear stresses, in Eq. (5), for cases in high latitudes.

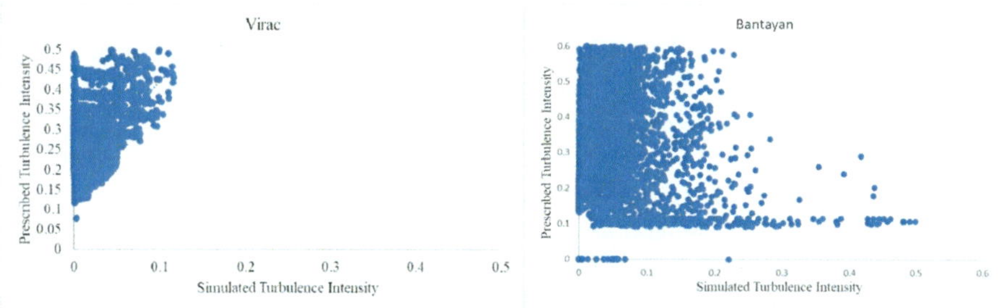

Fig. 14. Turbulence Intensities

4 Conclusions

The hybrid simulation model is tested over two cases: Typhoon Haiyan in Bantayan, Cebu, and Typhoon Nina in Virac, Catanduanes. For the case in Bantayan, Cebu, the hybrid simulation model simulated a peak gust of 66.26 m/s compared to the actual reading of 64.1 m/s. For the case of Virac, Catanduanes, the hybrid simulation model simulated a peak gust of 57.83 m/s compared to the actual reading of 54 m/s. Subsequent evaluation of the site-specific parameters, such as the gust factors, roughness length, and directional deviations, revealed the dynamic nature of the exposure parameters.

Same to the observation noted by Harper et al. (2010) for urban settings and clustered cases, this study has shown that the roughness lengths were not independent of the wind. The current assumptions for roughness lengths were, in fact, the results of the previous assumptions building over the other.

However, for isolated cases, the assumption for the exposure categories would hold if the development length criteria were satisfied. Considering this observation and past meteorological observations, wind codes should provide separate provisions for the cases involving urban settings and clustered cases.

This study showed in detail how the flow channeling, separation, and reattachment affected the roughness lengths. Establishing an empirical approach that would relate the topographic gradient (change in topography with respect to the direction of the wind), building envelopes, and wind interference (effects of buildings) can be related not only to the roughness lengths but also to the gust factors.

Since the roughness lengths established the local winds at a mean roof height from the winds at the reference height of 10 m, an overestimation of the roughness length can lead to the underestimation of the local winds. As winds turned out to be faster to develop from the ground, engineers might be designing structures at a local wind lower than that of the possible extreme event.

The inception of the friction velocity to approximate the shear stresses caused by the ground led to the current equations for the vertical wind profile. Roughness lengths mainly rely on the assumptions behind the boundary layer flow where only the ground was introducing shear stresses to the wind flow. In the presence of mid-rise buildings, the significant shear stresses from the building surfaces disrupt the boundary layer profile. Establishing the exposure categories on roughness lengths may not be enough to predict the actual local winds to be designed against, in this case, the presence of mid-rise buildings and, by extension, high-rise buildings.

The turbulence intensity is described as a function of height and roughness length by wind codes that came from a series of assumptions and idealizations. The simulation showed that Eq. (5) – which showed the turbulence intensity as a function of height and roughness length alone – did not hold in predicting turbulence intensities for urban settings, especially since it was established over cases of high latitudes. However, a correlation was observed in the isolated case (Virac). A revisit of the assumptions would be necessary to adjust certain parameters to match the observed scenario.

As previously discussed, the basic wind speed map, which was mainly derived using GEV analysis on the meteorological data, had two major flaws. The first one was that it failed to distinguish first the winds as cyclonic and non-cyclonic. If the winds were only non-cyclonic, then the GEV analysis on the station data would be enough since these involved winds of the mesoscale. But if the winds were cyclonic, then the GEV analysis should be done to determine the return periods of a storm of a particular strength instead. Through typhoon simulations that would follow, along with the validation, the data points produced would then be subjected to another GEV analysis to determine the return period of cyclonic winds at a certain strength.

The second one was that the extrapolation into making the design wind speed map neglected the different wind exposure of the weather stations. Accounting for the wind exposure was also vital in the typhoon simulations recommended to address the first flaw.

Acknowledgements. The authors would like to acknowledge the UP ERDT for funding and supporting this research. The authors would also like to acknowledge the DOST Project titled “Enhanced Severe Wind Vulnerability Curves of Key Building Types in the Philippines” for funding some of the field surveys.

References

Agar, J., Hernandez, J.: Site-specific Evaluation of Hazard and Exposure of Typhoons in the Philippines Using Tropical Cyclone Modeling and Large Eddy Simulation. MS Thesis. University of the Philippines, Diliman, Quezon City (2021)

Association of Structural Engineers of the Philippines, 2015. National Structural Code of the Philippines (2015)

Charnock, H.: Wind stress on a water surface. Qtly. J. Royal Met. Soc. **81**, 639–640 (1955)

Davenport, A.: The spectrum of horizontal gustiness near the ground in high winds. Qtly. J. Royal Met. Soc. **87**(372), 194–211 (1961)

Davenport, A., Isyumov, N.: The Application of Boundary Wind Tunnel to the Prediction of Wind Loading. Conference on Wind Effects on Buildings and Structures (1967)

De Leoz, T., Kaw, E., Qudilla, A., Valbuena, J., Garciano, L.: Updating the Wind Zone and Developing Contour Maps For the Philippines: Adaptation Strategies for Extreme Winds. Undergraduate Research. De La Salle University – Manila (2014)

Demirdzic, I.A.: A Finite Volume Method for Computation of Fluid Flow in Complex Geometries. Dissertation. Imperial College, University of London (1983)

Harper, B., Kepert, J., Ginger, J.: Guidelines for converting between various wind averaging periods in tropical cyclone conditions. World Meteorological Organization (2010)

Holland, G.: An analytic model of the wind and pressure profiles. Mon. Weather Rev. **108**(8), 1212–1218 (1980)

Huang, W., Zhou, H.: A hybrid numerical simulation method for typhoon wind field over complex terrain (2014). https://doi.org/10.12989/was.2014.18.5.000

Kolmogorov, A.: The local structure of turbulence in incompressible viscous fluids for very large Reynolds number. Dokl. Akad. Nauk SSR **30**, 9–13 (1941)

Lumley, J., Panofsky, H.: The Structure of atmospheric turbulence. Wiley Interscience New York (1964)

Powell, M., Houston, S.: Hurricane Andrew’s landfall in South Florida. Part II: surface wind fields. Weather Forecast. **11**, 304–328 (1996)

Powell, M., Houston, S., Reinhold, T.: Hurricane Andrew’s landfall in South Florida. Part I; standardizing measurements for documentation of surface wind fields. Weather Forecast. **11**, 304–328 (1996)

Richardson, L.F.: Weather Prediction by Numerical Process. Cambridge University Press, Cambridge (1922)
Schloemer, R.W.: Analysis and synthesis of hurricane wind patterns over Lake Okechobee. FL. Hydromet Rep. **31**, 49 (1954)
Smagorinsky, J.: General circulation experiments with the primitive equations. Mon. Weather Rev. **91**(3), 99–164 (1963)
Van der Hoven, I.: Power spectrum of horizontal wind speed in the frequency range from 0.0007 to 900 cycles per hour. J. Meteorology **14**, 160–164 (1957)
Wieringa, J., Davenport, A., Grimmond, C., Oke, T.: New revision of Davenport roughness classification. In: Proceedings of the 3rd European and African Conference on Wind Engineering, Eindhoven, The Netherlands, 2–6 July 2001, pp. 285–292 (2001)
You, D., Moin, P.: A dynamic global-coefficient subgrid-scale eddy viscosity model for large-eddy simulation in complex geometries. Center for Turbulence Research. Annual Research Briefs 2006, pp. 41–53 (2007)

A Multidisciplinary Approach to Optimize the Gust Loading on Groups of Wind Turbine Towers at the Quayside

Alessandro Giusti[1(✉)] and Alessio Torrielli[1,2]

[1] Siemens Gamesa Renewable Energy A/S, Brande, Denmark
alessandro.giusti@siemensgamesa.com, alt@rdas.dk
[2] R&D Test Systems A/S, Hinnerup, Denmark

Abstract. Steel towers of offshore wind turbines are usually assembled at the quayside before being loaded onto the installation vessel and taken to the offshore wind farm. Towers are usually erected in groups at close distance on gravity-based foundations, which should be verified against the relevant environmental loads, including the wind load. The peculiarities of the system together with the necessity of optimization to limit the costs without compromising the safety inspired the development of an advanced multidisciplinary approach for the assessment of gust buffeting loads. The methodology aims at refining the calculation and introducing tailored input parameters. This is achieved through local wind measurements, wind tunnel tests and full-scale measurements. In this paper the methodology is presented and compared with the typical approach provided by EN 1991–1-4.

Keywords: Gust Buffeting · Groups of Towers · Load Optimization

1 Introduction

The installation of an offshore wind farm passes through some preparatory activities carried out at the so-called pre-assembly harbor and aimed at preparing the components (e.g., towers, nacelles, blades) to the load-out onto the installation vessel in order to limit as much as possible offshore operations, which are usually costly and time consuming. Among the pre-assembly activities, wind turbine towers are typically erected at the quayside on gravity-based foundations, forming regular groups at close distance, which optimize crane operations and harbor layout (jack-up positions of the vessel). The foundation should be designed against environmental loads, which in Northern Europe are practically always related to the sole wind (earthquakes are relevant for the USA and APAC region markets). Furthermore, pre-assembly activities usually last 6–12 months and have therefore a temporary character, which can be exploited to reduce the reference return period used for the structural design (see e.g., [6, 9]).

The assessment of wind loads on vertical slender cylindrical structures is addressed by design standards (see e.g., [2, 8, 10]) under specific conditions, whose extension to the study of groups of towers requires assumptions that can sometimes lead to an

P. Schito and A. Zasso (Eds.): IN VENTO 2022, LNCE 461, pp. 175–186, 2024.
https://doi.org/10.1007/978-3-031-53059-3_15

underestimation of the loads. Among the wind-structure interaction phenomena, the gust buffeting response is typically based on the calculation of the equivalent static force, which passes through the definition of the so-called gust response factor [5], further developed over the years with analytical formulations covering also more than one structural mode (see e.g., [17]). In the framework of design standards, loads due to gust buffeting (e.g., base force and moment) are related to the so-called peak velocity pressure [8] or directly to the 3-s gust wind velocity [1]. In this context, Lidars seem very promising (see e.g., [11]) being an integrated system able to measure the instantaneous wind velocity at several heights, with the possibility to properly characterize the local turbulent wind velocity field in terms of both mean wind profile and peak (e.g., gust) wind velocities. The tower response is significantly driven by the aerodynamic coefficients (see e.g., [16]), which in the case of group arrangements also capture the interference of neighboring towers (a sort of group effect). These cases are not comprehensively addressed by design standards, and wind tunnel tests are usually used to achieve a satisfying level of knowledge (see e.g., [12, 13]). Moreover, a detailed assessment of wind statistics and the damping of the system, as a supplement of design standards, may increase the accuracy of the assessment of the tower response. How should one properly assess the gust buffeting response of groups of wind turbine towers at the quayside?

This paper presents a multidisciplinary approach for the assessment of gust buffeting loads on groups of wind turbine towers at the quayside, with the main purpose to increase the accuracy in the design of the tower foundations. The approach is introduced in Sect. 2 by a one-to-one comparison with the Eurocode methodology. The following sections address the main topics more in detail. Section 3 describes the importance of a proper characterization of the wind climate. Section 4 presents the use of wind tunnel tests for the assessment of the aerodynamic coefficients. Section 5 discusses the estimation of the total damping. Finally, Sect. 6 draws the conclusions and perspectives for further improvements.

2 Multidisciplinary Approach

The multidisciplinary approach presented in this paper aims at improving the aspects where the standards are incomplete or not sufficiently accurate. Siemens Gamesa Renewable Energy developed a methodology to achieve a satisfying level of accuracy for the assessment of gust buffeting loads, with particular focus on the loads (forces and moments) at the base of wind turbine towers arranged in groups, with the final purpose to have a reliable estimation of the environmental loads for the design of the temporary foundation at the quayside (see Sect. 1). As detailed in the following, the methodology uses local wind data, wind tunnel tests and full-scale measurements to achieve a satisfying level of knowledge that helps the designers to optimize the project without compromising the safety. For instance, the case of a row of two towers (see Sect. 2.2) is emblematic of the difference in the outcomes. Indeed, the base bending moment of the tower calculated with the proposed multidisciplinary approach is about 22.1 MNm compared to about 14.7 MNm given by an assessment based only on the Eurocode.

2.1 Flowchart for the Assessment of Gust Buffeting Response

Limiting the discussion to the European environment, without loss of generality, the Eurocode (EN) provides a useful guideline for the assessment of gust buffeting loads, which is here tailored and improved by the multidisciplinary approach (MA).

Geometry of the system. (EN, MA) It is an input data of the problem.

Reference wind velocity. (EN) National Annexes provide a description of the wind climate to be used for design purposes on a country level. They usually cover extended areas and oversee local peculiarities the pre-assembly areas may fall in. The use of design standards is usually conservative. (MA) Statistics of local wind data provide a detailed knowledge of the wind climate through the assessment of the parent distribution and extreme event distribution (including directional and seasonal factors). The latter permits to calculate the wind velocity corresponding to a given return period. At least 10 years of wind observations are needed for a reliable assessment of the extreme events.

Mean wind profile. (EN) The standard and National Annexes provide guidelines for the definition of the mean wind velocity profile for a given terrain category. (MA) A site-specific assessment of the mean wind profile could help the optimization of gust buffeting loads, but it is not common/cheap practice because multi-level wind velocity readings are needed.

Peak velocity pressure. (EN) The standard and National Annexes provide guidelines for the assessment of the peak velocity pressure for a given terrain category associated with a given intensity of turbulence. (MA) A site-specific assessment of the intensity of turbulence and peak velocity pressures could help the optimization of gust buffeting loads, but it is not common practice.

Aerodynamic coefficients. (EN) The standard provides guidelines for the assessment of force coefficients on parallel-sided cylinders, including corrections to consider the finite length of the cylinder (end-effect factor) and the proximity to other cylinders but only in row arrangements. However, the extension of the standard to tapered cylinders and/or cylinders arranged in multi-row groups is not straightforward and can lead to significant underestimations of the aerodynamic coefficients. (MA) Wind tunnel tests address specific cases, like towers with variable geometry or interferences with neighboring structures, which are not covered by the design standards. Moreover, specific wind conditions (e.g., wind profile and intensity of turbulence) can be modelled. Wind tunnel tests provide a comprehensive understanding of the aerodynamic coefficients for all directions of the wind rose and are usually focused on the resultant actions at the base of the tower, which drive the design of the foundation.

Structural damping. (EN) Table F.2 [8] suggests values of structural damping for several structural types, which however does not constitute a comprehensive list and are rather far to capture all the features of a given structural system. (MA) Full scale tests consider the total damping, given by the structure itself, internals and any added damper (e.g., tuned mass dampers and tuned liquid dampers inside the tower, or laminated elastomeric bearings at the tower-foundation interface).

Aerodynamic damping. (EN) The formulation in Annex F of the standard is approximated and in principle valid only for parallel-sided cylinders. (MA) The knowledge of the mean aerodynamic force distribution along the height of the towers (through wind tunnel tests) combined with the theory permits a detailed calculation of the expected aerodynamic damping.

Structural factor. (EN) The formulation in Sect. 6, Annex B and Annex C of the standard [8] considers only the correlation in space of the wind turbulence but not the correlation induced by the structure itself. (MA) The analysis of the maximum aerodynamic forces and moments recorded by wind tunnel tests (e.g., in accordance with the Davenport-Rice approach) permits to embed the size factor in the (maximum) aerodynamic coefficient.

Gust buffeting loads. (EN) The standard provides guidelines for the calculation of forces but not moments. (MA) Calculation of forces and moments.

2.2 Case Study

The MA is hereafter presented with the support of numerical examples, which provide the reader with a more comprehensive understanding of this alternative approach and the influence of the different parameters on the results. Multiple towers of height H and bottom diameter D_b having a cylindrical lower part ($0.586 \cdot H$) and a linearly tapered upper part ($0.414 \cdot H$) with top diameter of $0.694 \cdot D_b$ are considered. Towers are arranged (unless isolated) in groups with regular grid having a center-to-center distance of $1.333 \cdot D_b$ (see also [13]). Moreover, the Port of Rønne (Denmark) is considered as pre-assembly location (see Fig. 1).

Fig. 1. Port of Rønne on the Danish island of Bornholm (left); pre-assembly area (right), from Google Earth Imagery 2022 - Maxar Technologies.

3 Wind Climate

The wind-structure interaction depends, among others, on the peculiarities of the local wind climate (e.g., mean wind profile and turbulence), which is usually a key driver of the design, being the mean structural response related to the square of the mean velocity. Although design standards provide guidelines about the wind climate to be used for the design, typically through macro-areas for the basic wind velocity and terrain categories, a local characterization can improve the accuracy of the load assessment, which often results in a reduction of the gust buffeting load.

3.1 Basic Wind Velocity

The basic wind velocity, for a given return period, gives indication of the severity of the wind climate. Design standards typically provide a map of the country with the different basic wind velocities throughout the territory, which can be further corrected to consider the reference return period, the directionality of the wind and also the season, if the usage period of the structure is limited to a certain period of the year.

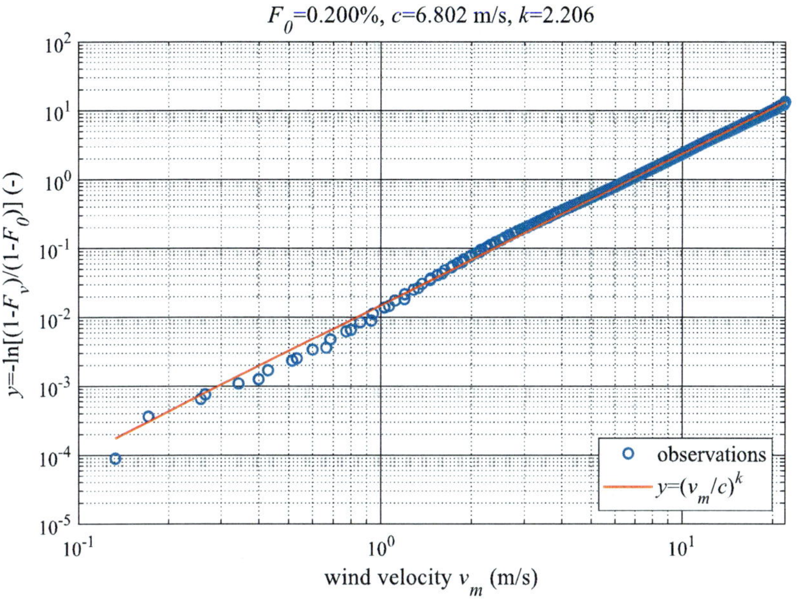

Fig. 2. Parent distribution of the mean wind velocity (10-min average) at 10 m above the sea level homogenized to Terrain Category II (F_v is the cumulative distribution function). Location: Rønne harbor (data curtesy of Port of Rønne).

Since design standards aim at covering huge areas with a limited number of rules, it is likely that the actual local wind climate may be milder than the one described by the standards being design standards generally conservative (or on the way around they should not give unsafe rules). Therefore, a local characterization of the wind climate in terms of basic wind velocity is an important step forward for the optimization of the design. Statistics of at least ten years [3] of local wind measurements are usually used in the assessment of the parent distribution (see Fig. 2) for fatigue verifications, and in the assessment of the extreme value distribution (see Fig. 3) for the definition of the basic wind velocity. Many methodologies have been developed over the years, depending also on the length and quality of the available dataset (see e.g., [3, 14, 18]). The 8-day maximum wind velocity (see Fig. 3) well fits the Gumbel probability distribution as long as an appropriate left-censorship of about 15 m/s is applied. This leads to a 50-year return period mean wind velocity of 23.85 m/s which is slightly lower than the value of 24 m/s recommended by [7].

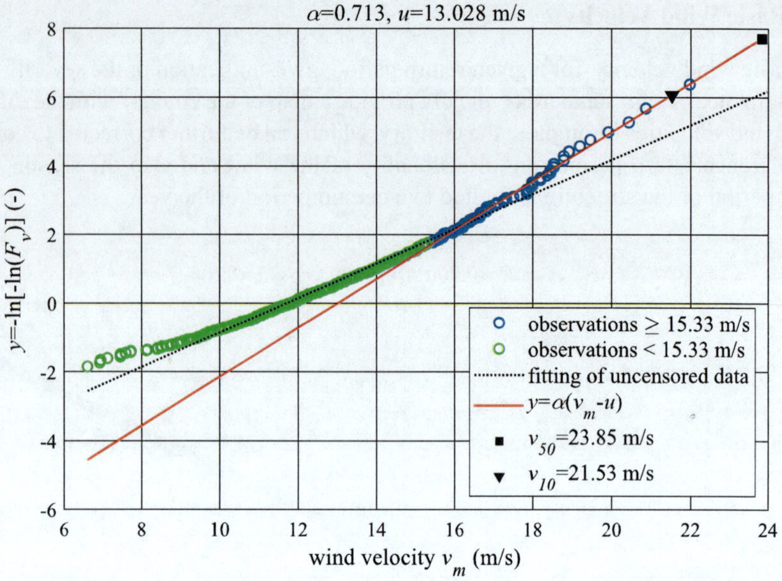

Fig. 3. Extreme value distribution of the mean wind velocity (10-min average) at 10 m above the sea level homogenized to Terrain Category II (F_V is the cumulative distribution function, v_{50} and v_{10} are respectively the 50-years and 10-years return period mean wind velocities). Location: Rønne harbor (data curtesy of Port of Rønne).

3.2 Mean Wind Profile and Peak Velocity Pressure

The mean wind profile is usually characterized following the indications of the local standards, e.g., by means of a logarithmic or a power-law relationship, namely:

$$v_m(z) = p \cdot \ln\left(\frac{z}{z_{ref}}\right) \tag{1}$$

$$v_m(z) = q \cdot \left(\frac{z}{z_{ref}}\right)^{\beta} \tag{2}$$

where p and q are dimensional constants related to the local climate, z_{ref} is the reference height, β is the exponent of the power-law. These relationships, prior a proper calibration of the constants, are used worldwide. However, they cannot capture possible local interferences, for instance related to neighboring structures and the orography in the surrounding, that may change entirely or partially the profile for some wind directions. Similarly, also the intensity of turbulence and peak velocity pressures given by reference standards cannot capture local singularities or peculiarities of the site (e.g., large storage yards in the pre-assembly harbor, where wind turbine components are parked in regular grids, see Fig. 1 right panel); therefore a local assessment may help to improve the load calculation and to optimize the design. However, the use of a local characterization (e.g., through Lidars) of the mean wind profile and peak wind velocities for pre-assembly operations in the wind energy sector has not been implemented yet as a standard procedure, because of costs and duration of the observation period.

4 Aerodynamic Coefficients

The aerodynamic force and moment coefficients are dimensionless quantities typically used for describing the forces and moments that a moving fluid exerts on elongated slender bodies immersed on it. They are typically referred to the mean (time average) response of the system, but it is also possible to introduce time dependent coefficients, related to the instantaneous response, and maximum coefficients, related to the peak response. Mean aerodynamic coefficients are hereafter considered, following the typical framework of design standards [8]. The peak response of the system can be calculated starting from the mean response by means of a gust effect factor, which considers both the turbulence of the wind and the dynamic properties of the system.

4.1 EN Design Standard

The Eurocode [8] addresses the aerodynamic response of chimneys starting from the aerodynamics of the circular cylinder (2D case) which is corrected afterwards to consider the finite length, namely (for an isolated tower):

$$c_f = c_{f,0} \cdot \psi_\lambda \tag{3}$$

$$c_{f,0} = \frac{0.11}{\left(Re/10^6\right)^{1.4}} \text{ or } c_{f,0} = 1.2 + \frac{0.18 \cdot \log(10 \cdot k/b)}{1 + 0.4 \cdot \log\left(Re/10^6\right)} \tag{4}$$

$$\psi_\lambda = 0.698 + 0.252 \cdot \left[\log(\lambda) - 1\right] \tag{5}$$

where $c_{f,0}$ is the force coefficient of cylinders without free-end flow (e.g., infinitely long cylinders), ψ_λ is the end-effect factor (Eq. (5) is valid for solid structures with slenderness greater than 10 and smaller than 100), k is the equivalent roughness, b is the width of structure, Re is the Reynolds number (the use of Eq. (4), left or right, depends on the value of Re), λ is the slenderness. It is stressed that the standard [8] is ambiguous in the definition of the force coefficient because it is not clearly stated if it is related to the cross section or to the whole structure. This ambiguity becomes particularly relevant in the case of structures with a variable diameter. Indeed, on the one hand, Eq. (3) may be applied at different cross-sections getting a force coefficient variable along the height, which is however rather far from the real distribution of the wind forces. On the other hand, the standard seems to prefer a global approach with the definition of a unique force coefficient for the whole structure. This leads to the definition of an equivalent diameter. There are some options for the definition of the equivalent diameter D_{eq} of the tower, either based on the shear force or bending moment, and including or not the mean wind profile, namely:

$$D_{eq} = \frac{1}{H} \cdot \int_0^H D(z) \cdot dz \tag{6}$$

$$D_{eq} = \frac{\int_0^H D(z) \cdot q_m(z) \cdot dz}{\int_0^H q_m(z) \cdot dz} \tag{7}$$

$$D_{eq} = \frac{2}{H^2} \cdot \int_0^H D(z) \cdot z \cdot dz \tag{8}$$

$$D_{eq} = \frac{\int_0^H D(z) \cdot q_m(z) \cdot z \cdot dz}{\int_0^H q_m(z) \cdot z \cdot dz} \tag{9}$$

$$q_m(z) = \frac{1}{2} \cdot \rho_{air} \cdot v_m^2(z) \tag{10}$$

where H is the height of the tower, D is the diameter, q_m is mean velocity pressure, ρ_{air} is the unit mass of the air, v_m is the mean wind velocity, z is the elevation along the axis with origin at the geometric center of the base of the tower and pointing vertically upwards. It is assumed that the base of the tower is at the ground level. Focusing on the case study, Eqs. (6), (7), (8) and (9) give respectively the results reported in Table 1; the smaller the equivalent diameter the smaller the force coefficient and the larger the structural factor $c_s c_d$. It is noted that in this case the choice of the equivalent diameter does not significantly affect neither the force coefficient (evaluated considering the Reynolds number given by the peak velocity at the top of the tower) nor the structural factor.

Table 1. Force coefficient and structural factor for different possible choices of the equivalent diameter of the structure.

Eq.	D_{eq} (D_b)	$c_{f,0}$ (-)	$c_s c_d$ ^) (-) Ann. C	$c_s c_d$ ^) (-) Ann. B
6	0.936	0.6622	1.0695	1.0157
7	0.923	0.6621	1.0701	1.0163
8	0.890	0.6617	1.0717	1.0179
9	0.881	0.6616	1.0721	1.0184

^) Tower with infinite base rotational stiffness (fixed support) and total damping of 2.5% log. dec.

The use of the standards becomes more complicated in the estimate of the aerodynamic coefficients in the case of groups of towers, passing though the correction of the coefficients of an isolated tower. Design standards usually cover a limited number of cases (as the line of towers [8]), whose extension to different arrangements can produce unquantifiable inaccuracies.

4.2 Wind Tunnel Tests

Static wind tunnel tests are often used for complex civil structures (e.g., bridges, tall buildings), which, because of their geometry and/or the presence of nearby structures, may be very sensitive to the wind action. Wind tunnel tests can provide a good understanding of the aerodynamic forces with some limitations associated with the specific characteristics of the wind tunnel, the necessity to operate with scaled models, and the general challenge to reproduce at the same time the mean wind profile, the intensity of turbulence, integral length scales and the Reynolds number. Focusing on the loads at the base of the tower, force coefficients c_F and moment coefficients c_M can be defined as follows (different normalizations can however be introduced):

$$c_F = \frac{F_b}{q_m(H) \cdot \int_0^H D(z) \cdot dz} \tag{11}$$

$$c_M = \frac{M_b}{q_m(H) \cdot \int_0^H D(z) \cdot z \cdot dz} \tag{12}$$

where F_b and M_b are respectively shear force and bending moment at the base of the tower. Aerodynamic coefficients of groups of towers depend, among others, on the geometry of the tower, the group configuration and the ratio of the center-to-center distance to the diameter (see e.g., [12, 13]). With reference to the case study, for instance, a group of four towers and a line of two towers have respectively a base moment coefficient up to about 15% and 30% larger than the case of an isolated tower. Wind tunnel tests on rigid models could also be used to get knowledge on the extreme response of the structure, including the size factor (aerodynamic admittance) but not the dynamic factor, reflecting the formulation in [8]. The latter would require aeroelastic tests.

5 Damping

The damping is the capability of the system to dissipate energy under dynamic loads. The total damping δ_{tot} of wind turbine towers is typically modelled as the sum of three main contributions [8]: a) the structural damping δ_s, related to the material and the structural type; b) the aerodynamic damping δ_a, related to the interaction of the system with the fluid flow; c) the damping δ_d due to special devices, e.g., sloshing and tuned mass dampers. In principle, swinging tower internals, like the high voltage cable, and special interfaces at the base connection, may introduce additional sources of damping; however, these additional contributions are traced to the global structural damping. The damping mitigates the dynamic response of the tower and is relevant for the gust buffeting and the other aeroelastic phenomena (e.g., among others, vortex induced vibration and galloping). The assessment of the available damping is crucial to achieve an optimum design.

5.1 Structural Damping

The structural damping of 1.2% logarithmic decrement assigned to unlined welded steel stacks (without external thermal insulation), i.e., the type of structure closest to the wind turbine tower in Table F.2 [8], underestimates the true overall structural damping. Full scale measurements of the tower vibrations have been performed for the identification of the structural damping of towers at the quayside. Multiple time series of strain in the shell at the base of the tower through Operational Modal Analysis resulted in the estimate of 3.6% logarithmic decrement of the structural damping.

5.2 Aerodynamic Damping

The Eurocode [8] proposes the following formula for the evaluation of the logarithmic decrement of the aerodynamic damping for the fundamental bending mode for along-wind vibrations:

$$\delta_a = \frac{c_f \cdot \rho_{air} \cdot v_m(z_s) \cdot b}{2 \cdot n_1 \cdot m_1} \int_0^H \varphi_1^2(z) \cdot dz \tag{13}$$

where c_f is the force coefficient, presumably the base force coefficient capturing the whole structure behavior, ρ_{air} is the unit mass of the air, v_m is the mean wind profile, z_s is the reference height, b is the diameter, n_1 is the first natural frequency of the structure, m_1 is the first modal mass, φ_1 is the first mode shape (see [8]). Equation (13) is a particular case of a more general formulation presented in [15], namely:

$$\delta_a = \frac{\rho_{air}}{2 \cdot n_1 \cdot m_1} \cdot \int_0^H D(z) \cdot c_d(z) \cdot v_m(z) \cdot \varphi_1^2(z) \cdot dz \tag{14}$$

where D is the diameter of the structure, c_d is the drag coefficient of the cross-section. In this case the global force coefficient in Eq. (13) is replaced by the sectional drag coefficient together with the use of the actual wind profile in place of $v_m(z_s)$. The use of Eq. (14) requires the knowledge of the variability of the drag coefficient along the height of the structure, which is not always available. It can however be approximated by a global aerodynamic coefficient, e.g., either the base force coefficient c_F or the base moment coefficient c_M. The use of the base force coefficient is conservative for aerodynamic damping evaluation of chimneys because it is usually smaller than the moment coefficient. Indeed, the point of application of the base shear force is above the geometric center of the reference area (see [10]), and the smaller the aerodynamic coefficient the smaller the damping. An underestimation of the damping of the system produces an overestimation of the base bending moment (see Fig. 4), and it is always a conservative choice, although less pronounced for high values of damping. Finally, with reference to the case study, the use of the procedure in Annex C [8] (as required by [7]) provides larger base bending moments than the procedure of Annex B [8] if the total damping is smaller than about 20% logarithmic decrement. This trend reverses for infinitely large damping, even though the deviation is below 2%, being the two estimates of the base bending moment due to gust buffeting equal to 13.95 MNm and 14.19 MNm respectively for Annex C and Annex B.

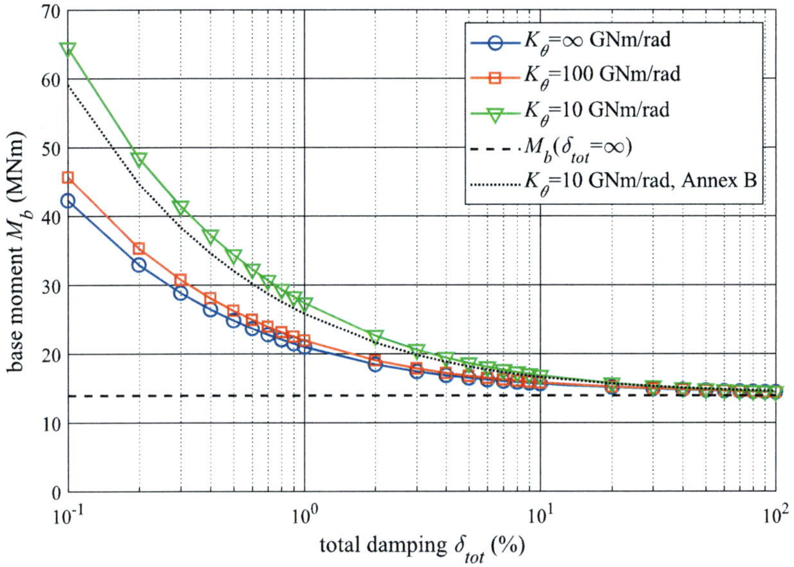

Fig. 4. Influence of the total damping (logarithmic decrement) on the base moment due to gust buffeting, for different values of the equivalent rotational stiffness K_θ of the tower-foundation connection (structural factor evaluated with Annex C [8]).

6 Conclusions and Perspectives

A MA for the optimization of gust buffeting loads on groups of offshore wind turbines towers is presented, along with a one-to-one comparison with the Eurocode [8], to point out the main weakness of the standard when it is applied to this type of structures. The MA intends to overcome the limitations of the Eurocode by means of statistics of the local wind climate, wind tunnel tests and full-scale measurements. A case study dealing with the evaluation of the gust buffeting response of a row of towers standing at the quayside in Port of Rønne illustrates the improvements of the MA and points out the key parameters driving the assessment of the gust-buffeting loads. In order of relevance, the mean wind velocity, being the mean load a function of wind velocity squared; the aerodynamic coefficients, being the load proportional to them; the total damping (aerodynamic and structural contributions), which affects the dynamic part of the structural response.

A possible improvement of the presented approach is the implementation of a more accurate aerodynamic admittance function pursued through full scale and wind tunnel tests. Indeed, the calculation of the gust buffeting load in [8] relies on a simplified model of aerodynamic admittance developed for thin sections [4], which relates the fluctuating wind force to the wind velocity only considering its horizonal and vertical correlation. The aerodynamics of groups of towers is expected to be affected also by the boundary layer separation and the resulting large-scale vortex shedding. A refined modeling of the aerodynamic admittance, considering also the correlation between wind pressures on windward and leeward surfaces of the towers may result in more accurate gust buffeting loads and so in the optimization of the design of the quayside tower foundations.

References

1. ASCE/SEI 7–22: Minimum design loads and associated criteria for buildings and other structures. American Society of Civil Engineers, Reston, Virginia (2022)
2. CICIND: Model code for steel chimneys, Revision 1 (1999)
3. Cook, N. J.: The designer's guide to wind loading of building structures. Part 1: Background, damage survey, wind data and structural classification. Building Research Establishment, Garston, Watford (1985)
4. Davenport, A.G.: The spectrum of horizontal gustiness near the ground in high winds. Q. J. R. Meteorol. Soc. **87**(372), 194–211 (1961)
5. Davenport, A.G.: Gust loading factors. J. Struct. Div. **93**(3), 11–34 (1967)
6. DNVGL-ST-N001: Marine operations and marine warranty (2016)
7. DS/EN 1991–1–4 DK NA. National annex to Eurocode 1: actions on structures. Part 1–4: general actions – wind actions (2015)
8. EN 1991–1–4. Eurocode 1: actions on structures. Part 1–4: general actions – wind actions (2007)
9. EN 1991–1–6. Eurocode 1: actions on structures. Part 1–6: general actions – actions during execution (2007)
10. ESDU 81017. Mean forces, pressures and moments for circular cylindrical structures: finite-length cylinders in uniform and shear flow (1981)
11. Kim, D., Kim, T., Oh, G., Huh, J., Ko, K.: A comparison of ground-based LiDAR and met mast wind measurements for wind resource assessment over various terrain conditions. J. Wind Eng. Ind. Aerodyn. **158**, 109–121 (2016)
12. Mannini, C., Massai, T., Giachetti, A., Giusti, A.: Aerodynamic loads on offshore wind turbine towers arranged in groups at the quayside. In: Proceedings of the 8th European and African Conference on Wind Engineering (EACWE 2022), Bucharest, Romania, in Press (2022)
13. Mannini, C., Massai, T., Giachetti, A., Giusti, A.: Aerodynamic loads on wind turbine towers arranged in groups with and without helical strakes. In: Proceedings of the 17th International Conference of the Italian Association for Wind Engineering - IN-VENTO 2022, Milano, Italy (2024)
14. Palutikof, J.P., Brabson, B.B., Lister, D.H., Adcock, S.T.: A review of methods to calculate extreme wind speeds. Meteorol. Appl. **6**(2), 119–132 (1999)
15. Repetto, M.P., Solari, G.: Dynamic alongwind fatigue of slender vertical structures. Eng. Struct. **23**(12), 1622–1633 (2001)
16. Simiu, E., Yeo, D.: Wind Effects on Structures: Modern Structural Design for Wind, 4th edn. John Wiley & Sons (2019)
17. Solari, G., Martin, P.: Gust buffeting and aerodynamic admittance of structures with arbitrary mode shapes. I: enhanced equivalent spectrum technique. J. Eng. Mech. **147**(1), 04020142 (2021)
18. Torrielli, A., Repetto, M.P., Solari, G.: Extreme wind speeds from long-term synthetic records. J. Wind Eng. Ind. Aerodyn. **115**, 22–38 (2013)

On the Modeling of Self-damping in Overhead Electrical Line Conductors Subject to Vortex-Induced Vibrations

Francesco Foti, Luca Martinelli(✉), Stefano Corazza, and Federico Perotti

Department of Civil and Environmental Engineering, Politecnico di Milano, Milan, Italy
luca.martinelli@polimi.it

Abstract. The paper reviews some recent results of the authors' research group related to the modeling of the hysteretic behaviour of metallic stranded cables and applies them to the definition of a theoretical model for the self-damping of a widespread typology of overhead electrical line conductors, namely the ACSR conductors. Closed-form expressions of the dissipated power per unit of length of the conductor, previously derived by the authors, are cast in a new form that allows for a more effective comparison with available experimental data of the literature.

Keywords: cable dynamics · self-damping · vortex induced vibrations

1 The Proposed Self-damping Model

Overhead electrical line conductors are prone to vortex-induced vibrations. The severity of vibration is typically evaluated through an application of the Energy Balance Principle (EBP): mono-modal vibrations are assumed and the balance between the input power provided by the wind and the power dissipated within the cable is imposed at each vibration frequency. As already argued in [3], the accuracy and reliability of the outcomes of the EBP can be strongly affected by the criteria adopted to define the input data, namely: the wind input power and the cable self-damping (i.e. the power dissipated within the oscillating cable).

The current practice to characterize the self-damping of overhead electrical line conductors relies upon an empirical approach. Mono-modal vibration tests are performed on laboratory spans and the power dissipated per unit of length of the cable (P_d) is measured for different values of the antinode vibration amplitude Y, vibration frequency f and axial force T. The experimental data are then interpolated through the empirical power law:

$$P_d = kY^l f^m T^{-n}, \tag{1}$$

where l, m and n are exponents with values that typically fall in the ranges: $l = 2 \div 2.5$; $m = 4 \div 6$; $n = 2 \div 2.8$ for the widespread class of Aluminium Conductors Steel Reinforced (ACSR) cables, while k is a proportionality coefficient.

P. Schito and A. Zasso (Eds.): IN VENTO 2022, LNCE 461, pp. 187–195, 2024.
https://doi.org/10.1007/978-3-031-53059-3_16

Whenever k is not defined from experimental data, it can be computed on the base of both the internal geometry and material properties of the cable using empirical equations such as:

$$k = D/\sqrt{\gamma RTS} \tag{2}$$

where D is the cable diameter that has to be expressed in mm, γ is the mass per unit length of the cable, expressed in kg/m, and RTS is the rated tensile strength of the cable expressed in kN. It is worth noting, however, that this definition of k leads to a dimensionally non-homogeneous equation. Typical values of k are in the range 1–2.5 for ACSR conductors.

Due to the power form of the empirical self-damping equation, even a small variation in the values of the exponents l, m and n can lead to large differences in the computed values of dissipated power. This, in turn, can significantly affect the reliability of the outcomes of the EBP method. This observation motivated the interest in developing a self-damping formulation based on a mechanical description of the hysteretic dissipation mechanism associated to the vortex-induced-vibration of overhead electrical conductors [6, 7]. Basing on the theoretical formulation in [6, 7], significant experimental results are herein analyzed adopting a new parametrization to provide a new insight into the physics behind the self-damping of cables.

2 The Proposed Self-damping Model

Modeling the bending behavior of conductors is an old topic that has attracted the interest of many researchers, see e.g. the review in [2].

Overhead electrical conductors are made of concentric layers of metallic helical wires; each layer is twisted around the preceding one. Whenever conductors are bent, an axial force gradient is generated in the wires and tries to makes the wires slide one with respect to the others. Friction forces between the wires are opposing these relative movements.

These frictional forces make the bending behavior of strands non-linear and non-holonomic, and characterize the strand bending behavior as one controlled by a hysteretic damping mechanism as the bending stiffness will change between an upper (EI_{max}) and a lower (EI_{min}) limit value. Bound values of the bending stiffness can be obtained, respectively, under the assumption of: (a) full-stick and (b) full-slip cross sectional behavior. In the first case, which leads to EI_{max}, the wires are ideally bounded together to prevent all relative displacements. In the second case, which leads to EI_{min}, the external surfaces of the wires are assumed to be perfectly lubricated, so each wire behaves as an independent curved thin rod.

The change in bending stiffness can be large, as the ratio EI_{min}/EI_{max} can be in the order of 1/100 for typical overhead lines cables such as Aluminium Conductors Steel Reinforced (ACSR) cables.

Several different analytical expressions, mostly based on the assumption of rigid contact interfaces between the wires (e.g. [4, 5]), have been proposed in the literature to calculate EI_{max} and EI_{min} on the basis of the sole knowledge of the material and geometrical properties of the wires.

Theoretical and experimental studies point out as calculated values of EI_{min} lead to a good prediction of the bending stiffness in the large curvature regime, e.g. strands wound around a pulley or strands subjected to large localized transverse loads, while calculated values of EI_{max} tend to overestimate the bending stiffness, even in the small curvature regime. This tendency has been recently recognized as a consequence of the deformability of the contact interfaces between the wires [1, 9], and more specifically, of the tangential contact compliance that allows for micro-slip (MS) phenomena between the wires even though contacting wires are macroscopically still sticking together.

An effective maximum value ($EI_{max,ef}$) of the cross sectional bending stiffness can be defined, for practical purposes, as [6, 7]:

$$EI_{max,ef} = \beta EI_{max} \tag{3}$$

where β is a stiffness reduction factor that for typical ACSR conductors can be in the range $\beta = 0.5 \div 0.95$.

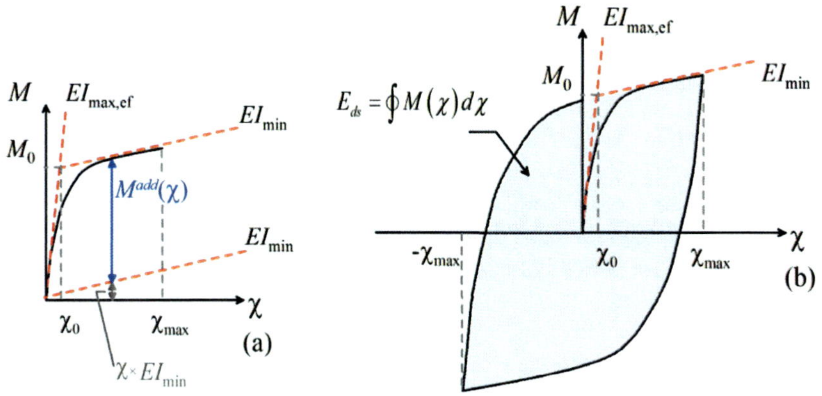

Fig. 1. Moment-curvature diagram. (a) Monotonic loading case. (b) Cyclic loading case.

The transition from small curvatures (full-stick regime) to large curvatures (full-slip regime) depends on the possible activation of relative displacements between the wires (gross-sliding (GS)) and is inherently non-linear and non-holonomic. As already formulated in several pioneering works of the 1930's–50's, later revisited and developed by (e.g. [4, 5, 8]), a widely accepted physical interpretation is that whenever a strand is bent, an axial force gradient is generated along the length of the wires, which gives the wires the tendency to slip relatively to the nearby ones. The axial force gradient is counteracted by the friction forces acting on the external surface of the wires. The friction forces depend on the geometry and material properties of the wires, on the friction coefficient of the surfaces in contact and on the internal contact pressure value. The contact pressure value increases for increasing values of the axial load acting on the strand as a consequence of the helicoidal shape of the wires that creates a clenching effect on the lower layers. As long as the friction forces are large enough to prevent the relative displacement between the wires, the cross section of the strand can be assumed to be in the full-stick state and only MS phenomena can occur.

The axial force gradient increases for increasing values of the bending curvature and can overcome the effect of the frictional tangential forces acting on the wire. For a given value of the strand curvature, hence, the wires of the strand start to slip relatively one to another (gross-sliding), and the cross sectional bending tangent stiffness gradually decreases approaching, for a sufficiently large enough bending curvature, the minimum value associated to the full-slip state in which all the wires are slipping. Figure 1 shows a typical moment-curvature diagram for a metallic strand under both monotonic (Fig. 1a) and cyclic (Fig. 1b) bending loading conditions.

An analytical expression of the moment-curvature (M-χ) diagram can be written in the form:

$$M = EI_{min}\chi + M^{add}(\chi) \tag{4}$$

The linear term, $EI_{min}\chi$, can be regarded as the (theoretical) response of the strand under the full-slip assumption while the non-linear second term, $M^{add}(\chi)$, is non-holonomic and accounts for the additional contribution to the bending moment of the strand coming from the axial force of each wire.

Several models have been proposed in the literature to describe the term $M^{add}(\chi)$ in Eq. (2), among which the one recently developed by the authors' research group [4, 5]. Based on this model, Foti and Martinelli [6, 7] have proposed a unified moment-curvature diagram for ACSR conductors in the non-dimensional plane (χ^*, M^*), where: $\chi^* = \chi/\chi_0$, $M^* = M/M_0$ and χ_0 and M_0 are, respectively, the coordinates in the (χ, M) plane of the transition point from one branch to the other of a bi-linear envelope of the moment-curvature diagram (see Fig. 1).

The curvature χ_0, which is typically associated to the onset of gross-sliding (GS) between the wires of the outermost layer of the strand and the wires of the penultimate layer, can be conventionally considered as the curvature value at which transition from MS to GS occurs between all the wires of the strand.

Based on theoretical developments, the following equation has been proposed in [6, 7] to calculate χ_0:

$$\chi_0 = c_0 \mu\, \eta \tag{5}$$

where μ is the interwire friction coefficient. An indicative range of values for μ can be $0.15 \div 0.7$, with lower values associated to new and well-lubricated strands and higher values associated to old or non-lubricated). η is the ratio between the tensile force in the cable and its Rated Tensile Strength (RTS) and c_0 is a coefficient depending only on the internal geometry of the strand. Typical values of c_0 are in the range $0.1 \div 0.3\ \text{m}^{-1}$ for ACSR conductors, see [7].

Using the model proposed in [6, 7] and briefly previously recalled, by assuming a mono-modal steady state vibration the conductor self damping can be calculated in closed form under both the assumption of MS and GS dissipation mechanism, yielding the two expressions:

$$P_d^{ms} = \alpha_{ms} Y^3 f^7 T^{-4}, \tag{6}$$

$$P_d^{gs} = \alpha_{gs} Y^2 f^5 T^{-2} \tag{7}$$

where the superscripts refer to the assumed dissipation model while α_{ms} and α_{gs} are two dimensional coefficients that depend on the geometry and material properties of the conductor.

Theoretical expression for α_{ms} and α_{gs} have been derived in [6, 7] and read:

$$\alpha_{ms} = \frac{128\pi^5\gamma^3 RTSEI_{max,ef}}{3c_0\mu} \tag{8}$$

$$\alpha_{gs} = 4\pi^4\gamma^2 EI_{max,ef} \tag{9}$$

3 Comparison Between Theoretical and Experimental Results

The predictions of the proposed self-damping models have been compared against the data-set of experimental data available in [3], obtained under different test conditions and for different conductor cross sections. Experimental results are typically reported by showing the self-damping, measured for a constant value of vibration frequency and axial force, as a function of the non-dimensional vibration amplitude Y/D. As an example, Fig. 2 reports measured values of self-damping for a Bersfort ACSR (diameter $D =$ 35.6 mm, $EI_{max} = 3750$ Nm^2, $RTS = 180$ kN, $c_0 = 0.139$ m^{-1}). Measurements were carried out through the Inverse Standing Wave Ratio (ISWR) technique and corrected to subtract the aerodynamic damping (see [3] for further details on the testing procedure).

In order to ease the comparison between experimental results and the predictions of the proposed theoretical models a novel reparametrization of Eqs. (6) and (7) is herein proposed.

Focusing on high-order vibration modes and consistently with the approach presented in [6, 7], the mode shapes of the conductor are described through the classic taut-string model, i.e.:

$$y = Y\sin(2\pi x/\lambda) \tag{10}$$

where $y(x)$ is the modal displacement of the cable centerline, $x \in [0, 1]$ is a spatial coordinate running over the span l of the cable, Y is the single-peak antinode vibration amplitude and λ is the wavelength of the mode:

$$l = 1/f\sqrt{(T/\gamma)} \tag{11}$$

where f is the vibration frequency, T is the cable axial force and γ the mass per unit of length. A non-dimensional wavelength ρ is defined as it follows:

$$\rho = \lambda/2l \tag{12}$$

By deriving with respect to the coordinate x, then, the modal rotations can be readily obtained from Eq. (10) as:

$$\varphi_{mod} = \Phi\cos(2\pi x/\lambda) \tag{13}$$

where $\Phi = 2\pi Y/\lambda$ is the rotation angle (in absolute value) of a node of the mode shape.

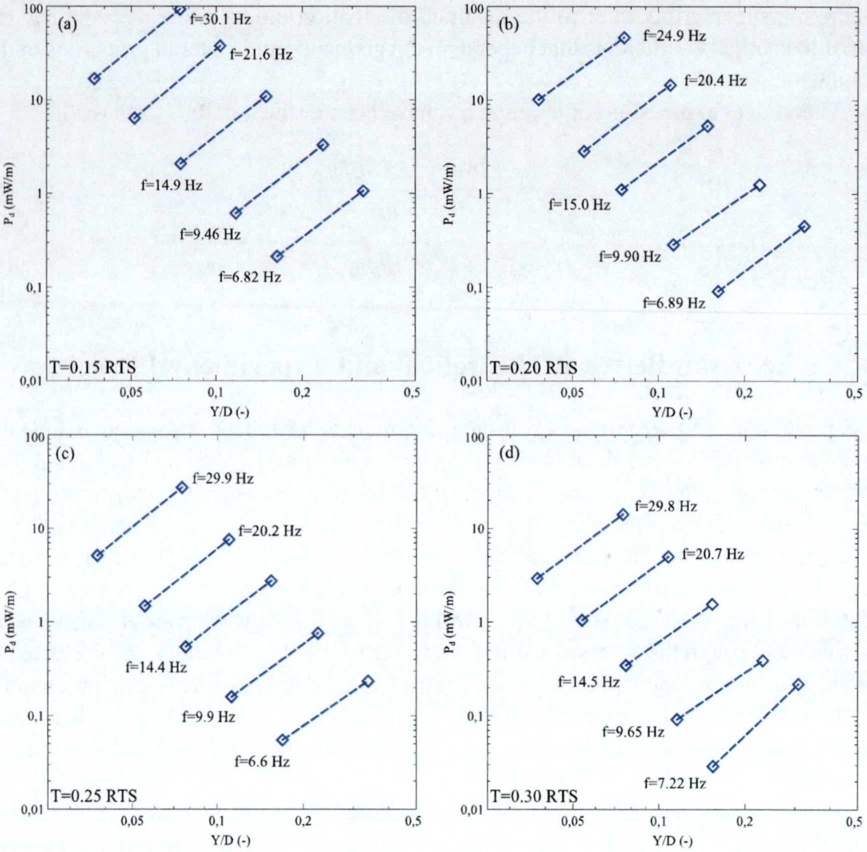

Fig. 2. ACSR Bersfort conductor. Dissipated power per unit of length vs. non-dimensional antinode amplitude. Experimental results are from [3]. The different figures correspond to different values of the axial force T. (a) $T = 0.15$ RTS. (b) $T = 0.20$ RTS. (c) $T = 0.25$ RTS. (d) $T = 0.30$ RTS.

By taking into account the above definitions, the predictions of the GS self-damping model can be restated, after some straightforward calculations, in the following non-dimensional form:

$$P_d^{gs}/(C \cdot 4\chi_0^2 l^2) = J \tag{14}$$

Equation (14) identifies a straight line with unitary slope in the non-dimensional plane (J, Π), where Π is the non-dimensional dissipated power due to self-damping of the cable:

$$\Pi = P_d/(C \cdot 4\chi_0^2 l^2) \tag{15}$$

and J is the non-dimensional variable:

$$J = \Phi^2/\rho^3 \tag{16}$$

Plotting experimental and theoretical results in the non-dimensional plane (J, Π) allows one to easily judge about the range of validity of the proposed models. Figure 3 shows a comparison between the experimental data and the theoretical predictions of the MS and GS models, obtained for a Bersfort ACSR. Calculations have been carried out by assuming $\beta = 0.5$ and a friction coefficient $\mu = 0.3$ in Eqs. (8) and (9). This last value is herein tentatively supposed to be representative of contact conditions for a greased new ACSR as the one tested in the laboratory. Further theoretical and experimental studies, however, should be devoted to a better characterization of the tribological properties of interwire contact interfaces in ACSR cables.

As it can be clearly appreciated from Fig. 3, all experimental data points tend to collapse on a single straight line, independent of the different testing conditions (i.e. different values of cable axial force, frequency and vibration amplitude). For large values of J (i.e. greater to about 0.005), the slope of the line interpolating experimental data is approximately equal to one predicted by the GS damping model. For smaller values of J, instead, the experimental data are in closer agreement with the theoretical predictions of the MS damping model rather than to the ones of the GS damping model.

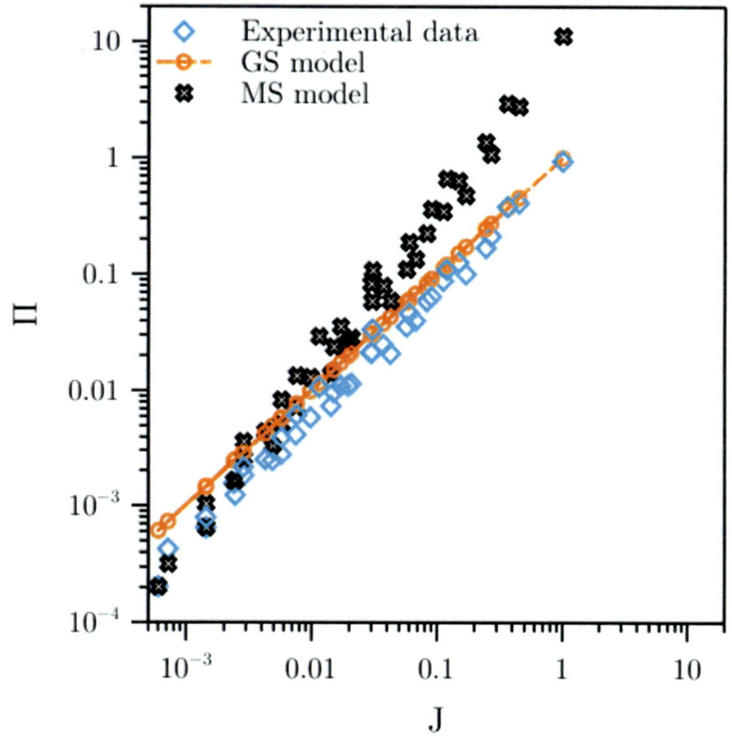

Fig. 3. ACSR Bersfort conductor. Non-dimensional dissipated power per unit of length. Comparison between experimental data (from [3]) and theoretical predictions of the Micro-Slip (MS – Eq. (6)) and Gross-Sliding (GS – Eq. (7)) models.

The comparison in Fig. 3 allows to appreciate as the conductor self damping predicted with the closed form equations reported in Eq. (6) and Eq. (7) give values close to those of the experimental data for a range of testing conditions typical of overhead conductors. Besides this first interesting result, the proposed reparametrization seems to confirm the occurrence of two different dissipation mechanisms, associated to different ranges of bending curvatures of the cable, which where only hypothesized in the previous studies of this research group. Further studies however are needed to precisely define the transition curvature between these dissipation regimes; this is the subject of an ongoing research.

4 Conclusions

Due to the peculiar internal structure of overhead electrical conductors, whenever they are bent, the wires they are formed of tend to slide relative one to another. Sliding of the wires is the physical mechanism providing power dissipation in the conductors subject to vortex-induced-vibrations. Starting from the theoretical description of the frictional interaction between the wires, two different dissipative phenomena have been identified in previous works of the research group, namely: the Micro-Slip (MS) and the Gross-Sliding (GS) dissipation mechanisms. These two mechanisms are here assumed to control, respectively, the small-amplitude and large-amplitude vibration regimes. Closed-form expressions of the dissipated power per unit of length of the conductor, previously derived by the authors, have been cast in a new form that allows for a more effective comparison with available experimental data of the literature.

The comparison with qualified experimental data allows one to appreciate the good predictive capabilities of the proposed self-damping models. Besides this first interesting result, the proposed reparametrization suggests the occurrence of two different dissipation mechanisms, associated to different ranges of bending curvatures of the cable. Further studies will be devoted to investigate the transition curvature between these dissipation regimes.

References

1. Baumann, R., Novak, P.: Efficient computation and experimental validation of ACSR overhead line conductors under tension and bending. CIGRE Sci. Eng. **9**, 5–16 (2017)
2. Cardou, A.: Taut helical strand bending stiffness, p. 9 (2006). www.utfscience.de
3. EPRI (Electric Research Power Institute): Transmission Line Reference Book: Wind-induced Conductor Motion, Palo Alto, U.S. (2006)
4. Foti, F., Martinelli, L.: Mechanical modeling of metallic strands subjected to tension, torsion and bending. Int. J. Solids Struct. **91**, 1–17 (2016)
5. Foti, F., Martinelli, L.: An analytical approach to model the hysteretic bending behavior of spiral strands. Appl. Math. Model. **40**, 6451–6467 (2016)
6. Foti, F., Martinelli, L.: A unified analytical model for the self-damping of stranded cables under aeolian vibrations. J. Wind Eng. Ind. Aerodyn. **176**, 225–238 (2018)
7. Foti, F., Martinelli, L.: An enhanced unified model for the self-damping of stranded cables under aeolian vibrations. J. Wind Eng. Ind. Aerodyn. **182**, 72–86 (2018)

8. Papailiou, K.O.: On the bending stiffness of transmission line conductors. IEEE Trans. Pow. Del. **12**, 1576–1588 (1997)
9. Paradis, J.P., Legeron, F.: Modelling of the free bending behavior of a multilayer cable taking into account the tangential contact compliance of contact interfaces. In: 9th International Symposium on Cable Dynamics, p. 8, Shanghai (2011)

Influence of Stochastic Load Perturbations on the Performance of a Torsional-Flutter Wind Harvester

Luca Caracoglia(✉)

Northeastern University, Boston, MA 02115, USA
lucac@coe.neu.edu

Abstract. Wind energy technologies are emerging as the need for clean energy resources has considerably increased in the last decade. Apart from horizontal-axis wind turbines, less efficient at the intermediate scales (or meso-scales) and for moderate wind speeds, a competitive alternative is necessary. This alternative is represented by simpler, smaller-size wind-based energy systems, activated by various aeroelastic phenomena. An ongoing investigation on the performance of an aeroelastic harvester, which can efficiently replace meso-scale wind turbines, is discussed. The apparatus exploits the torsional flutter of a streamlined "rigid" blade that rotates about a pre-set axis through a nonlinear torsional spring mechanism that activates post-critical behavior. A stochastic model of the apparatus, including output power estimation, has been derived to evaluate relevance of uncertain wind loads, induced by imperfect characterization and simplified load assumptions. Studies are carried out to examine mean square stability.

Keywords: Wind Energy Technology · Aeroelastic Harvester · Torsional Flutter · Random Loads · Stochastic Differential Equations

1 Introduction

Wind energy is rapidly emerging field because of the need for clean energy. Current technology advancements mainly focus on large-scale, horizontal-axis wind turbines that are less efficient at intermediate scales, i.e., the meso-scales. An interesting alternative is represented by simpler wind-based systems, triggered by various aeroelastic phenomena such as galloping [1], vortex induced vibration [2] and coupled flutter [3, 4]. Caracoglia [5] proposed the use of torsional flutter of a blade-airfoil in Fig. 1, simpler than coupled flutter and other aeroelastic phenomena, to produce energy. The pivot axis (O) in Fig. 1 coincides with the windward position close to stagnation point of the mean flow, from left to right in the figure.

Building on previous studies, a new stochastic model has been recently derived to investigate effects of random perturbations (i.e., load variability) on the performance of the apparatus. Parametric perturbations are employed to describe aeroelastic load variability as a first attempt to replicate imperfect fluid-structure interaction. Examples are three-dimensional flow effects, neglected in standard aeroelastic theory and nonlinear

P. Schito and A. Zasso (Eds.): IN VENTO 2022, LNCE 461, pp. 196–204, 2024.
https://doi.org/10.1007/978-3-031-53059-3_17

coupling. Stochastic differential equations are employed to study mean square stability (and output power) of the apparatus, conditional on the value of the wind speed that triggers the instability. This study also preliminarily evaluates output energy and identifies physical properties controlling output power.

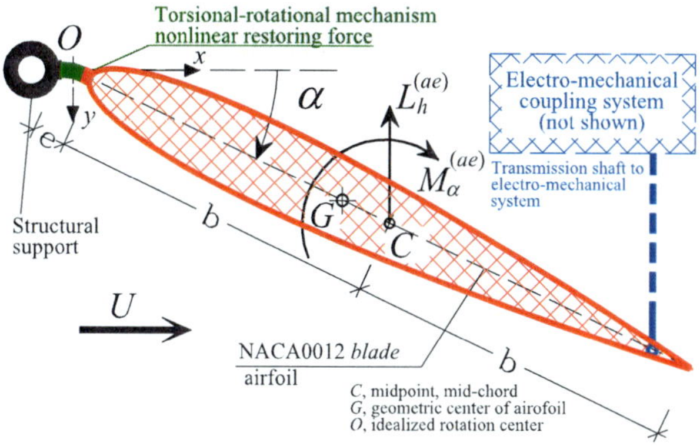

Fig. 1. Cross-sectional schematics of the apparatus and its components.

2 Methodology

A state-space model, based on stochastic differential equations, is employed to study the mean square stability of the apparatus as a function of mean flow speed U. In the reduced-order state-space model, aeroelastic torque is simulated using unsteady load formulation. The triggering mechanism depends on the following quantities: reduced frequency $k_\alpha = \omega_\alpha b/U$ with ω_α [rad/s] angular frequency of linear flapping motion, damping ratio ζ_α, generalized inertia ε and cubic stiffness κ. The mean flow speed is U The main, one-degree-of-freedom (1DOF) dynamic equation depends on flapping angle (torsional rotation) α and dimensionless time $\tau = t\omega_\alpha$:

$$
\begin{gathered}
\psi_0 \frac{\mathrm{d}^2\alpha}{\mathrm{d}\tau^2} + \left(\frac{1.5\varepsilon\eta_{3D}}{k_\alpha} + 2\zeta_\alpha\right)\frac{\mathrm{d}\alpha}{\mathrm{d}\tau} + (\alpha + \kappa\alpha^3) = \\
-\varepsilon\eta_{3D}k_\alpha^{-2}\left[\begin{array}{c}\Phi_0\left(\alpha + 1.5k_\alpha\dfrac{\mathrm{d}\alpha}{\mathrm{d}\tau}\right) \\ +1.5\left(\nu_{ae,1} + \nu_{ae,2}\right) + \mu_{ae,1} + \mu_{ae,2}\end{array}\right] - \Psi_\iota
\end{gathered} \tag{1}
$$

The right-hand side of Eq. (1) shows, within square brackets, the "indicial function" (Wagner) formulation with $\nu_{ae,1}$, $\nu_{ae,2}$, $\mu_{ae,1}$ and $\mu_{ae,2}$ being aeroelastic states; the formulation is corrected for three-dimensional flow effects by quasi-static, simplified correction function $\eta_{3D} = \mathrm{AR}/(\mathrm{AR}+2)$ that depends on aspect ratio $\mathrm{AR} = \ell/b$ [blade-airfoil's chord half-length b in Fig. 1 *vs.* depth of the apparatus ℓ]. Quantity $\Phi_0 = 0.5$ denotes the unit-step Wagner function at time $\tau = 0$; parameter $\psi_0 = (1 + 9/8\varepsilon\eta_{3D})$

accounts for non-circulatory flow load effects ("added mass"). Derivation of the equations of motion and aeroelastic states is omitted for the sake of brevity but is available from a previous work [5].

The mass moment of inertia of the apparatus, rotating about O, is $I_{0\alpha}$. On the left-hand side of Eq. (1) an important feature is the nonlinear cubic stiffness term ($\kappa\alpha^3$). The cubic term ensures that the apparatus can efficiently work by achieving a stable post-critical vibration regime. The nonlinear torsional spring mechanism is schematically indicated in Fig. 1 only. The plan is to design a translation spring connected to the chord axis of the rigid apparatus (dashed line in Fig. 1) with cubic properties. This may be achieved, for example, by exploiting the nonlinear geometric properties of a "loose", non-taut cable, aptly designed; an example of this nonlinear stiffness feature has been proposed for the design of "nonlinear energy sinks" [6] that have been used for the passive control of various structures.

On the right-hand side of Eq. (1), $\iota(\tau)$ is a normalized output current. Coupling with an eddy-current power circuit is possible through $\Psi = 4b^2(\Phi_{\text{e.m.c.}})^2/(\omega_\alpha I_{0\alpha} R_C)$, dimensionless electro-mechanical coupling coefficient. $\Phi_{\text{e.m.c}}$ is the electro-mechanical coupling coefficient [newton/ampère] due to magnetic induction and interaction of the moving coil with the moving shaft, schematically indicated in Fig. 1. The current ι is found from the relationship $I(\tau) = \iota(\tau)[\Phi_{\text{e.m.c.}} 2b\omega_\alpha / R_C]$, with $I(\tau)$ being the current generated by the eddy-current circuit [in Ampères] and R_C the resistance of the circuit. The properties of the secondary power circuit are found following the steps indicated by Kwon *et al.* [7], i.e., through Faraday's law and induced output current (ι in dimensionless units). Furthermore, the dimensionless equation of the power circuit is derived as $\mathrm{d}\iota/\mathrm{d}\tau = \lambda_{RL}(\mathrm{d}\alpha/\mathrm{d}\tau - \iota)$ [5]. The quantity $\lambda_{RL} = R_C/(\omega_\alpha L_C)$ is defined as "generalized impedance of the power circuit" with L_C inductance [Henries]; λ_{RL} must be selected to avoid that the electro-magnetic apparatus behaves as a damper that suppresses the flapping [5].

The Itô-type differential equation is derived from Eq. (1), after perturbing the Wagner function needed for airfoil flutter in time domain τ [8], defined as $\Phi(\tau) = c_0 - c_1\exp[-d_1\tau/k_\alpha] - c_2\exp[-d_2\tau/k_\alpha]$; with the constant parameters being derived for a NACA0012 cross section as [9]: $c_0 = 1.0$, $c_1 = 0.165$, $c_2 = 0.335$, $d_1 = 0.0455$ and $d_2 = 0.3$.

In this study, random load perturbation is obtained by modifying the reference coefficient d_2 as $d_2 = 0.3 + \Delta_{d2}(\tau)$ with $\Delta_{d2}(\tau)$ being a zero-mean Gaussian white noise of standard deviation σ_{d2}. The reason for exclusively considering randomness in the d_2 parameter is because the second exponential term $\{c_2\exp[-d_2\tau/k_\alpha]\}$ mainly controls the rapidly varying load variations through $\Phi(\tau)$ function and is mostly responsible for the memory effects. The standard deviation σ_{d2} is selected to replicate a coefficient of variation equal to about 20%, plausible with this type of randomness. Although other perturbations can readily be included in the formulation, they are not considered herein for the sake of simplicity. Further experimental verification would be necessary to better determine their variability.

The randomization $\Delta_{d2}(\tau)$ enables the subsequent derivation of the system of equations below,

$$\mathrm{d}\mathbf{w}_{\text{em}} = \mathbf{q}_{\text{NL},\Delta}(\mathbf{w}_{\text{em}})\mathrm{d}\tau + \sqrt{2\pi}\,\mathbf{Q}_{\text{L},\Delta}\mathbf{w}_{\text{em}}\mathrm{d}B(\tau) \tag{2}$$

where $\mathbf{W}_{\text{em}}$ is the state vector, $\mathbf{q}_{\text{NL},\Delta}$ is a nonlinear drift vector, $\mathbf{Q}_{\text{L},\Delta}$ is a diffusion matrix, $B(\tau)$ is a scalar Wiener noise of dimensionless time τ that is used to represent the randomness of $\Delta_{d2}(\tau)$. Turbulence effects are neglected.

The state vector has seven variables and includes physical states (rotation and its derivative), aeroelastic states and dimensionless current $\iota(\tau)$; $\mathbf{W}_{\text{em}}$ reads

$$\mathbf{W}_{em}(\tau) = \left[\alpha(\tau), \mathrm{d}\alpha(\tau)/\mathrm{d}\tau, \nu_{ae,1}(\tau), \nu_{ae,2}(\tau), \mu_{ae,1}(\tau), \mu_{ae,2}(\tau), \iota(\tau)\right]^T \tag{3}$$

The non-zero elements of the 7-by-7 diffusion matrix $\mathbf{Q}_{\text{L},\Delta}$ in Eq. (2) are:

$$\left(\mathbf{Q}_{\text{L},\Delta}\right)_{4,2} = \left(\mathbf{Q}_{\text{L},\Delta}\right)_{6,1} = \sigma_{d2} k_\alpha^{-1} c_2, \quad \left(\mathbf{Q}_{\text{L},\Delta}\right)_{6,6} = -\sigma_{d2} k_\alpha^{-1} \tag{4}$$

Coefficients in Eq. (4) are constant and depend on the random load perturbation $\Delta_{d2}(\tau)$ through standard deviation σ_{d2}. The drift function in Eq. (2) is nonlinear:

$$\mathbf{q}_{\text{NL},\Delta}(\mathbf{W}_{\text{em}}) = \begin{Bmatrix} W_{\text{em},2} \\ 8/[8 + 9\varepsilon\eta_{3D}]\widetilde{\Pi}_{\mathbf{W}_{\text{em}}}(\mathbf{W}_{\text{em}}) \\ d_1\left[c_1 W_{\text{em},2} - k_\alpha^{-1} W_{\text{em},3}\right] \\ \left[d_{2,m} + \pi\sigma_{d2}^2\right]\left[c_2 W_{\text{em},2} - k_\alpha^{-1} W_{\text{em},4}\right] \\ k_\alpha^{-1} d_1\left[c_1 W_{\text{em},1} - W_{\text{em},5}\right] \\ k_\alpha^{-1}\left[\overline{d}_2 + \pi\sigma_{d2}^2\right]\left[c_2 W_{\text{em},1} - W_{\text{em},6}\right] \\ \lambda_{RL}\left[W_{\text{em},2} - W_{\text{em},7}\right] \end{Bmatrix} \tag{5}$$

Equation (5) is presented as a 7-by-1 functional $\mathbf{q}_{\text{NL},\Delta}(\mathbf{W}_{\text{em}})$; the last equation also describes electro mechanical coupling through $W_{\text{em},2} = \mathrm{d}\alpha/\mathrm{d}\tau$ and $W_{\text{em},7} = \iota(\tau)$. The nonlinear function of the second – row element is:

$$\begin{aligned} \widetilde{\Pi}_{\mathbf{W}_{\text{em}}}(\mathbf{W}_{\text{em}}) = & -\begin{pmatrix} W_{\text{em},1} \\ +\kappa W_{\text{em},1}^3 \end{pmatrix} - \left(\frac{3\varepsilon\eta_{3D}}{2k_\alpha} + 2\zeta_\alpha\right) W_{\text{em},2} \\ & - \frac{\varepsilon\eta_{3D}}{k_\alpha^2}\left[\begin{matrix} \Phi_0\left(W_{\text{em},1} + \dfrac{k_\alpha W_{\text{em},2}}{2/3}\right) + \\ \dfrac{W_{\text{em},3} + W_{\text{em},4}}{2/3} + W_{\text{em},5} + W_{\text{em},6} \end{matrix}\right] - \Psi W_{\text{em},7} \end{aligned} \tag{6}$$

In Eqs. (4) and (5) the Wong and Zakai [10] correction terms are included. Finally, Eq. (2) is combined with Eqs. (4–6) and solved numerically. The Euler numerical integration scheme [11] is employed to solve Eq. (2) several times, i.e., by defining a "population of realizations" and collecting each solution or $\mathbf{W}_{\text{em}}$ as a function of time. It is noted that initial random vector $\mathbf{W}^{(0)}{}_{\text{em}}$ at $\tau = 0$ is also needed; initial conditions are imposed by assuming random initial rotation different from zero, expressed as a random, Gaussian, scalar angle perturbation $W_{\text{em},1}$ with given properties and identically zero initial conditions for other states, i.e., $\mathbf{W}^{(0)}{}_{\text{em}} = [\alpha_0, 0, \ldots, 0]^T$. The ensemble of the solutions is collected and later utilized to examine stability.

The second Moment Lyapunov Exponent (MLE) $\Lambda_{\Xi}(2)$ is used to study the dynamic stability. The second-moment Lyapunov exponent [12] examines the propensity of a stochastic dynamic system to become unstable in terms of mean squares (variances and co-variances). This is scalar quantity that measures the rate of change of the "slow time dynamics" and evaluates the state as time tends to infinity; in other words, it can be assimilated to the total damping of the system at steady state after an initial, transitory stage.

A negative value of $\Lambda_{\Xi}(2)$ is needed to ensure mean-square stability; therefore, $\Lambda_{\Xi}(2) > 0$ yields an apparatus that produces energy. The stability limit varies as a function of wind speed. By varying U, both incipient and post-critical conditions are evaluated. Relevance is attributed to the output current $\iota = w_{\mathrm{em},7}$ that is extracted from the secondary system and, consequently the output power.

The $\Lambda_{\Xi}(2)$ quantity is found by solving Eq. (2) using numerical integration; $\Lambda_{\Xi}(2)$ is approximated as:

$$\Lambda_{\Xi}(2) \approx \log_e\left(\mathrm{E}\left[\|\Xi\|^2\right]\right) \Big/ \tau_j \tag{7}$$

where $\Xi(\tau) = [\alpha(\tau),\mathrm{d}\alpha(\tau)/\mathrm{d}\tau]^T$ is a suitable, dynamic sub-vector of $\mathbf{W}_{\mathrm{em}}$ and τ_j a discrete time, needed in conjunction with step-by-step integration. The time index j is taken sufficiently large, i.e., tending to infinity. Discussion on the validity of Eq. (7) to approximate the limit may be found in a previous study [13]. Accuracy in the calculation of Eq. (7) is achieved by reducing the time step for numerical integration of Eq. (2) and selecting a sufficiently large sample of solutions to derive expectations in Eq. (7).

3 Results

Numerical solution of the stochastic Eq. (2) in a post-critical regime is studied. The main properties of the apparatus, selected in this study, are: damping ratio ζ_α, electro-mechanical coupling coefficient Ψ, generalized impedance λ_{RL} and aspect ratio AR. The following reference quantities are set: $\zeta_\alpha = 0.25\%$, $\Psi = 0.01$, $\lambda_{RL} = 0.75$, AR = 10, and $\kappa = 100$ in dimensionless units.

Three basic configurations of the apparatus are investigated:

- *Type 0* with $\omega_\alpha/2\pi = 0.25$ Hz, $b = 0.25$ m, $I_{0\alpha}/\ell = 20$ kg-m^2/m;
- *Type 1* with $\omega_\alpha/2\pi = 0.20$ Hz, $b = 0.25$ m, $I_{0\alpha}/\ell = 40$ kg-m^2/m;
- *Type* 2 with $\omega_\alpha/2\pi = 0.10$ Hz, $b = 0.50$ m, $I_{0\alpha}/\ell = 300$ kg-m^2/m.

These configurations are derived to achieve minimum operational conditions at mean flow speed between 10 m/s and 15 m/s [5, 13] and, at the same time, to be relatively compact (small b) and have the potential for practical implementation and installation on building roofs.

Numerical investigations examine both a variation in the random load perturbation $\Delta_{d2}(\tau)$, with zero mean and standard deviation σ_{d2} variable between 0.07 and 0.10. Moreover, because of the perceived nonlinear uncertainty perturbation in Eq. (6), effects of the initial random conditions are evaluated, i.e., an initial "flapping" angle [α_0] with zero mean and standard deviation equal to both $\sigma_{\alpha 0} = 1.8$ deg (small deviations) and $\sigma_{\alpha 0}$ = 18.0 deg (large deviations). An initial angle α_0 is always needed to trigger instability and cause the system to store energy.

Figure 2 illustrates an example of the numerical results for the three basic apparatuses with AR = 10; this large AR value is considered to simulate negligible three-dimensional flow effects. In this figure, the designation "Type" is abbreviated as "Ty.", with 0,1,2 indicating the numeral identifier.

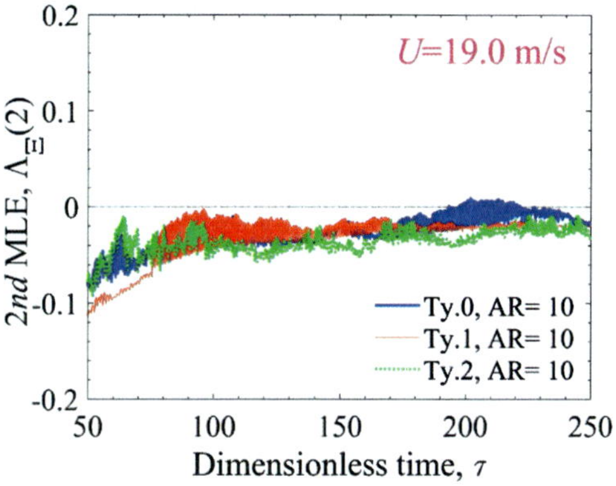

Fig. 2. Mean-square dynamic stability analysis via 2nd MLE [$\Lambda_{\Xi}(2)$] at flow speed $U = 19.0$ m/s for an apparatus with aspect ratio AR = 10, $\sigma_{d2} = 0.07$ and $\sigma_{\alpha 0} = 1.8$ deg (small initial amplitude).

Figure 2 demonstrates that, as a first example, at the stated wind speed $\Lambda_{\Xi}(2)$ of all three systems are tending to an unstable asymptote, with $\Lambda_{\Xi}(2) > 0$, and can trigger energy conversion. In fact, the trends of $\Lambda_{\Xi}(2)$ found by Eq. (7) numerically exhibit a progressively positive value despite an initial negative trend of Type 2. The introduction of randomness in the aeroelastic load is considered, with Δ_{d2} having moderate variability and a standard deviation $\sigma_{d2} = 0.07$. The flow speed at which the system transitions from stable to unstable is above $U = 15$ m/s, which is a plausible estimate of the stability threshold [5] in the absence of random aeroelastic load, suggesting that load variability has a stabilizing effect. This remark indicates that variability in the loads produces a delay in the onset of the flutter vibration and may be detrimental to the efficiency of the apparatus.

Figure 3 illustrates the stability analysis of the Type-2 apparatus with AR = 10 but considers the influence of initial conditions on the stability. In this case, the objective is to study potential nonlinear effects in Eq. (6) (cubic nonlinearity in state variable $W_{\mathrm{em},1}$) through initial state α_0 perturbation. Large initial amplitudes are imposed with zero-mean α_0 but standard deviation $\sigma_{\alpha 0} = 18.0$ deg, possibly unrealistic since the vibration is evaluated at angles close to the stall for this type of airfoils and for which the aeroelastic load through Wagner function becomes unrealistic and a different formulation is needed [14]. The stability limit is examined in Fig. 3 at a mean flow speed U lower than the previous figure.

The trend exhibited by $\Lambda_{\Xi}(2)$ at large τ in Fig. 3 appears to be negative, at least for Type-2 apparatus, already at $U = 17.8$ m/s; the situation persists for $U > 17.8$ m/s. This observation suggests that, contrary to Fig. 2, stability is negatively affected by larger-amplitude initial conditions and that larger α_0 would be needed to ensure that the apparatus is active and converts flow kinetic energy into electrical energy.

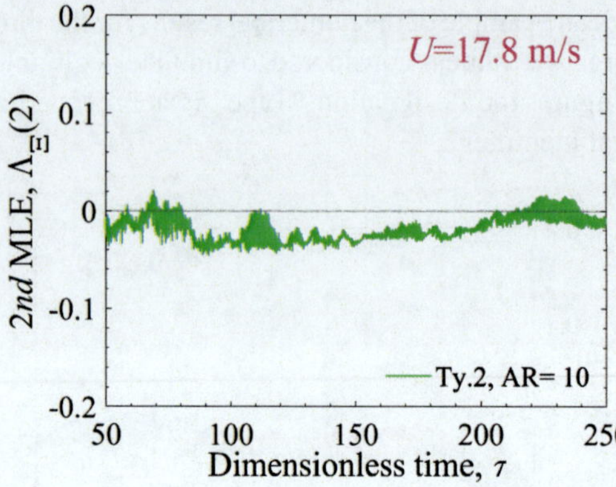

Fig. 3. Mean-square dynamic stability analysis via 2nd MLE [$\Lambda_{\Xi}(2)$] at flow speed $U = 17.8$ m/s for Type-2 apparatus with aspect ratio AR = 10, $\sigma_{d2} = 0.07$ and $\sigma_{\alpha 0} = 18.0$ deg (large initial amplitude).

Figure 4 depicts the time evolution of $\Lambda_{\Xi}(2)$ for all three apparatuses for small initial flapping ($\sigma_{\alpha 0} = 1.8$ deg) at triggering, when the random perturbation of the load parameter Δ_{d2} has a standard deviation 1.41 times larger than the first scenario (Fig. 2). The stability is studied at $U = 18.8$ m/s very close to $U = 19.0$ m/s in Fig. 2.

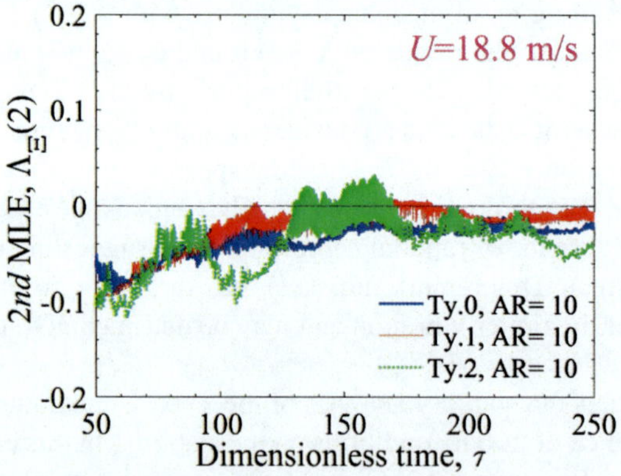

Fig. 4. Mean-square dynamic stability analysis via 2nd MLE [$\Lambda_{\Xi}(2)$] at flow speed $U = 18.8$ m/s for an apparatus with aspect ratio AR = 10, $\sigma_{d2} = 0.10$ and $\sigma_{\alpha 0} = 1.8$ deg.

The trends of in Fig. 4 are very similar to the ones observed earlier. More pronounced fluctuations of $\Lambda_{\Xi}(2)$ about the zero horizontal axis are noted, prior to the asymptotic limit. This result indicates a possibly small influence of σ_{d2} on the overall stability. Nevertheless, numerical integration issues cannot be excluded; they may be related to the selection of the sample size (200 realizations) used to solve Eq. (2) (time step $\Delta\tau =$ 1E-4) and to find the MLE by Eq. (7) and ensemble averaging.

In any case, the choice of the time "limit" $\tau = \tau_j = 250$ seems appropriate since it coincides with 40 periods of torsional oscillation at angular frequency ω_α, approximately coincident with flutter angular frequency. This aspect is examined in Fig. 5, where the estimation of Eq. (7) is extended to $\tau = \tau_j = 500$; the same asymptotic trends are already discernible for all the three apparatuses at about $\tau = 200$, excluding the need for continuing the numerical integration beyond the value used in Fig. 4 and in the previous cases.

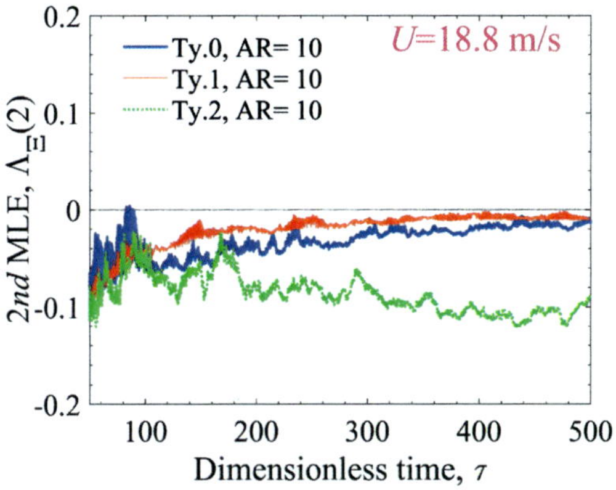

Fig. 5. Extended-time mean-square stability analysis [$\Lambda_{\Xi}(2)$] at flow speed $U = 18.8$ m/s for an apparatus with aspect ratio AR = 10, $\sigma_{d2} = 0.10$ and $\sigma_{\alpha 0} = 1.8$ deg.

4 Conclusions

Numerical study results indicate that introduction of variability in the aeroelastic load may be detrimental and may postpone flutter occurrence, requiring a mean flow speed of magnitude 20% larger than the scenario in the absence of load modeling uncertainty. This result requires further investigation also because the estimation of MLE by ensemble averaging is affected by the choice of the time step used for integration of the stochastic differential equation [Eq. (2)]. Additional studies should be conducted to confirm the influence of uncertainty in the aeroelastic loads, as a function of the standard deviation of the random parameters in the Wanger function. The final goal of this research is the identification and exclusion of low-efficiency operational regimes by rigorous probabilistic analysis. Anticipated results will open new avenues that will be employed in conjunction with future wind tunnel verification.

Acknowledgments. This material is based in part upon work supported by the National Science Foundation (NSF) of the United States of America, Award CMMI-2020063. Any opinions, comments or conclusions are those of the author and do not necessarily reflect the views of the NSF or any other organization.

References

1. Abdelkefi, A., Nayfeh, A.H., Hajj, M.R.: Design of piezoaeroelastic energy harvesters. Nonlinear Dyn. **68**(4), 519–530 (2012)

2. Bernitsas, M.M., Raghavan, K., Ben-Simon, Y., Garcia, E.M.H.: VIVACE (Vortex Induced Vibration Aquatic Clean Energy): a new concept in generation of clean and renewable energy from fluid flow. J. Offshore Mech. Arct. Eng. **130**(4), 041101 (2008)
3. Shimizu, E., Isogai, K., Obayashi, S.: Multiobjective design study of a flapping wind power generator. J. Fluids Eng. ASME **130**(2), 021104 (2008)
4. Pigolotti, L., Mannini, C., Bartoli, G., Thiele, K.: Critical and post-critical behaviour of two-degree-of-freedom flutter-based generators. J. Sound Vib. **404**, 116–140 (2017)
5. Caracoglia, L.: Modeling the coupled electro-mechanical response of a torsional-flutter-based wind harvester with a focus on energy efficiency examination. J. Wind Eng. Ind. Aerodyn. **174**, 437–450 (2018)
6. Vakakis, A.F., Gendelman, O.V., Bergman, L.A., McFarland, D.M., Kerschen, G., Lee, Y.S.: Nonlinear targeted energy transfer in mechanical and structural systems (Volumes I and II). Springer Science, New York, New York, USA (2008). https://doi.org/10.1007/978-1-4020-9130-8
7. Kwon, S.-D., Park, J., Law, K.: Electromagnetic energy harvester with repulsively stacked multilayer magnets for low frequency vibrations. Smart Mater. Struct. **22**(5), 055007 (2013)
8. Bisplinghoff, R.L., Ashley, H., Halfman, R.L.: Aeroelasticity. Dover Publications Inc., Mineola, NY, USA (1955)
9. Jones, R.T.: The unsteady lift of a finite wing. Technical Note 682. NACA (1939)
10. Wong, E., Zakai, M.: On the relation between ordinary and stochastic differential equations. Int. J. Eng. Sci. **3**, 213–229 (1965)
11. Kloeden, P.E., Platen, E., Schurz, H.: Numerical Solution of Stochastic Differential Equations through Computer Experiments. Springer-Verlag, Berlin-Heidelberg, Germany (1994)
12. Xie, W.-C.: Dynamic Stability of Structures. Cambridge University Press, New York, NY, USA (2006)
13. Caracoglia, L.: An Euler-Monte Carlo algorithm assessing Moment Lyapunov Exponents for stochastic bridge flutter predictions. Comput. Struct. **122**, 65–77 (2013)
14. Dunn, P., Dugundji, J.: Nonlinear stall flutter and divergence analysis of cantilevered graphite/epoxy wings. AIAA J. **30**(1), 153–162 (1992)

Free-Stream Turbulence Effects on a Square Cylinder with a Forebody Screen at a Small Distance

Andrea Giachetti(✉), Gianni Bartoli, and Claudio Mannini

CRIACIV/DICEA University of Florence, Florence, Italy
{andrea.giachetti,gianni.bartoli,claudio.mannini}@unifi.it

Abstract. This work shows that the presence of a forebody solid screen fixed at a relatively small distance can significantly influence the global aerodynamic behavior of a square cylinder, even for a flow perpendicular to the screen and in presence of free-stream turbulence. The present investigation represents an extension of previous two-dimensional experimental and numerical studies conducted by the same Authors on rectangular and square cylinders in smooth flow. Experiments in the wind tunnel were carried out by reproducing two different homogeneous isotropic turbulent flows and setting different distances between the screen and the front face of the prism. The pressure measurements revealed in turbulent flow qualitatively the same behaviour observed in smooth flow, though the effects of the screen are less pronounced. In particular, a pressure reduction immediately behind the front corners, a fast recover of it moving towards the back corners, and a strong variation of the lateral force coefficient in the direction of stability are the most apparent features.

Keywords: Square Cylinder · Screen · Bluff body · Turbulent Flow · Wind Tunnel Tests

1 Introduction

Permeable building envelopes are often used in modern buildings. Despite the literature concerning wind effects on such façade typologies (*e.g.*, [1]), in certain configurations, the flow mechanisms acting on the system composed by the building and the permeable façade are not fully understood [2]. In addition to studies about standard permeable building envelopes (*e.g.*, [3]), innovative solutions have been recently investigated showing that some permeable building envelopes can improve the building response under the wind action [4] or allow the installation of energy-harvesting devices [5]. Given the presence of an internal cavity connected to the exterior, an interaction between internal and external flows can occur, and the aerodynamics of the original building can strongly be affected by a permeable building envelope [6].

The complexity of the problem and the lack of studies stress the need for a research effort on this topic, even starting from basic geometries. However, to the Authors' best knowledge, the literature concerning bluff bodies equipped with a screen (*e.g.*, [7, 8])

P. Schito and A. Zasso (Eds.): IN VENTO 2022, LNCE 461, pp. 205–213, 2024.
https://doi.org/10.1007/978-3-031-53059-3_18

is mainly focused on the drag reduction of the overall system, and the screen distances considered are comparable with the characteristic dimension of the body cross section, instead of those typical of a permeable building envelope.

A series of two-dimensional exploratory studies were performed by the Authors in the last years [9, 10]. Experimental tests and numerical simulations were performed on a 2:3 rectangular and a square prism with a forebody screen at small distances (between *D*/10 and *D*/40, where *D* is the characteristic cross-flow dimension). Tests were carried out for different angles of attack. The screen was solid, so the internal cavity was connected to the exterior through the lateral sides only. The main campaign was devoted to the comprehension of possible screen effects in smooth flow (namely, the approaching flow longitudinal turbulence intensity, I_u, was lower than 1%). It was observed that the presence of the screen affected the global aerodynamics of the body if the cavity was not vertically sealed (thus allowing the connection between the openings behind the opposite front corners). In this condition, for a flow perpendicular to the screen, an oscillating flow was found to occur inside the cavity, even for the smallest gap width tested. The influence of the screen on the pressure distribution over the cylinder's faces was apparent. Nevertheless, when the flow direction was varied, it was also possible to appreciate the influence of the screen on bulk quantities, such as the slope of the lateral force coefficient. The screen caused a remarkable reduction of this parameter, up to make the cylinder stable against galloping according to the quasi-steady theory.

Then, the Authors extended the study to the three-dimensional case of a 1:1:5 tall building, immersed in an atmospheric boundary layer, equipped with the same type of solid, laterally opened, screen [6–11]. Considering gap widths equal to 1/40 and 1/20 of the cross-section side length, the mean pressures on the lateral sides of the building and the slope of the lateral force coefficient were remarkably affected by the screen, while other aerodynamic quantities, like the drag force and the Strouhal frequency, did not change much when the screen was installed. Thus, the results revealed a screen-induced aerodynamic interference less pronounced but similar to that observed in the previously mentioned two-dimensional studies.

The current work aims to bridge the gap between the two-dimensional study in smooth flow and the three-dimensional investigation in atmospheric boundary layer flow. Indeed, a two-dimensional square cylinder with a forebody solid screen is tested in homogeneous isotropic turbulent flows. Three screen distances were tested (*D*/10, *D*/20 and *D*/40), and the wind direction was varied in the range $\pm 5°$ around the condition of flow perpendicular to the screen.

2 Experimental Set up

The experimental campaign was carried out in the CRIACIV (Inter-University Research Centre on Building Aerodynamics and Wind Engineering) wind tunnel. The model cross-flow section depth (*D*) was equal to 0.12 m (which corresponds to a blockage ratio of 7.5%) and the model length was 1.24 m. Circular end plates with a radius of about 3.5*D* were employed to confine the flow. The sectional model was mounted horizontally in the test chamber through a computer-controlled rotating system able to set the desired angle of attack with high accuracy (Fig. 1). The screen was obtained from 1 mm-thick stainless-steel foils. Steel and plastic spacers were used to create gap widths equal to $D/10$, $D/20$ and $D/40$.

44 pressure taps were installed around the middle section of the model, but it was not possible to install pressure taps around the screen due to its limited thickness. The pressure data were recorded with two 32-channels miniaturized piezoelectric scanners and the Pressure Systems' Digital Pressure Compensation (PSI DTC) Initium system, at a sampling rate of 500 Hz. The pressure signals were transferred to the scanners through 500 mm long Teflon tubes with an internal diameter of 0.8 mm, which were split at a distance of 130 mm from the scanner to insert a 10 mm long damper with an internal diameter of 0.3 mm. Thanks to this configuration of the tubes, no significant signal distortion was observed in the frequency range of interest.

The characteristics of the incoming flow were determined through measurements at a sampling rate of 2000 Hz with a Dantec single-component hot-wire probe 55P11 connected to a Dantec CTA 56C01 module and a 56C17 bridge. The mean flow speed was monitored with a Prandtl tube placed upstream of the model and connected to a high-accuracy low-differential pressure transducer by Setra (AccuSense model ASL). Prior to model installation in the wind tunnel, the conversion coefficient between the mean dynamic pressure registered by the reference Prandtl tube and that measured at the midspan of the future model location was determined for each wind speed and free-stream turbulence considered.

The cross-section side length D was employed as reference dimension to evaluate the Reynolds number and to normalize the aerodynamic coefficients. Reynolds numbers between 1.4×10^5 and 2.0×10^5 were tested.

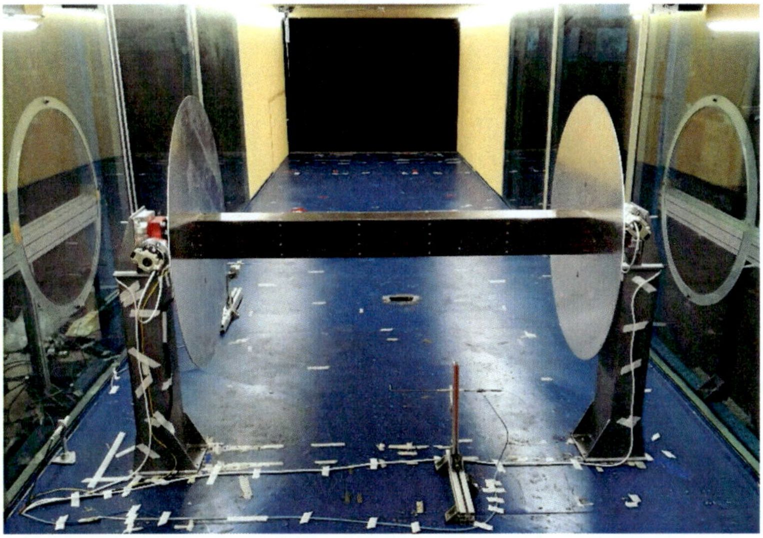

Fig. 1. A picture of the section model in the wind tunnel without turbulence generators.

The smooth flow in the wind tunnel presented a turbulence intensity lower than 1%, while two different turbulent flow conditions (hereafter referred to as Turb_1 and

Fig. 2. Grid used to generate turbulence.

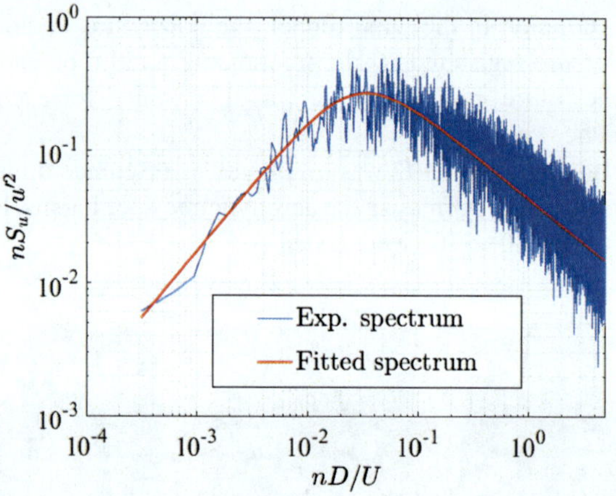

Fig. 3. Normalized power spectral density of longitudinal velocity fluctuations for Turb_1 flow.

Table 1. Main characteristics of flows reproduced in the current study.

Flow	x [m]	I_u [%]	L_{ux} [m]
Smooth	–	< 1	–
Turb_1	5.0	13	0.231
Turb_2	7.0	10	0.254

Turb_2) were obtained through a wooden grid characterized by a mesh of 550 mm × 550 mm (see Fig. 2) and slats of 140 mm (having a thickness in the streamwise direction of 25 mm), installed in two different upwind positions (see x values in Table 1). These approaching flows were characterized by longitudinal integral length scales (L_{ux}) of about 1.9D and 2.1D, and by longitudinal turbulence intensities (I_u) of about 13% and 10%, respectively for Turb_1 and Turb_2 (see Fig. 3 for an example of the energy content of longitudinal flow velocity fluctuations). The turbulent flow was uniform along the model. An attempt to consider a two-dimensional square cylinder immersed in a turbulent flow has already been made in [9], but unfortunately the flow generated through a grid was not uniform along the sectional model (an issue discussed also in [12]), and the results were considered only from the qualitative standpoint.

3 Results

Firstly, the mean pressure coefficients obtained in smooth and turbulent flow around the square section without the screen were compared with some data available in the literature (Fig. 4). No blockage corrections were adopted. The current results in smooth flow are very close to those measured by Bearman and Obasaju [13] and Lander et al. [14]. In turbulent flow, the variability of the results in the literature is more pronounced, but the current data for the Turb_2 flow well retrace those of Lander et al. [14] on the lateral face of the cylinder and those of Lee [15] on the rear face. In contrast, lower pressures are encountered on the side face for the Turb_2 flow. The larger discrepancy in turbulent flow between the results in the literature can probably be ascribed to the significantly different integral length scales in the various cases. The major difference between the results in smooth and in turbulent flow, is the increase of pressure on the lateral sides near the back corners and on the rear side in the second case (responsible for the drag coefficient reduction in Table 2).

When the screen is present, the results obtained in smooth flow for a wind perpendicular to the screen present the same general pattern observed in [10] for a smaller blockage ratio, though with some slight differences in the numerical values. In addition, the following features can be noticed in Fig. 5a: (i) a pressure reduction due to the presence of the screen on the lateral sides behind the separation point; (ii) a pressure jump between the first point on the lateral sides behind the front corners and the pressure in the cavity; and (iii) a mean pressure recovery on the lateral sides towards the trailing edge, indicating the aerodynamic interference caused by the screen.

Figure 5b and Fig. 5c shows the mean pressure coefficients respectively for the approaching flows Turb_1 and Turb_2. Generally, the turbulent flow tends to attenuate the screen effects, partly because, as previously noted, the pressure recovery on the rear portion of the lateral sides occurs even without the screen. However, the pressure reduction on the lateral side behind the separation point occurs in turbulent flow too (and, in general, the pressure is lower for Turb_1 flow, which is characterized by a higher turbulent intensity). In contrast, in both turbulent flow conditions, the screen effects on the rear portion of the lateral sides are nearly negligible for a gap width of $D/40$. The jump between the pressure measured on the lateral side, in correspondence of the front corner, and the mean cavity pressure is also reduced in turbulent flow. Finally, in correspondence of the first pressure taps in the cavity, for the gap widths of $D/20$ and $D/10$, the local effect associated with the separation of the flow drawn into the cavity are still visible.

Table 2 reports the global aerodynamic coefficients obtained by pressure integration. The values for the smooth flow case are slightly different from those reported in

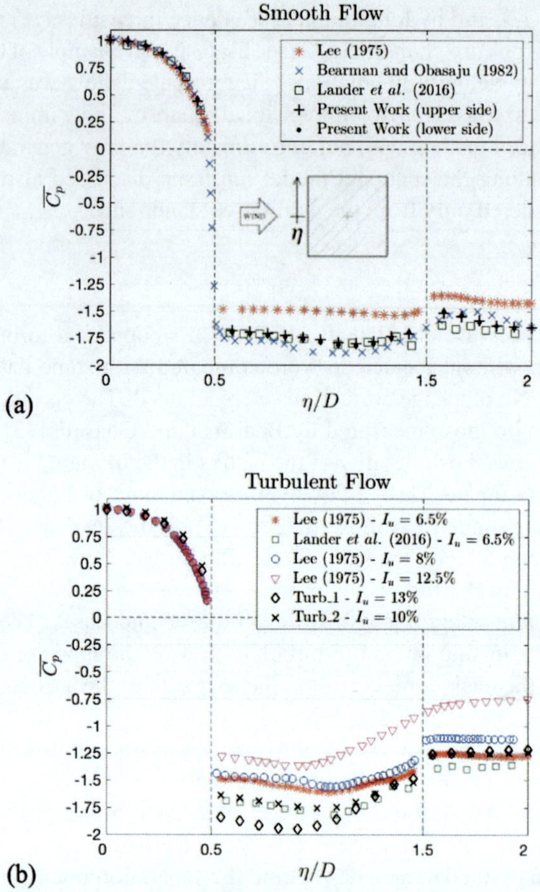

Fig. 4. Mean pressure coefficients measured in smooth (a) and turbulent (b) flow around the square cylinder. Data are taken from [13, 14] and [15].

Table 2. Global aerodynamic coefficients obtained from pressure integration. "NoSc" denotes the configuration without the screen.

	Smooth Flow				Turbulent Flow – Turb_1				Turbulent Flow – Turb_2			
	NoSc	$D/40$	$D/20$	$D/10$	NoSc	$D/40$	$D/20$	$D/10$	NoSc	$D/40$	$D/20$	$D/10$
C_D	2.33	–	–	–	1.97	–	–	–	2.03	–	–	–
C_{RBD}	1.58	1.64	1.66	1.67	1.24	1.26	1.26	1.26	1.26	1.28	1.27	1.28
St	0.13	0.12	0.12	0.12	0.13	0.13	0.13	0.13	0.13	0.13	0.13	0.13
C_L'	1.46	1.53	1.49	1.43	0.94	0.98	0.99	0.97	0.99	1.05	1.05	1.03
$dC_{Fy}/d\alpha$	2.96	0.96	−1.74	−3.79	2.10	0.28	−1.07	−2.39	3.08	1.18	−0.31	−1.91

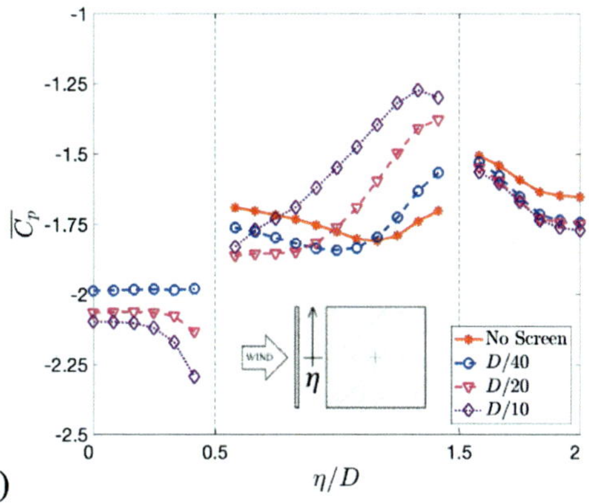

(a)

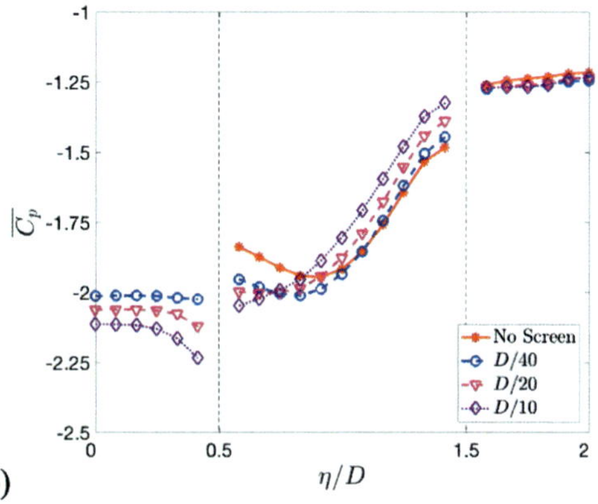

(b)

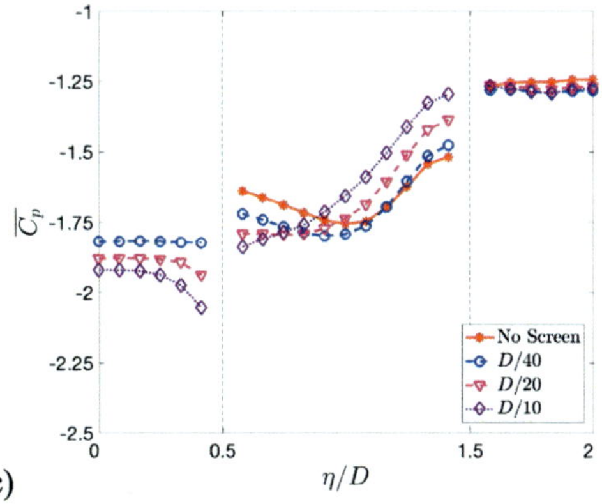

(c)

Fig. 5. Mean pressure coefficients around the square cylinder for a wind perpendicular to the screen: smooth flow (a), Turb_1 flow (b), and Turb_2 flow (c).

[8], probably because of the previously mentioned higher blockage ratio. Generally, in turbulent flow the variations of such integral quantities due to the presence of the screen are comparable with those measured in smooth flow, except for the integration

of pressures on the rear side (*i.e.*, the rear-body drag coefficient, C_{RBD}), which exhibits negligible variations, as confirmed by the mean pressure distributions (Fig. 5). The frequency of vortex shedding (expressed through the Strouhal number, *St*) varies no more than 3%, and the lift coefficient standard deviation ($C_L{}'$) exhibits variations comparable with those observed in smooth flow (in particular, the lift oscillation is slightly larger for the less turbulent flow, namely Turb_2 condition). The slope in the origin of the lateral force coefficient, $dC_{Fy}/d\alpha$, shows the strongest sensitivity to the presence and distance of the screen both in smooth and in turbulent flow. Indeed, the square cylinder without the screen is always highly prone to galloping. It becomes either slightly or moderately unstable (for Turb_1 and Turb_2 flows, respectively) for a screen distance of *D*/40, and it finally results stable for *D*/20 and *D*/10.

4 Conclusions

A turbulent approaching flow tends to attenuate the effects caused by a forebody solid screen fixed on a side of a square cylinder at a small distance. Nevertheless, a reduction of pressure behind the front corners, the characteristic pressure pattern close to the end of the cavity, and typical variations of some bulk aerodynamic quantities (in particular, the reduction of the standard deviation of the lateral force coefficient and the change of its slope in the direction of stability) still occur.

The current study also shows that the partial abatement of the influence of the screen on a three-dimensional building exposed to a turbulent boundary layer wind, as observed in a previous work conducted by the Authors [6], can partly be ascribed to the turbulent flow.

References

1. Gerhardt, H.J., Janser, F.: Wind loads on wind permeable facades. J. Wind Eng. Ind. Aerodyn. **53**, 37–48 (1994)
2. Manara, G.: Some topics about wind engineering that curtain walling design might be longing for in standards. In: Ricciardelli, F., Avossa, A.M. (eds.) Proceedings of the XV Conference of the Italian Association for Wind Engineering IN-VENTO2018, Napoli, pp. 17–31 (2019)
3. Lou, W., Huang, M., Zhang, M., Lin, N.: Experimental and zonal modeling for wind pressures on double-skin facades of a tall building. Energy Build. **54**, 179–191 (2012)
4. Hu, G., Hassanli, S., Kwok, K.C.S., Tse, K.T.: Wind-induced responses of a tall building with a double-skin façade system. J. Wind Eng. Ind. Aerodyn. **168**, 91–100 (2017)
5. Hassanli, S., Hu, G., Kwok, K.C.S., Fletcher, D.F.: Utilizing cavity flow within double skin façade for wind energy harvesting in buildings. J. Wind Eng. Ind. Aerodyn. **167**, 114–127 (2017)
6. Giachetti, A., Bartoli, G., Mannini, C.: Aerodynamics of a tall building equipped with a solid screen close to the façade. J. Wind Eng. Ind. Aerodyn. **225**, 104999 (2022)
7. Koenig, K., Roshko, A.: Experimental study of geometrical effects on the drag and flow field of two bluff bodies separated by a gap. J. Fluid Mech. **156**, 167–204 (1985)
8. Cooper, K.R.: The use of a forebody plate to reduce the drag and to improve the aerodynamic stability of a cylinder of square cross-section. J. Wind Eng. Ind. Aerodyn. **28**, 271–280 (1988)
9. Giachetti, A.: Wind effects on permeable building envelopes: a two-dimensional exploratory study. PhD Thesis, University of Florence – TU Braunschweig (2018)

10. Giachetti, A., Bartoli, G., Mannini, C.: Two-dimensional study of a rectangular cylinder with a forebody airtight screen at a small distance. J. Wind Eng. Ind. Aerodyn. **189**, 1–11 (2019)
11. Giachetti, A., Bartoli, G., Mannini, C.: Wind effects on permeable building envelopes of tall buildings: main issues and potentialities. CTBUH J. 2019 (3) (2019)
12. Bearman, P.W., Morel, T.: Effect of the free stream turbulence on the flow around bluff bodies. Prog. Aerosp. Sci. **20**(2–3), 97–123 (1983)
13. Bearman, P.W., Obasaju, E.D.: An experimental study of pressure fluctuations on fixed and oscillating square-section cylinders. J. Fluid Mech. **119**, 297–321 (1982)
14. Lander, D.C., Letchford, C.W., Amitay, M., Kopp, G.A.: The influence of the bluff body shear layers on the wake of a square prism in a turbulent flow. Physical Review Fluids **1**, 044406 (2016)
15. Lee, B.E.: The effect of turbulence on the surface pressure field of a square prism. J. Fluid Mech. **69**, 263–282 (1975)

Can Wind Lidars Be Used to Calibrate Mean Wind Profiles?

Vincenzo Sepe[1], Alberto M. Avossa[2(✉)], Fabio Rizzo[3], and Francesco Ricciardelli[2]

[1] University "G. d'Annunzio" of Chieti-Pescara, Pescara, Italy
vincenzo.sepe@unich.it
[2] University of Campania "Luigi Vanvitelli", Aversa, Italy
{albertomaria.avossa,friccia}@unicampania.it
[3] Polytechnic University of Bari, Bari, Italy
fabio.rizzo@poliba.it

Abstract. This paper explores to which extent Wind Lidars can be used to calibrate mean wind profiles, also in relation to the characteristics of the available measurements. Starting from measured wind speeds in the range 50 m to 250 m from the ground obtained in a LIDAR experimental campaign, pseudo-experimental wind speed profiles are generated by adding a zero-mean Gaussian noise to the logarithmic law; this is representative of realistic measurement errors. Then, a least-square fitting procedure is applied to identify the roughness length and the zero-plane displacement. The procedure is applied to two different mean wind profiles, corresponding to densely built areas featuring mid-rise and low-rise buildings, respectively. The results show increasing scatter in the identified parameters of the log-law with increasing height of the lowest measurement point and with decreasing number of measurement points. This suggests that to improve accuracy of the identified profile at low heights, an additional point of measurement is needed, lower than those allowed by available LIDAR technology, or that a very large number of measured profiles must be available.

Keywords: Wind LIDAR · Mean Wind Profile · Wind Models

1 Introduction

Measurement of the wind velocity field at a given site can be useful for defining and validating analytical and numerical models for wind actions on structures. To this aim, the use of Doppler Lidar (Light Detection and Ranging, LIDAR) technology has become popular in the last decades [1–6], in spite of the inaccuracies associated with non-synchronous measurement and space averaging. LIDARs in fact allow a relatively easy acquisition of wind velocities at different heights, up to several hundred meters above the ground; however, they are unable to measure at low heights, as they have a lower measurement threshold at about 30 m or even more.

On the other hand, the gradient of the mean wind speed is more pronounced near the ground, therefore the parameters that characterize analytical models are deeply influenced by the wind speed at low heights; this in particular in the urban environment, where the wind flow near the ground interacts with the buildings.

P. Schito and A. Zasso (Eds.): IN VENTO 2022, LNCE 461, pp. 214–226, 2024.
https://doi.org/10.1007/978-3-031-53059-3_19

This paper explores to which extent LIDARs can be used to calibrate mean wind profiles in the urban environment, also in relation to the characteristics of the available measurements.

2 Methodology

It is widely accepted that in the urban environment any analytical law for the mean wind speed profile has to include the zero-plane displacement, i.e. the level below which the flow is nearly blocked by the obstructing buildings. As a part of a wider research on the effectiveness of the LIDAR technology for the experimental characterization of wind fields [7], this paper describes numerical analyses devoted to investigate how identification of the parameters z_0 (roughness length) and d (zero-plane displacement) of the logarithmic law is affected by the characteristics of the available measurements.

Assuming a synoptic, neutral wind regime, for different values of z_0 and d a set of vertical profiles of the 10-min averaged wind speed is generated, corresponding to moderately strong winds ($v_{ref} \geq 12$ m/s). The profiles are affected by numerical random noise, whose magnitude is kept in a range representative of the typical accuracy of LIDAR measurements. Reference profiles for the generation of pseudo-experimental data are those obtained in a LIDAR experimental campaign [7] performed by some of the authors and corresponding to neutral stratification. Wind speeds were measured in the range 50 m to 250 m from the ground, therefore in the analyses presented here v_{ref} is set equal to the wind speed at 50 m of elevation. Pseudo-experimental mean wind speed profiles are then assumed as input data for a least-square fitting procedure, that allows to identify the parameters z_0 and d of the logarithmic law.

3 Numerical Investigation

Two cases are examined, corresponding to pairs of parameters z_0 and d representative of 30° sectors centred around 120° and 270° oncoming wind directions, as identified in [7]. Namely, the results reported in Sect. 3.1 are based on a set of pseudo-experimental profiles generated by adding a zero-mean Gaussian noise to the logarithmic law with $z_0 = 1.65$ m and $d = 17.50$ m (urban areas with mid-rise buildings), whereas those of Sect. 3.2 are based on a reference logarithmic profile with $z_0 = 0.50$ m and $d = 4.50$ m (urban areas with low-rise buildings). For the purpose of the analyses, the logarithmic profile is expressed through the following equation:

$$v(z) = v_{ref} \frac{log[(z - d)/z_0]}{log\left[\left(z_{ref} - d\right)/z_0\right]} \tag{1}$$

3.1 Urban Areas with Mid-Rise Buildings

As a first sample case, a reference logarithmic profile $v_0(z)$ with $v_{ref} = 12.7$ m/s at $z_{ref} = 50$ m, $z_0 = 1.65$ m and $d = 17.50$ m (Eq. 1) is considered, representative of the parameters identified from measurements downwind a densely built urban area during the LIDAR experimental campaign described in [7]. The reference profile v_0 (z) was generated with a vertical step $dz = 1$ m between the elevation $z = d + z_0$, where $v = 0$ according to Eq. 1, and $z = 200$ m, the latter being the maximum height of validity for Eq. (1) in EN1991–1-4 [8]. Two sets of one hundred pseudo-experimental profiles were then generated by adding to the value $v_0(z)$ of the logarithmic profile of Eq. (1) a zero-mean Gaussian-distributed noise with a standard deviation σ_{noise} $(z) = 0.02 \cdot v_0(z)$ and σ_{noise} $(z) = 0.04 \cdot v_0(z)$, respectively.

For the two levels of noise, Fig. 1 shows the exact logarithmic profile (red line), a sample pseudo-experimental profile (black line), as well as the envelope of the whole set of 100 pseudo-experimental profiles (green stars). For each pseudo-experimental profile, the parameters z_0 and d were identified by means of a least-square error minimization procedure, assuming the availability of measured data according to different scenarios. These are defined by the elevation z_1 of the lowest measurement point (variable between around 23 m and 60 m above the ground) and by the number of equally spaced measurement points n_p between z_1 and the maximum elevation $z_{max} = 200$ m above the ground ($n_p = 15$ or $n_p = 8$).

For a noise standard deviation σ_{noise} $(z) = 0.02 \cdot v_0(z)$, Fig. 2 shows the identified pairs of parameters z_0 and d for each pseudo-experimental profile (green stars) when the number of available measurement points is $n_p = 15$, while Fig. 3 shows the identified parameters when $n_p = 8$.

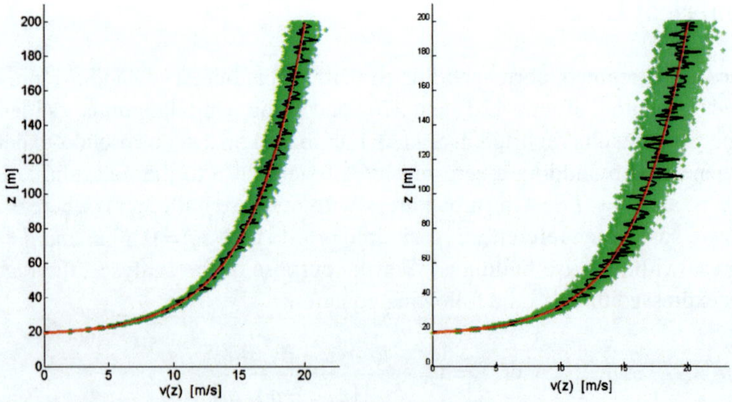

Fig. 1. Reference logarithmic profile $v_0(z)$ (red line), sample pseudo-experimental profile (black line) and envelope of 100 pseudo-experimental profiles (green stars); left: σ_{noise} $(z) = 0.02 \cdot v_0(z)$; right: σ_{noise} $(z) = 0.04 \cdot v_0(z)$. (Color figure online)

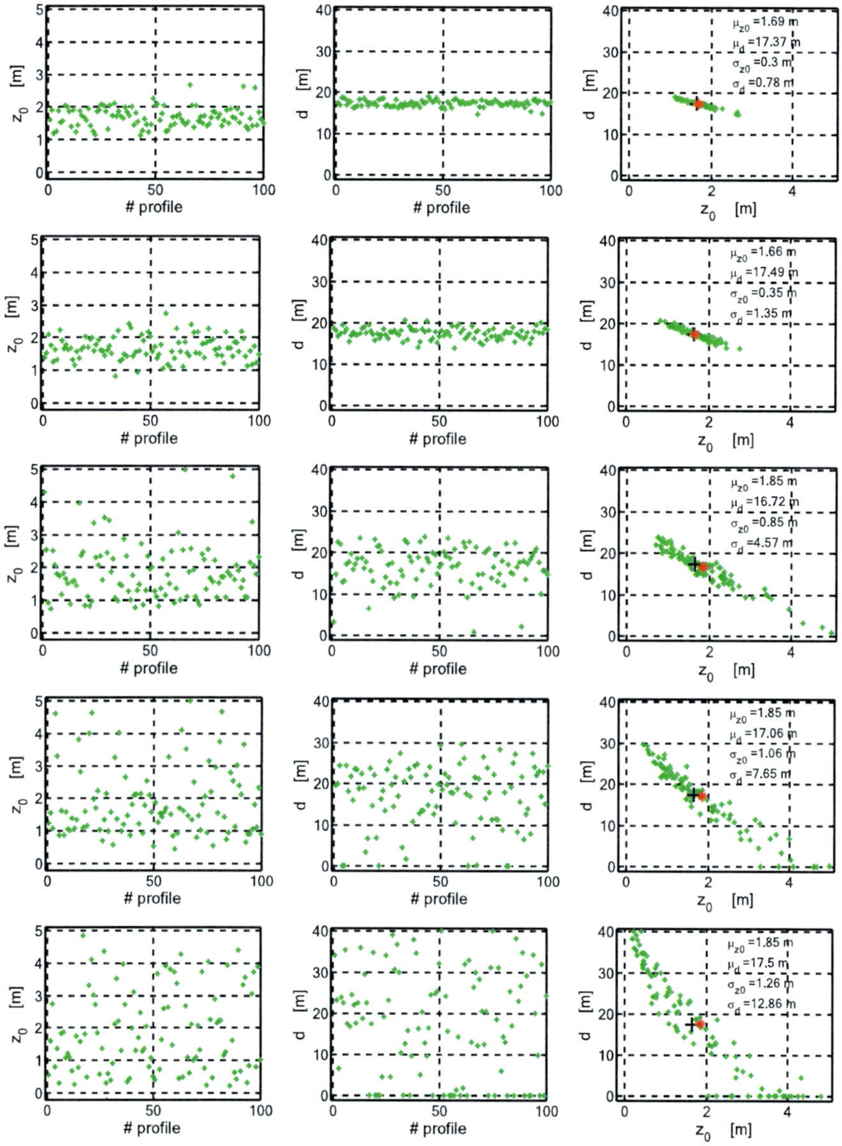

Fig. 2. Identified values of z_0 and d (green stars) for different values of the lowest available measurement point z_1; $n_p = 15$, $\sigma_{noise}(z) = 0.02 \cdot v_0(z)$; red stars and black crosses in the right column correspond to mean identified values and to the target values, respectively; 1st to 5th row: $\delta = (z_1\text{-}d) = 6$ m, 10 m, 20 m, 28 m, and 43 m. (Color figure online)

In both cases five different elevations of the lowest measurement point have been considered, i.e. z_1 equal to 23 m, 27 m, 37 m, 45 m and 60 m, corresponding to a distance $\delta = (z_1 - d)$ of about 6 m, 10 m, 20 m, 28 m, and 43 m, respectively, above the assumed zero-plane displacement height *d*.

The results in Figs. 2 and 3 show that for a given number of measurement points (either 15 or 8) identification of the parameters z_0 and d is strongly influenced by the elevation z_1 of the lowest measurement point. For the lowest value of z_1, corresponding to $z_1 - d = 6$ m (first row of Fig. 2 and Fig. 3), the 100 pairs of identified parameters (z_0, d) have only a moderate scatter around their average values $\mu_{z0} = 1.69$ m and $\mu_d = 17.37$ m ($n_p = 15$) and $\mu_{z0} = 1.71$ m and $\mu_d = 17.37$ m ($n_p = 8$), which are indeed very close to the values $z_0 = 1.65$ m and $d = 17.5$ m. When increasing z_1 (2nd to 5th row of Fig. 2 and Fig. 3) the scatter of identified parameters increases significantly, but the difference between their average values and the target parameters varies only moderately. In particular, for the largest value of z_1, corresponding to $\delta = (z_1 - d) = 43$ m, $\mu_{z0} = 1.85$ m and $\mu_d = 17.50$ m when $n_p = 15$, and $\mu_{z0} = 2.10$ m and $\mu_d = 16.43$ m when $n_p = 8$. It is concluded that availability of a large number of measured profiles can compensate for the large uncertainty, and average identified parameters are well representative of the target values.

The conclusion above is shown in a graphical form in Fig. 4 and Fig. 5 where, for the two cases of $n_p = 15$ and $n_p = 8$, μ_{z0} and μ_d are shown as a function of z_1 (left figures), together with the target value and with the values $\mu_{z0} \pm \sigma_{z0}$ and $\mu_d \pm \sigma_d$. Right figures show the coefficient of variation $CV_{z0} = \sigma_{z0} / \mu_{z0}$ and $CV_d = \sigma_d / \mu_d$ of the identified parameters, again as a function of z_1.

To better explain this circumstance, for a selected pseudo-experimental profile with $\sigma_{noise}(z) = 0.02 \cdot v_0(z)$ Fig. 6 and Fig. 7 show the logarithmic profiles identified from 15 equally spaced measurement points (red stars), and for five different values of z_1. It can be observed that (Fig. 7) all profiles identified (red lines) by means of the same number of wind speed data ($n_p = 15$) are very close to each other at higher elevations, while they differ from one another at lower elevations. It is concluded that as a consequence, when only one or few measured profiles are available, measurement at a low elevation is needed to calibrate the logarithmic profile parameters z_0 and d in a way such to accurately reproduce wind speeds also at low elevations.

The scatter of the identified pairs of parameters (z_0, d) increases when pseudo-experimental profiles are generated with a larger noise $\sigma_{noise}(z) = 0.04 \cdot v_0(z)$, as reported in Fig. 8 (right column) for the case of $n_p = 15$ measurement points. As a comparison, the left and central columns of Fig. 8 report the identified pairs of parameters for $\sigma_{noise}(z) = 0.02 \cdot v_0(z)$ for $n_p = 15$ and for $n_p = 8$ already shown in Fig. 2 and Fig. 3, respectively. It is also observed that for large values of z_1 the capacity of identifying the target parameters reduces with increasing noise in the pseudo-experimental profiles (compare left and right columns in Fig. 8).

For $\sigma_{noise}(z) = 0.04 \cdot v_0(z)$, still representing a realistic measurement error at the highest elevations (see [7]), the standard deviation of the identified pseudo-experimental profiles becomes very large, with coefficients of variations equal to 0.8 for $z_1 = 35$ m (for CV_{z0}) and $z_1 = 45$ m (for CV_d), as shown in Fig. 9. However, the availability of a

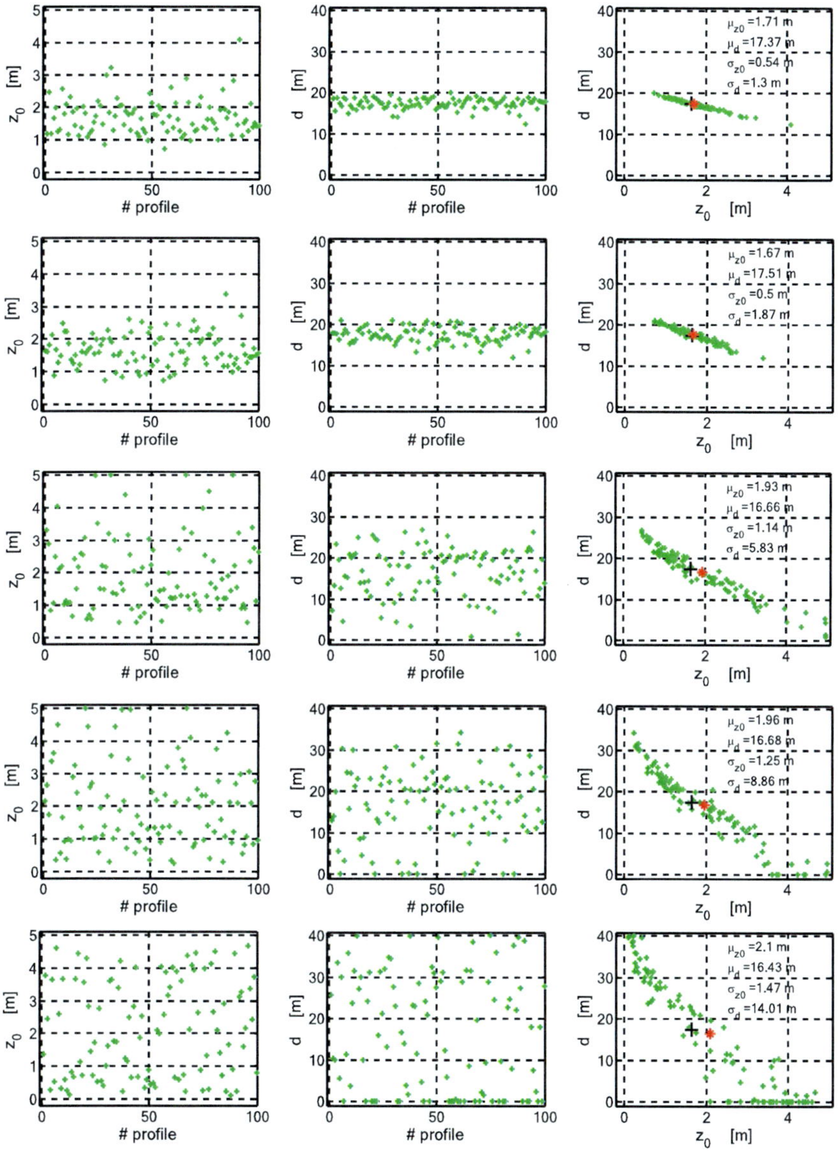

Fig. 3. Identified values of z_0 and d (green stars) for different values of the lowest available measurement point z_1; $n_p = 8$, $\sigma_{noise}\ (z) = 0.02{\cdot}v_0(z)$; red stars and black crosses in the right column correspond to mean identified values and to the target values, respectively; 1st to 5th row: $\delta = (z_1\text{-}d) = 6$ m, 10 m, 20 m, 28 m, and 43 m. (Color figure online)

large number of measured profiles (100 in this paper) still allows a reasonably accurate estimate of the target parameters of the profile.

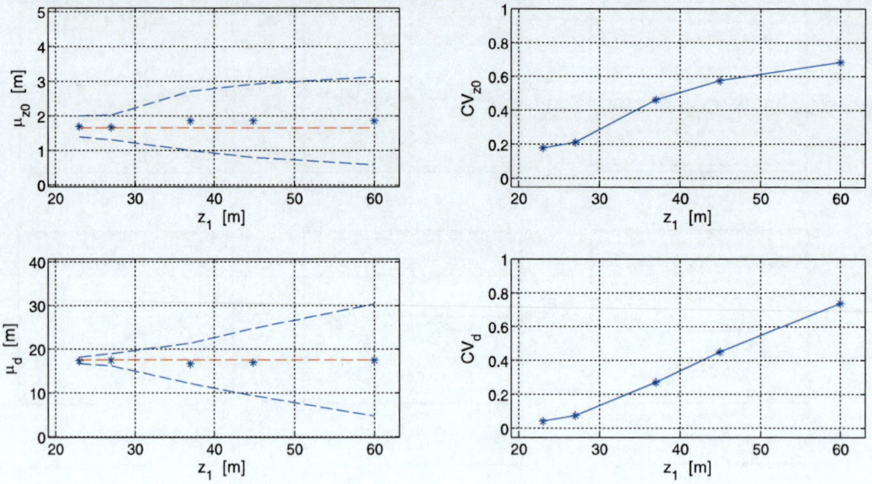

Fig. 4. Mean values of identified parameters (blue stars) and mean values ± standard deviation (blue dashed lines); red dashed lines show the target values (left column). Coefficient of variation of the identified parameters (right column). $n_p = 15$, $\sigma_{noise}(z) = 0.02 \cdot v_0(z)$. (Color figure online)

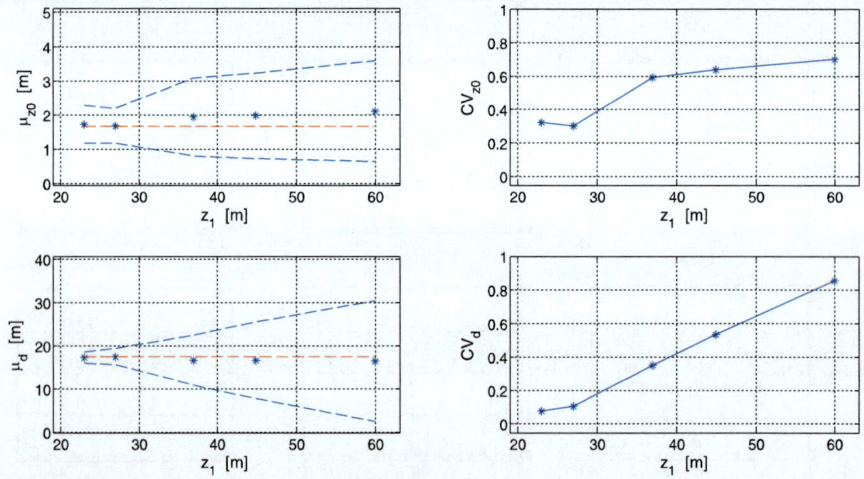

Fig. 5. Mean values of identified parameters (blue stars) and mean values ± standard deviation (blue dashed lines); red dashed lines show the target values (left column). Coefficient of variation of the identified parameters (right column). $n_p = 8$, $\sigma_{noise}(z) = 0.02 \cdot v_0(z)$. (Color figure online)

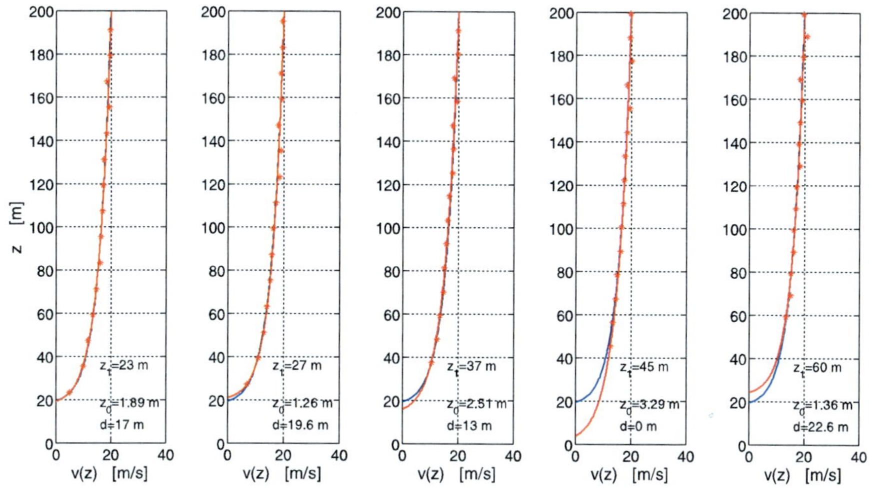

Fig. 6. Target (blue line) and selected identified (red lines) profiles for five different values of z_1; from left to right: $\delta = (z_1 - d) = 6$ m, 10 m, 20 m, 28 m, and 43 m. $n_p = 15$, $\sigma_{noise}(z) = 0.02 \cdot v_0(z)$. (Color figure online)

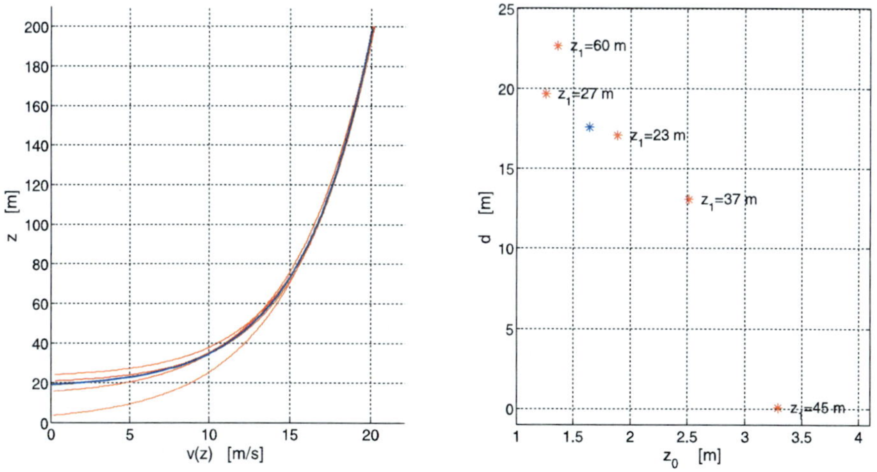

Fig. 7. Reference profile (blue line) and identified profiles (red lines) of Fig. 6 (left). Identified pairs of parameters (red stars) and reference values (blue star) (right). (Color figure online)

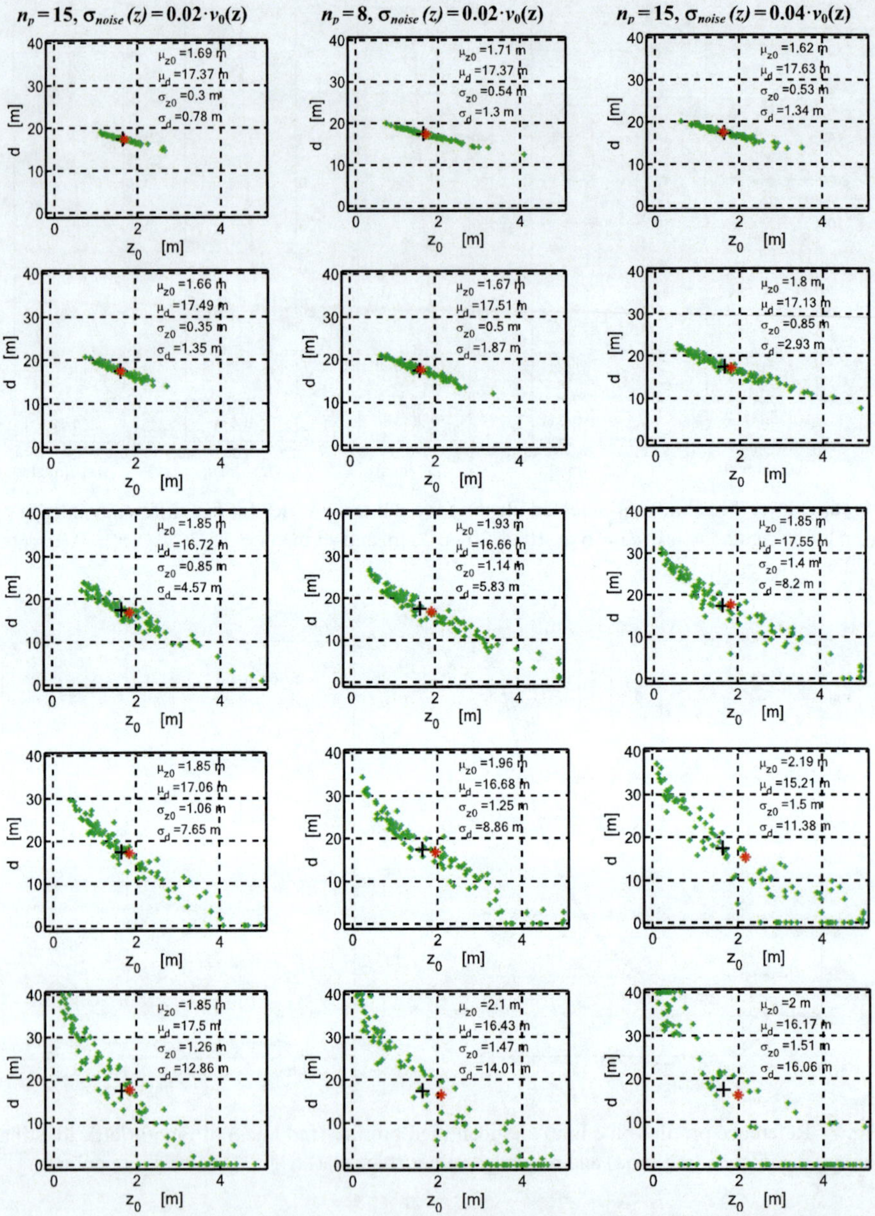

Fig. 8. Identified parameters z_0 and d (green stars) for different values of z_1; red stars and black crosses correspond to mean and target values, respectively. 1st to 5th row: $\delta = (z_1\text{-}d) = 6$ m, 10 m, 20 m, 28 m, and 43 m. 1st column: $n_p = 15$, $\sigma_{noise}(z) = 0.02 \cdot v_0(z)$; 2nd column: $n_p = 8$, $\sigma_{noise}(z) = 0.02 \cdot v_0(z)$; 3rd column: $n_p = 15$, $\sigma_{noise}(z) = 0.04 \cdot v_0(z)$. (Color figure online)

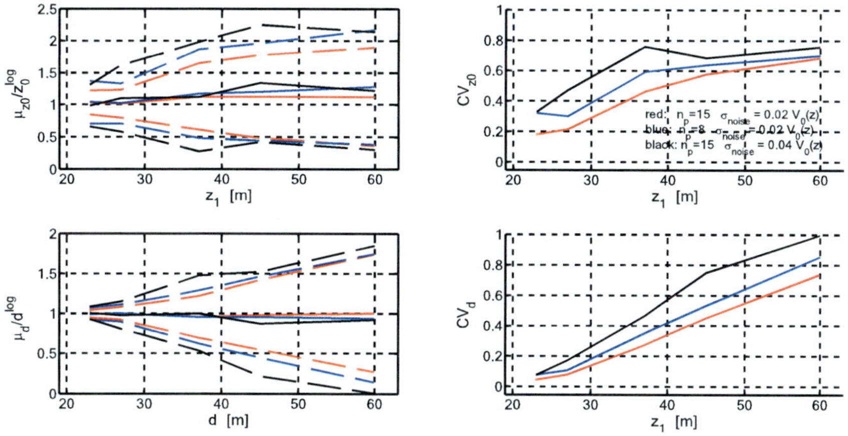

Fig. 9. Mean values of identified parameters (solid lines) and mean ± standard deviation (dashed lines) normalized with respect to the target values (left column). Coefficients of variation (right column). Red lines: $n_p = 15$, $\sigma_{noise}(z) = 0.02 \cdot v_0(z)$; blue lines: $n_p = 8$, $\sigma_{noise}(z) = 0.02 \cdot v_0(z)$; black lines: $n_p = 15$, $\sigma_{noise}(z) = 0.04 \cdot v_0(z)$. (Color figure online)

3.2 Urban Areas with Low-Rise Buildings

As a second sample case, a reference logarithmic profile $v_0(z)$ with $v_{ref} = 12$ m/s at $z_{ref} = 50$ m, $d = 4.50$ m and $z_0 = 0.50$ m (Eq. 1) was considered, which is representative of less urbanised area, as compared to sample case of Sect. 3.1. Similarly to Sect. 3.1, the reference mean wind profile $v_0(z)$ was generated with a vertical step $dz = 1$m between the elevation $z = d + z_0$ and $z = 200$ m, and a set of one hundred pseudo-experimental profiles was then generated by adding to $v_0(z)$ a zero-mean Gaussian-distributed noise with a standard deviation $\sigma_{noise}(z) = 0.02 \cdot v_0(z)$.

Following the same format as that of Fig. 2 and Fig. 3, Fig. 10 shows the results of the identified pairs of parameters z_0 and d for each pseudo-experimental profile (green stars) assuming that $n_p = 15$ measurement points are available and for four different elevations z_1 of the lowest measurement point. Namely, z_1 equal to 10 m, 14 m, 24 m, and 32 m, corresponding to a distance $\delta = (z_1 - d)$ of about 6 m, 10 m, 20 m, and 28m, respectively, above the assumed zero-plane displacement height d. Also for this second sample case the results in Fig. 10 confirm that the identification of the parameters z_0 and d is strongly influenced by the elevation z_1 of the lowest measurement point, above all for the identification of d. Namely, while the coefficient of variation of the identified roughness length z_0 (Fig. 11) shows a trend similar to the sample case discussed in Sect. 3.1, with values around 0.4 for elevations z_1 of about 10–20 m from ground, much larger coefficients of variation are observed in this second sample case for the identification of the zero-plane displacement height d, with values $CV_d \cong 1$ for elevations z_1 around 30 m from ground. It is confirmed that also in this case the availability of a large number of experimental profiles allows a rather accurate identification of the target parameters.

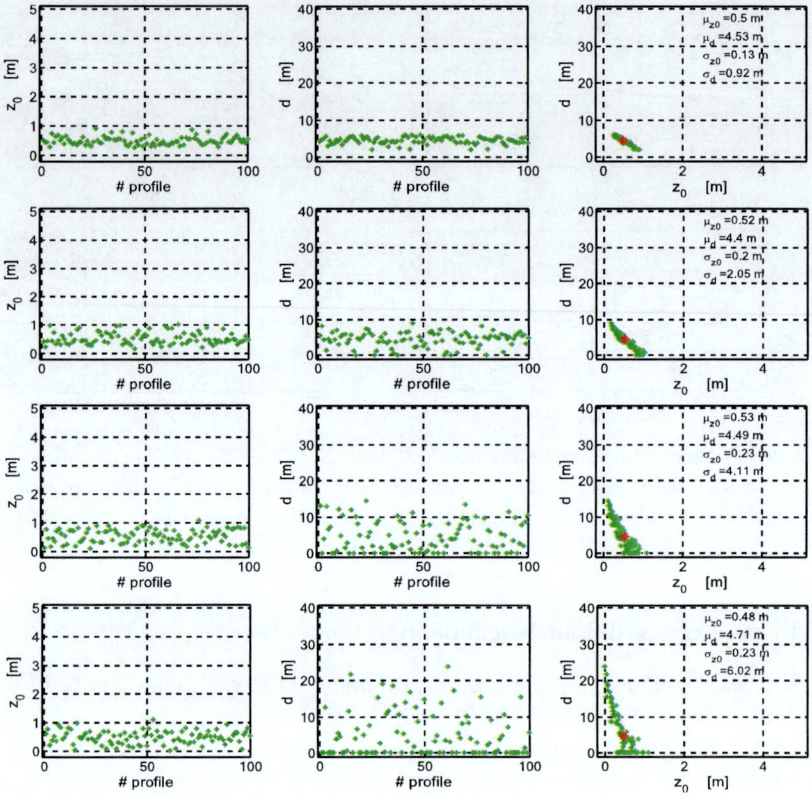

Fig. 10. Identified values of z_0 and d (green stars) for different values of z_1; $n_p = 15$, $\sigma_{noise}(z) = 0.02 \cdot v_0(z)$; red stars and black crosses in the right column correspond to mean identified values and to the target values, respectively; 1st to 4th row: $\delta = (z_1\text{-}d) = 6$ m, 10 m, 20 m and 28 m. (Color figure online)

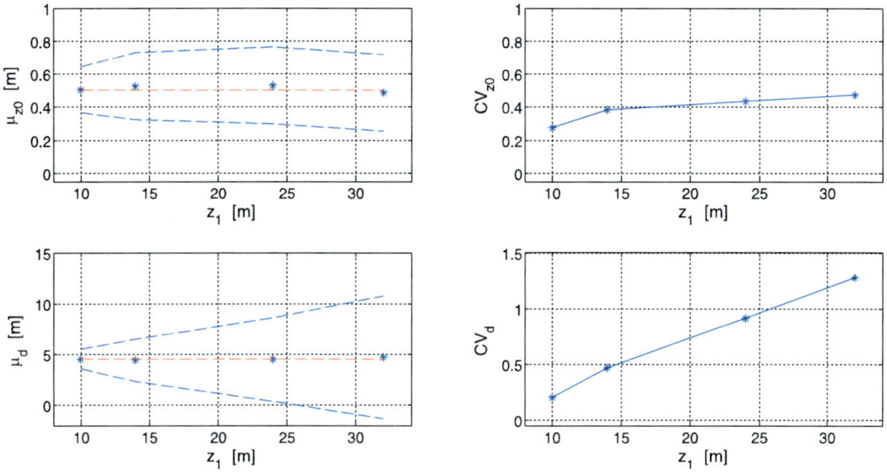

Fig. 11. Mean values of identified parameters (blue stars) and mean values ± standard deviation (blue dashed lines); red dashed lines show the target values (left column). Coefficient of variation of the identified parameters (right column). $n_p = 15$, $\sigma_{noise}(z) = 0.02 \cdot v_0(z)$. (Color figure online)

4 Concluding Remarks

Based on the results of a numerical investigation, this paper explores to which extent Wind Lidars can be used in the urban context to calibrate mean wind profiles, depending on the available measurements. With no loss of generality, the logarithmic law was taken as representative of the wind speed profile and the uncertainty related to the identification of the parameters z_0 (roughness length) and d (zero-plane displacement) is discussed as a function of the number of measuring points, of the measuring error and of the lowest available measuring point. It is concluded that when only one or few measured profiles are available, then a measurement point at low elevation is required to accurately identify the target logarithmic profile. However, when a large sample of measured profiles is available, then the average identified profile parameters rather nicely converges to the target values.

References

1. Demartino, C., Avossa, A.M., Ricciardelli, F., Calidonna, C.R.: Wind profiles identification using wind LIDARS: an application to the area of Lametia Terme. In: Proceedings of 7th European and African Conference on Wind Engineering, EACWE, 4–7 July 2017, Liège, Belgium (2017)
2. Drew, D.R., Barlow, J.F., Lane, S.E.: Observations of wind speed profiles over Greater London, UK, using a Doppler Lidar. J. Wind Eng. Ind. Aerodyn. **121**, 98–105 (2013)
3. Kent, C.W., Grimmond, C.S.B., Gatey, D., Barlow, J.F.: Assessing methods to extrapolate the vertical wind-speed profile from surface observations in a city centre during strong winds. J. Wind Eng. Ind. Aerodyn. **173**, 100–111 (2018)
4. Lane, S.E., Barlow, J.F., Wood, C.R.: An assessment of a three-beam Doppler Lidar wind profiling method for use in urban areas. J. Wind Eng. Ind. Aerodyn. **119**, 53–59 (2013)

5. Lim, K.E.W., Watkins, S., Clothier, R., Ladani, R., Mohamed, A., Palmer, J.L.: Full-scale flow measurement on a tall building with a continuous-wave Doppler Lidar anemometer. J. Wind Eng. Ind. Aerodyn. **154**, 69–75 (2016)
6. Wood, C.R., et al.: Wind observations above an urban river using a new Lidar technique, scintillometry and anemometry. Sci. Total. Environ. **442**, 527–533 (2013)
7. Sepe, V., Rizzo, F., Ricciardelli, F., Avossa, A. M.: Characterization of mean wind profiles and surface roughness assessment from wind LIDAR measurements. In: Ricciardelli, F., Avossa, A.M. (eds.) Proceedings of the XV Conference of the Italian Association for Wind Engineering. Lecture Notes in Civil Engineering, vol. 27, pp. 689–702. Springer, Cham (2019). https://doi.org/10.1007/978-3-030-12815-9_52
8. CEN: Eurocode 1: Action on structures – Part 1–4: General actions – Wind actions, EN1991–1–4:2005, Comité Européen de Normalisation (2005)

Numerical Investigation of a TLP Wind Turbine Under Wind and Wave Loads

M. Vardaroglu[1,3](✉), Z. Gao[2,4], A. M. Avossa[1], and Francesco Ricciardelli[1]

[1] Department of Engineering, University of Campania "Luigi Vanvitelli", Aversa, Italy
mustafa.vardaroglu@unicampania.it

[2] Department of Marine Technology, Faculty of Engineering, Norwegian University of Science and Technology, Trondheim, Norway

[3] Power & New Energy, Worley Europe, London, UK

[4] School of Naval Architecture, Ocean and Civil Engineering, Shanghai Jiao Tong University, Shanghai, China

Abstract. Research on offshore wind technology has increased in the last two decades following progress and cost reduction in turbine design and operation, and the increased demand for renewable energy production. Coupled dynamic response of a floating offshore wind turbine under simultaneous wind and wave loading can be investigated by experimental and computational methods. In the scope of this paper, a numerical study is conducted on a floating offshore wind turbine. A standard baseline wind turbine supported by a Tension Leg Platform is investigated under simultaneous wind and wave loading. For comparison, cases with only wind loading and only wave loading were also simulated to identify the dominating contributions in different design situations. Larger platform surge response is observed due to the resonance triggered by the turbulent wind at the platform surge natural frequency. Pitch motions are large at the wave frequencies and at the resonant frequency, mainly due to first-order wave loads. Aerodynamic damping associated with the rotation of the rotor resulted in decreased platform fluctuating pitch response. Results of this study are the basis of a sensitivity study aiming at improving the MIT/NREL TLP wind turbine design.

Keywords: Floating Offshore Wind Turbines · Tension Leg Platforms · Resonant dynamic response · Wind Turbulence · Aerodynamic Damping

1 Introduction

Offshore wind energy has been rapidly developed in the last two decades, with aim of achieving the net zero targets in carbon emission and fight against global warming. Decreasing costs in floating offshore wind energy technology makes it possible to generate energy in deep water areas, where wind speeds are constant and relatively higher than onshore or close to shore. Tension Leg Platform (TLP) wind turbines achieve stability by using the tension in the mooring lines connecting the floating platform to the seabed. TLP wind turbines exhibit smaller heave and pitch motions, compared to alternative platform types such as spar buoys and semi-submersibles.

P. Schito and A. Zasso (Eds.): IN VENTO 2022, LNCE 461, pp. 227–238, 2024.
https://doi.org/10.1007/978-3-031-53059-3_20

In this study, coupled response of a TLP wind turbine is discussed under simultaneous wind and wave loading over numerical analyses. Fully coupled motion response analyses are conducted in the time domain by using the open-source Fatigue, Aerodynamics, Structures, and Turbulence code OpenFAST [1]. The numerical model has been calibrated and validated in [2] under regular and irregular waves. Hydrodynamic damping in the numerical model was calibrated by using scaled physical model decay tests. A modified version of the MIT NREL TLP design [3] and its mooring system, supporting the NREL 5 MW standard baseline turbine [4], was used to this purpose, which is termed as DHI TLP wind turbine. The effect of wind excitation on the response of the numerical model turbine is investigated by comparing wave loading and operational wind conditions. The importance of aerodynamic damping is highlighted through comparisons between operational and parked turbine cases, with the floater subjected to irregular wave loading. Platform motions, tendon tensions, tower base bending moments, and blade root moments from the numerical analyses are discussed by means of their statistics and Power Spectral Density Functions (PSDFs).

2 Numerical Model

2.1 TLP Floating Wind Turbine

The structure consists of a floating cylinder, the tendons, and the wind energy generator. Figure 1 shows the numerical model, where the global coordinate system and system degrees of freedom are indicated. The structure is anchored to the 200 m deep seabed by four groups of tendons. Horizontal cylinders (pontoons) connect the floating platform to the tendons. The wind turbine tower is fixed to the floating platform at still water level. The basic features of the model are shown in Table 1.

Table 1. Principal features of the model.

Feature	Value	Unit
Platform radius	9.00	m
Platform height	47.89	m
Pontoon length	18.00	m
Nacelle height	90.00	m
Total mass	9806	tons
Water displacement	12696	m^3
Unstretched line length	151.58	m

The numerical model is based on the readily available MIT/NREL TLP standard baseline design [3]. Vardaroglu et al. [2] calibrated the hydrodynamic damping in the numerical model according to a scaled physical model, which was tested in the offshore wave basin of the Danish Hydraulic Institute in 2012 [5, 6]. Vardaroglu et al. [2] calculated the natural frequencies of the platform relevant Degrees of Freedom (DoF) and tower bending modes by conducting free decay analyses. Selected values are listed in Table 2.

Table 2. Natural frequencies of the model.

Mode	Value [Hz]
Platform Surge	0.015
Platform Pitch	0.200
Platform Heave	0.364
Tower Fore-Aft 1st bending	0.605

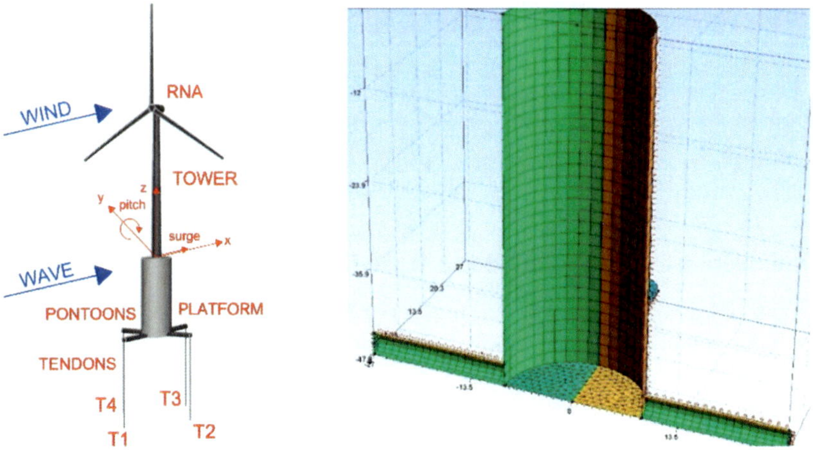

Fig. 1. TLP model (left, Edited from [2]) and hydrodynamic analysis model mesh (right, Source [2]).

2.2 Numerical Modelling Approach

Motion response analyses of the TLP wind turbine were conducted in the time-domain using the open-source numerical tool OpenFAST. The hydrodynamic problem was solved with frequency-domain computations by using the open-source Boundary Element Method (BEM) tool NEMOH [7]. For this purpose, half of the platform is modeled since the incident waves are orthogonal to the structure and the structure is axisymmetric (see Fig. 1). Mean-drift loads on the platform are also considered in hydrodynamic calculations. Resultant hydrodynamic coefficients and wave-induced loads are calculated for a wave frequency interval, which are used as input to the HydroDyn module in OpenFAST (see Fig. 2). For the hydrodynamic model see [2].

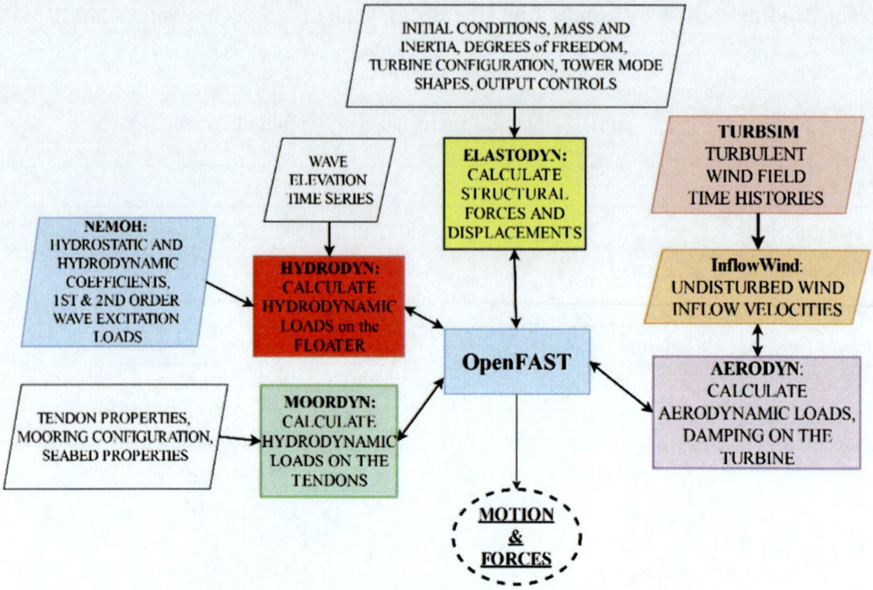

Fig. 2. OpenFAST modules and organization scheme [8].

2.3 Simulation of Wind-Induced and Wave-Induced Loads

The main objective of this study is to observe the response of the TLP wind turbine under wind-induced loads. Accordingly, turbulent wind field time series were generated according to the Kaimal spectrum using the open-source turbulent-wind simulator, TURBSIM. Turbsim is a stochastic, full-field, turbulent-wind simulator that uses a statistical model to numerically simulate time series of three-component wind speed vectors at points in a two-dimensional vertical rectangular grid that is fixed in space [9]. The diameter of the NREL 5MW baseline turbine is 126 m. For this study, turbulent wind fields were generated in a 150 m square domain, featuring a mesh of 33 x 33 nodes. Hub height of the NREL 5MW turbine is at 90 m and located in the middle of the generated turbulent wind field. A constant time step is considered of 0.05 s for all the generated wind fields. Spatial coherence parameters were selected according to IEC 61400-1 [10]. The aerodynamic forces on the blades and on the tower of a wind turbine are affected by the turbulent wind field [11]. OpenFAST processes the turbulent wind field with the InflowWind [12] module, where a uniform and constant wind field can also be utilized by considering a uniform and constant wind speed along the vertical axis.

The aerodynamic problem is solved according to Blade-Element Momentum (BEM) approach within the AeroDyn module. Corresponding wind- and wave-induced loads and responses of the turbine, tower, platform and tendons are calculated by using a combined time-domain approach in OpenFAST. Aerodynamic drag loads on the tower proportional to the square of the incoming wind speed are also taken into account, together with the tower shadow effect.

Dynamic response analyses were conducted on the numerical model for wave loading (LC1), simultaneous wave and turbulent wind loading (LC2), and simultaneous wave and uniform and constant wind conditions (LC3). Uniform and constant wind loading is not realistic but helps to show the effects of turbulent wind conditions clearly. Furthermore, to better isolate the effects of wind-induced loads, analyses under solely wind excitation are conducted considering turbulent wind (LC4) and uniform and constant wind conditions (LC5). Besides, due to the match of platform pitch natural frequency (0.200 Hz) and rotor rotation frequency (0.202 Hz), an additional case (LC6) is simulated, considering also the tower shadow. All load cases are listed in Table 3.

Table 3. Load Cases.

Load Case	Composition
LC1	Wave only
LC2	Wave & Turbulent wind
LC3	Wave & Uniform and constant wind
LC4	Turbulent wind only
LC5	Uniform and constant wind only
LC6	Turbulent wind only (tower shadow activated)

In the scope of this paper, only rated wind speed (11.4 m/s) is considered, which is the wind speed when blade pitch control starts to activate. The corresponding angular speed of the rotor is 12.1 rpm (0.202 Hz.) Wind-induced loads on the tower are calculated based on the tower diameter and drag coefficient. Drag coefficient is chosen as 0.6 from the experimental study of Roshko [13] and accepted as constant through the tower height. Blade airfoil aerodynamics are modeled as steady, where static airfoil data, i.e., lift-force and drag-force are used directly to calculate nodal loads [14]. In LC6, cyclic loading with the rotational frequency of the rotor (1P) forces are expected to occur on the turbine due to the rotor mass imbalance and aerodynamic imbalance under turbulent wind excitation and tower shadow effect. For this purpose, Beddoes-Leishman unsteady model is also utilized in the blade airfoil aerodynamics model in AeoroDyn. The duration of the analyses is 3780 s. The first 600 s of each response time series are interpreted as a transition period and neglected in the data processing. NREL 5MW turbine power curve, the PSDF of the turbulent wind speed horizontal component, and the PSDF of the wave elevation time series are given in Fig. 3. The wave elevation times series is obtained from the physical model measurements discussed by Vardaroglu et al. [2], which was generated according to the JONSWAP wave spectrum. The significant wave height is 4.00 m, and the peak spectral period is 10.1 s (0.099 Hz.). The same wave record is used for LC1, LC2, and LC3.

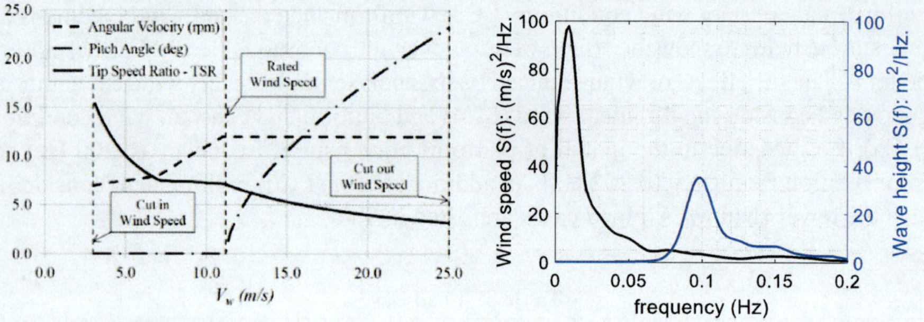

Fig. 3. NREL 5 MW turbine power curve (left, Source: [15]). Turbulent wind speed and incident wave elevation PSDs (right).

3 Results

In this section, the dynamic response of the TLP wind turbine numerical model is discussed. Time histories, PSDFs and statistics of the platform surge, heave, and pitch, tendon tensions, tower base bending moment in the fore-aft direction, and rotor blade root moments are evaluated. For this purpose, frequency domain analyses were conducted by using the MATLAB toolbox Wave Analysis for Fatigue and Oceanography (WAFO) [16]. Figure 4 shows a comparison of the platform surge dynamic response. The platform surge DoF (see Table 2) is excited at 0.015 Hz by the turbulent wind under LC2, LC4, and LC6 (black, red and blue dashed lines). Accordingly, the low-frequency component of the response is amplified under resonance. The low-frequency response of wave and turbulent wind (LC2) is relatively smaller compared to the condition with only turbulent wind due to the quadratic damping provided by the incident wave kinematics. Furthermore, simultaneous uniform and constant wind and wave loading (LC3) result in a slightly lower response than wave loading alone (LC1). This should not be confused with aerodynamic damping, which is due to the high-velocity conditions under resonance. Aerodynamic damping results from the interaction between nacelle motions and the rotor thrust. The fluctuations in the nacelle horizontal velocity provoke changes in the flow through the rotor, leading to oscillations in the rotor speed and thrust [17]. The lower response under uniform and constant wind loading (LC3) is expected to occur due to the additional damping generated by the gyroscopic effect of the rotating rotor blades. A shift is expected to occur in the natural frequency due to additional damping resulting from the rotation of the blades; such shift is not observed in the 0.015 Hz surge response in Fig. 4; it is indeed visible in Fig. 5, showing a shift in the peak spectral value of platform heave response from 0.012 Hz to 0.014 Hz, and in platform yaw response from 0.002 Hz to 0.014 Hz. Same observations are also reported by Bahramiasl et al. [18] from physical tests conducted on a scaled model TLP wind turbine. The response component around the peak wave frequency (0.099 Hz) is dominated by the wave loading and constant in the three different loading scenarios LC1, LC2 and LC3. Uniform and constant wind only loading case (LC5) resulted in constant platform surge response.

Figure 6 shows the comparison of platform pitch PSDFs for the different load conditions. Peaks are observed at 0.099 Hz under LC1, LC2, and LC3 (black lines), given by

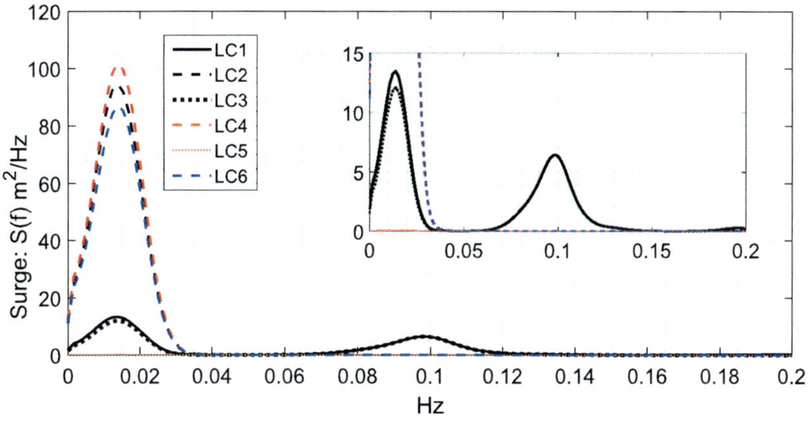

Fig. 4. Platform surge PSDF and close-up view.

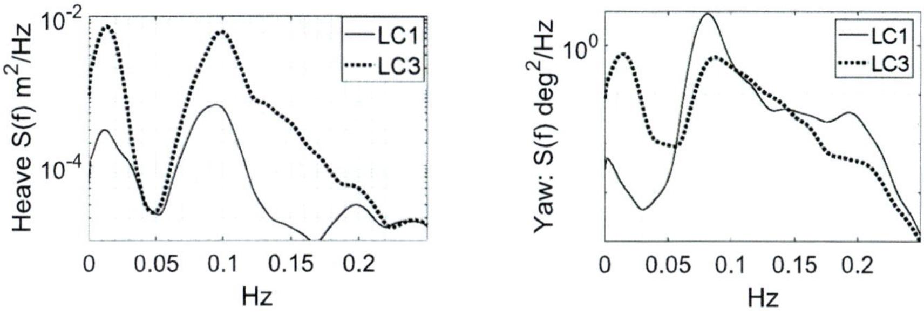

Fig. 5. Platform heave (left) and yaw PSD (right) PSDFs.

wave loading at the peak wave frequency. On the other hand, a higher peak is observed close to the platform pitch natural frequency (0.200 Hz, see Table 2) under the wave loading case (LC1), which is the resonance response triggered by the irregular wave loading components. The resonance response is reduced by the aerodynamic damping generated in LC2 and LC3. The dynamics of the platform pitch response is governed by the wave loads on the platform. Accordingly, no response is observed under wind-only loading cases (red and blue lines: LC4, LC5, and LC6) after 0.1 Hz (see also Fig. 3). Furthermore, under turbulent wind loading conditions (LC2, LC4 and LC6), a small peak is visible at 0.01081 Hz, which is the peak frequency of the wind speed spectrum. This is the low-frequency surge component of the platform pitch response, which is amplified by the turbulent wind. Uniform and constant wind-only loading (LC5) resulted in constant response.

The PSDFs of the tower base bending moment and the related time histories are given in Fig. 7 and Fig. 8, respectively. Figure 7 shows that the response is governed by the platform pitch natural frequency component in the wave-only loading case LC1. Similar to Fig. 4, a small peak is visible around the platform surge natural frequency, which is excited by the turbulent wind loading. On the other hand, simulations under operational conditions (LC2 and LC3) resulted in smaller responses around the platform pitch natural frequency compared to the wave-only loading case. The relatively small response observed in operational conditions are due to the aerodynamic damping generated by the rotor rotating at 0.202 Hz. The aerodynamic damping occurs under large-velocity response, which is the result of the resonance triggered by the irregular waves incoming with platform pitch natural frequency. The time histories of the tower base bending moment presented in Fig. 8 confirm the results of Fig. 7 with larger oscillations around the platform pitch natural frequency (0.200 Hz) in the wave-only loading case (black solid line). It should be noted that turbulent wind-only loading (LC4) resulted in slightly larger dynamic response compared to the uniform and constant wind-only loading case (LC5), due to the deviations in the wind-induced loads in LC4 (see Fig. 7).

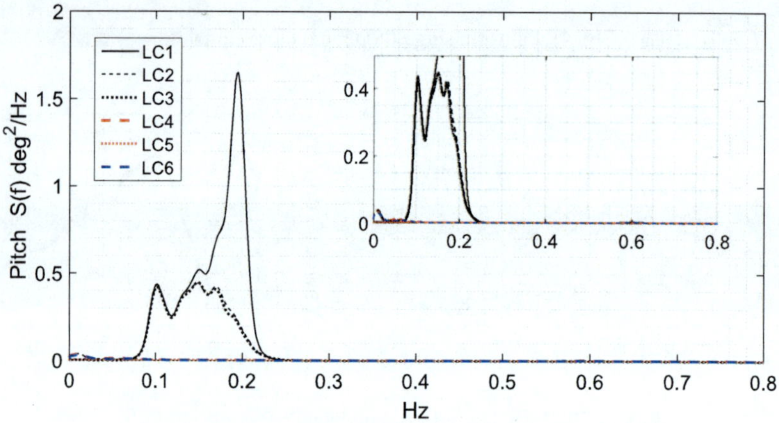

Fig. 6. Platform pitch PSD and close-up view.

Figure 6 and Fig. 7 show the resonance at the platform pitch natural frequency caused by the irregular waves. Furthermore, Table 2 shows that the rotor rotating frequency (0.202 Hz) matches the platform pitch natural frequency (0.200 Hz). It is also observed that rotor blade passing frequency (3P) matches the tower fore-aft bending natural frequency. Consequently, each time a blade passes the turbine tower, aerodynamic forces are expected to drop in the blade [3]. Accordingly, rotor rotation frequency-induced loads (1P and 3P) are expected to occur due to a possibly unbalanced rotor (such as mass) or turbulent winds and the tower shadow effect [19]. These loads are visible in blade root reactions rather than in the integrated effects such as tower bending moment and platform pitch. Figure 9 shows blade root flap-wise moment PSDFs. Larger response is observed under turbulent wind loading cases (dashed lines, LC2, LC4 and LC6). Furthermore, unsteady blade airfoil aerodynamics (blue dashed line, LC6) resulted in larger values compared to the turbulent wind-only loading case (red dashed line, LC4).

Windward (T1) and leeward (T3) tendon tension is governed by platform pitch response. Accordingly, the response spectra replicate the same shape of the platform pitch response PSDFs and are not shown for brevity. Besides, response statistics, i.e. mean values and standard deviations of each response, are presented in Table 4 and

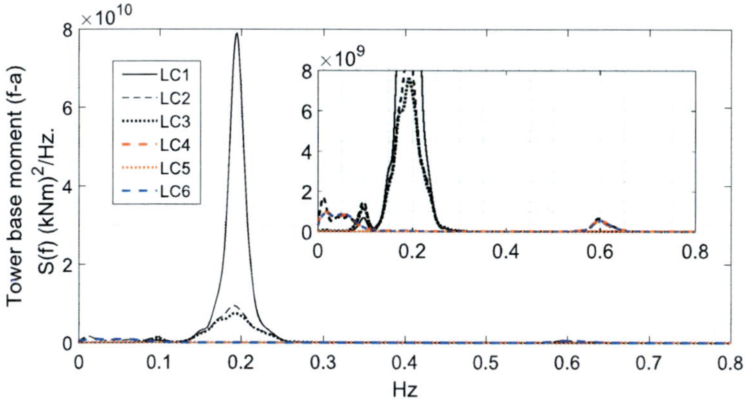

Fig. 7. Tower base bending moment (fore-aft) PSD and close-up view.

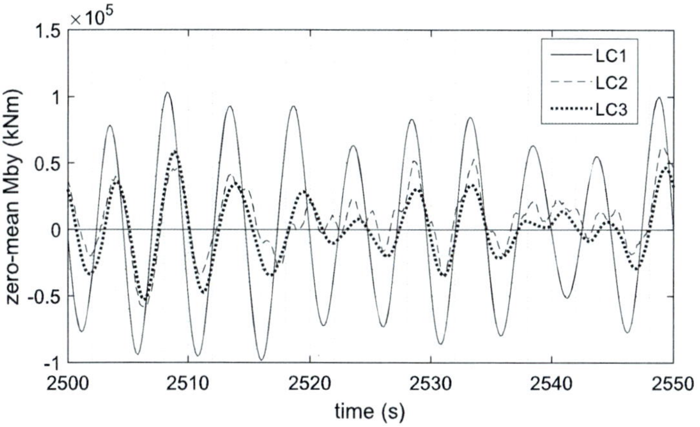

Fig. 8. Tower base bending moment (fore-aft) zero-mean time histories.

Table 5, respectively. According to Table 4, mean values of platform surge and pitch, windward tendon tension (T1), tower-base bending moment in fore-aft direction (Mbyy), and blade root moment in flap-wise direction (RootMyb1) are larger in the presence of wind excitation. On the other hand, the mean tension in the leeward line (T3) decreases due to the increased tension in the windward line. Table 5 confirms the observations drawn from Fig. 4 and Fig. 6. Platform surge dynamic response is the largest under simultaneous wave and turbulent wind excitation (LC2) because of the resonance triggered by the incoming turbulent wind. Platform pitch deviations are dominated by wave loading and decreases in rotating rotor conditions. Dynamic response of the windward and leeward tendons are similar to each other in the presence of wave loading, as seen in Table 5. Similar to the platform pitch dynamics, smaller tendon tension fluctuations are observed in rotor operating conditions compared to wave only loading case, LC1.

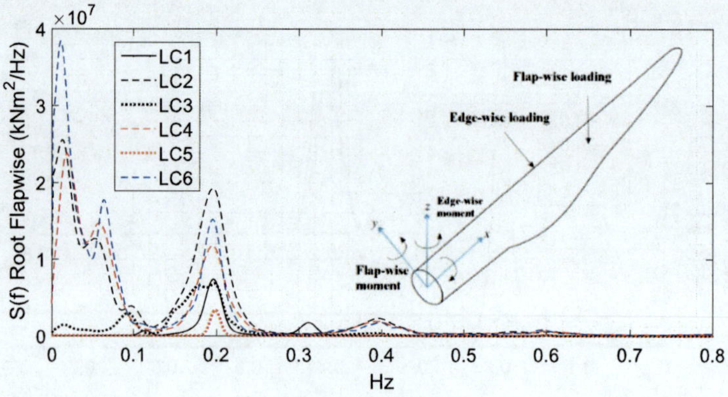

Fig. 9. Blade root flap-wise moment PSDFs and definition of the blade forces and moments (Edited from [20]).

Table 4. Mean values of the relevant response parameters.

Case	Surge (m)	Pitch (deg)	T1 (kN)	T3 (kN)	Mbyy (kNM)	RootMyb1 (kNm)
LC1	0.29	0.00	7902	7854	−1412	574
LC2	3.24	0.20	9327	6500	53650	8160
LC3	4.02	0.25	9672	6187	65829	9867
LC4	3.10	0.20	9325	6488	55319	8203
LC5	3.79	0.25	9636	6198	66766	9986
LC6	3.01	0.20	9287	6519	54117	8167

Table 5. Standard deviation of the relevant response parameters.

Case	Surge (m)	Pitch (deg)	T1 (kN)	T3 (kN)	Mbyy (kNM)	RootMyb1 (kNm)
LC1	0.63	0.28	1964	1956	46940	535
LC2	1.32	0.21	1451	1438	25571	1627
LC3	0.61	0.20	1411	1401	22455	779
LC4	1.30	0.04	270	209	10142	1476
LC5	0.03	0.00	8	8	445	240
LC6	1.20	0.04	277	210	11267	1598

4 Conclusions

In this paper, the dynamic response of a TLP wind turbine is evaluated by numerical simulations under simultaneous wind and wave induced loads. Results obtained for rated wind speed conditions show that incident waves excite the resonant responses at the platform surge and pitch natural frequencies. Furthermore, statistical evaluation of the numerical model responses resulted in larger mean values of platform surge, leeward tendon tensions under the rotating rotor (operational) conditions compared to the wave-only case. On the contrary, evaluation of standard deviations and power spectral density plots resulted in smaller pitch and related response dynamics under the rotating rotor (operational) conditions compared to the wave-only case. Finally, the comparison of the responses between uniform and constant wind and turbulent wind loading conditions reveals the importance of turbulent wind loading on the resonant response of the TLP wind turbine, which should be considered in the design of the structure. Due to the tendon stiffness difference compared to the original design (MIT NREL TLP), the platform pitch natural frequency and rotor rotation frequency match, and the tower fore-aft first bending mode natural frequency and rotor blade passing frequency of the TLP wind turbine also match in this study. These coincidences resulted in increased loads at the blade root and should be avoided. A broader response assessment of the structure (not only at rated wind speed but also cut-off, extreme wind speeds) under turbulent wind conditions is in progress, which aims on enhancing the standard baseline design of the MIT/NREL TLP wind turbine.

Acknowledgements. The experimental results used in this research were produced within a project funded under the Integrated Infrastructure Initiative HYDRALAB IV of EU FP7. Release of the data is acknowledged.

References

1. NREL: OpenFAST v3.5.1 Documentation. https://openfast.readthedocs.io/en/main/. Last accessed 15 Jan 2024
2. Vardaroglu, M., Gao, Z., Avossa, A.M. and Ricciardelli, F.: Validation of a TLP wind turbine numerical model against model-scale tests under regular and irregular waves. Ocean Eng. **256**, 111491 (2022)
3. Matha, D.: Model development and loads analysis of an offshore wind turbine on a tension leg platform, with a comparison to other floating turbine concepts. Master's thesis. University of Colorado-Boulder (2009)
4. Jonkman, J., Butterfield, S., Musial, W., Scott, G.: Definition of a 5-MW Reference Wind Turbine for Offshore System Development. Technical report NREL/EL-500- 38060. National Renewable Energy Laboratory (2009)
5. Armenio, E., D'Alessandro, F: Dynamic response of floating offshore wind turbines under random waves and wind action HyIV – DHI - 01 Offshore wave basin, DHI Data Storage Report. Denmark, Copenhagen (2013)
6. Tomasicchio, G.R., et al.: Physical model experiments on floating off-shore wind turbines. In: Proceedings of the HYDRALAB IV Joint User Meeting, Lisbon (2014)
7. Babarit, A., Delhommeau, G.: Theoretical and numerical aspects of the open source BEM solver NEMOH. In: 11th European Wave and Tidal Energy Conference, EWTEC2015 (2015)

8. Vardaroglu, M.: Dynamic Behavior of Tension Leg Platform Floating Wind Turbines. Ph. D. dissertation, Universita della Campania" Luigi Vanvitelli", Aversa (2021)
9. Jonkman, B.J. and Buhl Jr, M.L.: TurbSim user's guide (No. NREL/TP-500-39797). National Renewable Energy Lab. (NREL), Golden, CO, USA (2006)
10. International Electrotechnical Commission (IEC): Wind turbines – Part 1: Design requirements. IEC 61400-1, third edition (2005)
11. Avossa, A.M., Demartino, C., Ricciardelli, F.: Assessment of the peak response of a 5MW HAWT under combined wind and seismic induced loads. Open Const. Build. Technol. J. **11**, 441–457 (2018). https://doi.org/10.2174/1874836801711010441
12. NREL: InflflowWind Users Guide and Theory Manual. NREL (Veranst.). https://openfast.readthedocs.io/en/main/source/user/inflowwind/index.html. Last accessed 15 Jan 2024
13. Roshko, A.: Experiments on the flow past a circular cylinder at very high Reynolds number. J. Fluid Mech. **10**(3), 345–356 (1961)
14. Jonkman, J.M., Hayman, G.J., Jonkman, B.J., Damiani, R.R., Murray, R.E.: AeroDyn v15 user's guide and theory manual. NREL Draft Report, p. 46 (2015)
15. Avossa, A.M., Demartino, C., Contestabile, P., Vicinanza, D., Ricciardelli, F.: Some results on the vulnerability assessment of HAWTs subjected to wind and seismic actions. Sustainability **9**(9), 1525 (2017). https://www.mdpi.com/2071-1050/9/9/1525
16. Brodtkorb, P.A., Johannesson, P., Lindgren, G., Rychlik, I., Ryden, J., Sjö, E.: Wafo-a matlab toolbox for analysis of random waves and loads. In: Proceedings of the 10th International Offshore and Polar Engineering Conference, International Society of Offshore and Polar Engineers (2010)
17. Souza, C.E.S., Hegseth, J.M., Bachynski, E.E.: Frequency-dependent aerodynamic damping and inertia in linearized dynamic analysis of floating wind turbines. In Journal of Physics: Conference Series, vol. 1452, no. 1, p. 012040. IOP Publishing (2020)
18. Bahramiasl, S., Abbaspour, M., Karimirad, M.: Experimental study on gyroscopic effect of rotating rotor and wind heading angle on floating wind turbine responses. Int. J. Environ. Sci. Technol. **15**(12), 2531–2544 (2018)
19. Laugesen, R., Hansen, A.M.: Experimental study of the dynamic response of the DTU 10 MW wind turbine on a tension leg platform. In DTU Wind Energy. Phys. Conf. Ser. **753**, 092007 (2015)
20. Cho, S., Bachynski, E.E., Nejad, A.R., Gao, Z., Moan, T.: Numerical modeling of the hydraulic blade pitch actuator in a spar-type floating wind turbine considering fault conditions and their effects on global dynamic responses. Wind Energy **23**(2), 370–390 (2020)

Improving Long-Span Bridge Flutter Reliability Through Gyroscopic Stabilizer, Considering Random Aeroelastic Loads

Gian Felice Giaccu[1(✉)] and Luca Caracoglia[2]

[1] University of Sassari, Alghero, Italy
gf.giaccu@uniss.it
[2] Northeastern University, Boston, MA 02115, USA
lucac@coe.neu.edu

Abstract. Long span suspension bridges are challenging structures, generally sensitive to wind effects and to wind-induced instabilities. Torsional rotation of the bridge deck plays an important role for the dynamic stability of the structure. This study uses a gyroscopic device as a stabilizer of the long span bridge. The proposed apparatus has been modeled as a lumped 3DOF system installed inside a specific section of the bridge deck. To assess effectiveness in increasing the critical flutter wind speed and performance, a numerical study is conducted on a benchmark long-span cable-supported bridge as a function of apparatus' gyricity. Deck aeroelastic loads are randomly perturbed to simulate modeling simplifications and measurement errors. Monte Carlo simulations are used to predict flutter probability and efficiency of the stabilizer, within a practical operational range, in the presence of random aeroelastic loads.

Keywords: long- span cable-supported bridges · flutter control · gyroscopic stabilizer

1 Introduction

The last decades have witnessed significant steps in the evolution of long-span, cable-supported bridges. The need for longer and longer spans necessitates the study of new practices to increase the performance of these structures under wind loads. Flutter instability occurs when a bridge is exposed to a wind speed above a certain critical threshold. Above this threshold vertical and torsional deck vibrations couple together [1].

Gyroscopic devices have routinely been employed in mechanical engineering [2] for vibration suppression but more seldom employed in structural engineering [3–5]. This type of devices is proposed herein to improve bridge reliability against flutter. The main component of the system is a rotating mass with a horizontal-vector angular momentum $\boldsymbol{\Omega}$ parallel to the section-model of the deck. The mass is connected to the deck by a rotational spring and is schematically depicted in Fig. 1 [6]. The system allows relative rotations between the deck cross section and the rotating mass on the bridge deck, vertical plane. The gyroscopic system reacts to the torsional vibrations of

P. Schito and A. Zasso (Eds.): IN VENTO 2022, LNCE 461, pp. 239–247, 2024.
https://doi.org/10.1007/978-3-031-53059-3_21

the deck, coupling torsional vibrations with the other degrees of freedom of the structure and transferring the energy from the deck torsional mode to other frequencies of the deck modes. A mathematical model of the single-unit gyroscopic device has been derived and used herein.

The proposed device is effective in suppressing flutter. Robustness against variations of its aeroelastic properties is examined by perturbation of the Scanlan derivatives, simulating modeling simplifications and measurement errors.

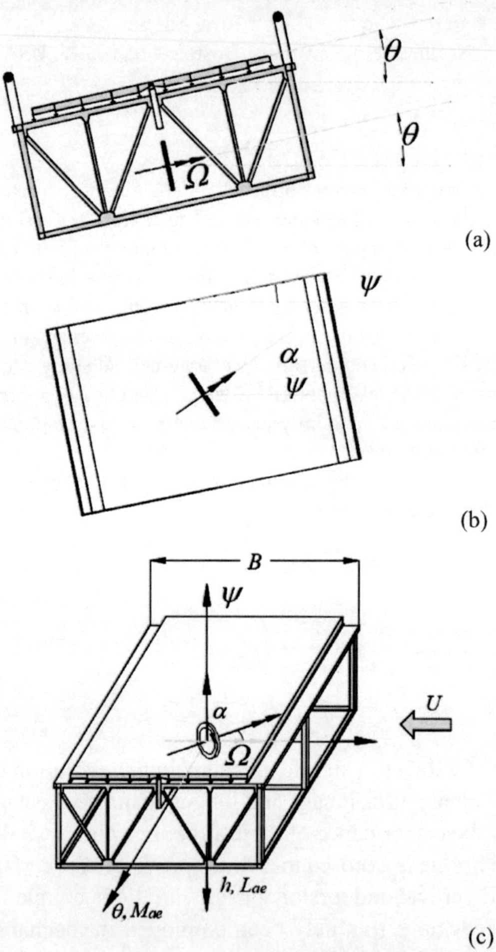

Fig. 1. Simplified model of the bridge deck (illustrative example): (a) cross-sectional view, (b) top view, (c) 3D lateral view with aeroelastic loads [6]

2 Theoretical Background

2.1 Dynamic Formulation of the Gyroscopic Device

The gyroscopic dynamics is modeled as a system with lumped, mass moment of inertia and three rotational DOFs (degrees of freedom) [7]. Figure 1 illustrates the rotating mass model geometry. The quantities $J_{\Omega,11} = J_{\Omega,22} = J_\alpha$, and $J_{\Omega,33} = J_{\Omega,11} + J_{\Omega,22} = J_{\Omega,p}$ represent the moment of inertia of the rotating mass with respect to the reference axes 1–1 and 2–2; $J_{\Omega,p} = M_{\Omega,TOT}\rho_\Omega^2$ is the polar moment of inertia with respect to the rotating axis 3–3, where ρ_Ω is the radius of gyration of the gyroscopic stabilizer and $M_{\Omega,TOT}$ the total mass of the gyroscope. The gyricity of the gyroscopic device can be expressed as $\boldsymbol{\Omega} = J_{\Omega,p}\boldsymbol{\omega}$ where $\boldsymbol{\omega}$ is the angular velocity vector of the gyroscope. The model equations, with three degrees of freedom (DOFs) θ, ψ and α, are:

$$\begin{aligned} (J_\theta + J_\alpha)\ddot{\theta} + c_\theta\dot{\theta} + k_\theta\theta &= -\Omega(\dot{\psi} + \dot{\alpha}) \\ (J_\psi + J_\alpha)\ddot{\psi} + c_\psi\dot{\psi} + k_\psi\psi &= \Omega\dot{\theta} \\ J_\alpha(\ddot{\alpha} + \ddot{\psi}) + c_\alpha\dot{\alpha} + k_\alpha\alpha &= \Omega\dot{\theta} \end{aligned} \tag{1}$$

The benchmark bridge in Fig. 1 is the Golden Gate Bridge [8]. Aeroelastic deck loads are described by Scanlan derivatives [1]. Scanlan derivatives are randomly perturbed about their reference values to account for wind load uncertainties (e.g., modeling simplifications, measurement errors, etc.) The critical flutter speed is evaluated by multimode approach through Monte Carlo sampling, to determine the empirical flutter probability and evaluate the effectiveness of the stabilizer from a reliability perspective.

2.2 Two-Mode Bridge Deck Generalized Formulation for Aeroelastic Instability

The dynamic deck response of a long-span bridge due to wind excitation was modeled after the multi-mode approach, for example discussed in [9, 10]. Self-excited forces per unit deck length (Fig. 2), based on the representation proposed by Scanlan [1], are a function of $h(x,t)$ heaving displacements and $\alpha(x,t)$ torsional rotations of the generic deck cross section, evaluated along the longitudinal bridge axis of coordinate x and time t. Wind load simulations are restricted to vertical force (L_{ae}) and torsional moment (M_{ae}) per unit span length for mean wind direction horizontal and orthogonal to the bridge deck (usually the most dangerous one). Horizontal lateral sway vibration and corresponding loads are neglected as they are usually secondary in the case of Golden Gate Bridge flutter. Aeroelastic forces per unit deck length are simulated [Fig. 2(c)] as [1]:

$$L_{ae}(x,t) = 0.5\rho U^2 B \begin{bmatrix} KH_1^*\dot{h}(x,t)/U + KH_2^*B\dot{\theta}(x,t)/U \\ +K^2H_3^*\theta(x,t) + K^2H_4^*h(x,t)/U \end{bmatrix} \tag{2}$$

$$M_{ae}(x,t) = 0.5\rho U^2 B^2 \begin{bmatrix} KA_1^*\dot{h}(x,t)/U + KA_2^*B\dot{\theta}(x,t)/U \\ +K^2A_3^*\theta(x,t) + K^2A_4^*h(x,t)/U \end{bmatrix} \tag{3}$$

In Eqs. (2) and (3), B is the bridge deck width, U is the horizontal, mean wind speed, orthogonal to the bridge axis, ρ is the air density, $K = \omega B/U$ is the reduced frequency, $\dot{h} = \mathrm{d}h/\mathrm{d}t$ and $\dot{\theta} = \mathrm{d}\theta/\mathrm{d}t$ are velocities of the deck. Equations (2–3) are valid for simple harmonic deck motion at incipient flutter, with ω being the circular vibration frequency [4]. The Scanlan derivatives are $H^*_{1,\ldots,4}$ and $A^*_{1,\ldots,4}$. The reduced frequency K is proportional to the reciprocal of the reduced velocity, $U_r = U/(nB)$, with $n = \omega/(2\pi)$ in Hz. Effects of loading due to wind turbulence are not considered.

The dynamic response of the bridge deck, accounting for gyroscopic forces introduced by the device and distributed aeroelastic loads, can be obtained by adaptation of Eqs. (1) and Eqs. (2–3); inertial and internal restoring force and damping effects are decomposed by modal expansion about the still-air structural modes [6]. The model becomes:

$$m_v\ddot{h} + c_v\dot{h} + k_v h = L_{ae}(x,t), \tag{4}$$

$$[J_\theta + J_\alpha\delta(x - x_\Omega)]\ddot{\theta} + c_\theta\dot{\theta} + k_\theta\theta = -\Omega\delta(x - x_\Omega)\big(\dot{\psi} + \dot{\alpha}\big) + M_{ae}(x,t), \tag{5}$$

$$(J_\psi + J_\alpha)\ddot{\psi}_\Omega + c_\psi\dot{\psi}_\Omega + k_\psi\psi_\Omega = \Omega\dot{\theta}(x_\Omega, t), \tag{6}$$

$$J_\alpha\big(\ddot{\alpha} + \ddot{\psi}_\Omega\big) + c_\alpha\dot{\alpha} + k_\alpha\alpha = \Omega\dot{\theta}(x_\Omega, t). \tag{7}$$

Equation (4) is the continuous bending or vertical equilibrium equation of the moving bridge deck. Equation (5) is the corresponding continuous torsional equilibrium equation; the gyroscope is inserted at deck coordinate x_Ω along the longitudinal bridge axis, and δ is the Dirac function. Equation (6) enforces the rotational equilibrium of the gyroscope and its local interaction with the rotational velocity of the bridge deck $\dot{\theta}(x_\Omega, t)$. In the gyroscope formulation the effect of the dynamic rotation on the horizontal plane of the deck is considered in Eq. (6) but has little effect due to the relative, local bending rigidity of the girders about vertical rotation axis ($k_\psi\psi_\Omega$) in comparison with the torsional restoring effect of the gyroscopic device ($k_\alpha\alpha$). The quantity ψ_Ω designates the rotation of the deck at x_Ω about a vertical axis [approximately equal to the first derivative of the lateral sway mode shape $p(x,t)$]. Although included in the formulation, previous remarks lead to $\psi_\Omega \approx \dot{\psi}_\Omega \approx \ddot{\psi}_\Omega \approx 0$.

Aeroelastic instability is solved through the method described, for example, in Ref. [9]. After modal expansion, a four-equation generalized homogeneous system is derived, rewritten as a function of dimensionless time $s = tU/B$ and $\xi_g\prime = \mathrm{d}\xi_g/\mathrm{d}s$ and $\xi_g'' = \mathrm{d}^2\xi_g/\mathrm{d}s^2$ of the g-th generalized modal amplitude. Equations (4–5) become:

$$\xi\prime\prime_{v1} + 2\zeta_{v1}K_{v1}\xi\prime_{v1} + K_{v1}{}^2\xi_{v1} = 0.5\rho B^4\ell J_{v1}{}^{-1}\begin{bmatrix} KH_1^*G_{v1,v1}\xi\prime_{v1} \\ +KH_2^*G_{v1,t1}\xi\prime_{t1} \\ +K^2H_3^*G_{v1,t1}\xi_{t1}+ \\ K^2H_4^*G_{v1,v1}\xi_{v1} \end{bmatrix}, \tag{8}$$

$$\big[1 + J_\alpha J_{t1}^{-1}\theta_{t1}(x_\Omega)\big]\xi''_{t1} + 2\zeta_{t1}K_{t1}\xi\prime_{v1} + K_{t1}{}^2\xi_{t1} =$$

$$+0.5\rho B^4\ell J_{t1}{}^{-1}\begin{bmatrix} KA_1^*G_{v1,t1}\xi\prime_{v1} \\ +KA_2^*G_{t1,t1}\xi\prime_{t1} \\ +K^2A_3^*G_{t1,t1}\xi_{t1} \\ +K^2A_4^*G_{v1,t1}\xi_{v1} \end{bmatrix} - J_{t1}{}^{-1}\Omega_j^*K_\alpha\theta_{t1}(x_\Omega)\alpha\prime; \tag{9}$$

Equations (8–9) are the main model equations with reduced "frequencies" of the gyroscope $K_\alpha = \omega_\alpha B/U$, $\Omega B/U = \Omega_J^* \omega_\alpha B/U = \Omega_J^* K_\alpha$, and $\dot{\theta}(x_\Omega, t) = \dot{\xi}_{t1}\theta_{t1}(x_\Omega)$. We note that the scalar gyricity Ω_J^* is dimensionless. Furthermore, in the previous equations modal damping ratios of modes $v1$ and $t1$ are denoted by ζ_{v1} and ζ_{t1}. The reduced frequencies of the two deck modes are indicated as $K_{v1} = \omega_{v1}B/U$ and $K_{t1} = \omega_{t1}B/U$, with ω_{v1} and ω_{t1} being the corresponding circular frequencies [rad/s].

In Eqs. (8–9) dimensionless modal integrals [9] are used: $G_{v1,v1} = \int h_{v1}^2(x)\mathrm{d}x/\ell$, $G_{t1,t1} = \int \theta_{v1}^2(x)\mathrm{d}x/\ell$, $G_{v1,t1} = \int h_{v1}(x)\theta_{t1}(x)\mathrm{d}x/\ell$; ; integrations are performed over the bridge longitudinal (central) span length ℓ.

Generalized modal inertias of the two modes, $v1$ and $t1$, are evaluated as $J_{v1} \approx m_v \ell B^2 G_{v1,v1}$ and $J_{t1} \approx J_\theta \ell G_{t1,t1}$ with m_v and J_θ being the equivalent mass and mass moment of inertia per unit span of the moving deck. The mode shapes $h_{v1}(x)$ and $\theta_{t1}(x)$ are simplified as skew-symmetric, harmonic of wavelength $\ell/2$, with deck modes purely vertical bending and torsional.

Aeroelastic instability is solved after combining the previous equations and using dimensionless Fourier analysis, e.g., $\overline{\xi}_{v1}(K) = \int_0^\infty \xi_{v1}(s)\mathrm{e}^{-\hat{i}Ks}\mathrm{d}s$ of ξ_{v1}, and $\overline{\alpha}(K) = \int_0^\infty \alpha(s)\mathrm{e}^{-\hat{i}Ks}\mathrm{d}s$ of the relative gyroscope rotation; the imaginary unit is $\hat{i} = \sqrt{-1}$.

Incipient flutter, equivalent to simple harmonic motion, is imposed [1]. Complex generalized coordinates are expressed in vector form as $\overline{\boldsymbol{\xi}}_\Omega = \{\overline{\xi}_{v1}, \overline{\xi}_{v1}, \overline{\alpha}\}^T$, after simplification by setting $\psi_\Omega \approx \dot{\psi}_\Omega \approx \ddot{\psi}_\Omega \approx 0$; the vector $\overline{\boldsymbol{\xi}}_\Omega$ includes both the modal coordinates of modes $v1$ and $t1$ as well as the relative gyroscope rotation variable and the local deck rotation about vertical axis. The last variable is included for completeness, even though it has negligible relevance on the flutter solution.

The critical condition was determined by re-writing the system in the Fourier domain, and in homogenous form, as $\mathbf{E}_\Omega(K, \chi)\overline{\boldsymbol{\xi}}_\Omega = 0$. The elements of the three-by-three complex matrix $\mathbf{E}_\Omega$ are, with frequency ratio $\chi = K/K_{t1}$ and dimensionless groups $q_{v1} = 0.5\rho\ell B^4/J_{v1}$ and $q_{t1} = 0.5\rho\ell B^4/J_{t1}$:

$$\{\mathbf{E}_\Omega\}_{1,1} = -\chi^2 + (K_{v1}/K_{t1})^2 - q_{v1}\chi^2 H_4^*(K)G_{v1,v1} + \hat{i}\Big(2\zeta_{v1}K_{v1}/K_{t1}\chi - q_{v1}\chi^2 H_1^*(K)G_{v1,v1}\Big), \tag{10a}$$

$$\{\mathbf{E}_\Omega\}_{1,2} = -q_{v1}\chi^2\Big(H_3^*(K)G_{v1,t1} + \hat{i}H_2^*(K)G_{v1,t1}\Big), \tag{10b}$$

$$\{\mathbf{E}_\Omega\}_{1,3} = 0, \tag{10c}$$

$$\{\mathbf{E}_\Omega\}_{2,1} = -q_{t1}\chi^2\Big(A_4^*(K)G_{v1,t1} + \hat{i}A_1^*(K)G_{v1,t1}\Big), \tag{10d}$$

$$\{\mathbf{E}_\Omega\}_{2,2} = -\Big[1 + J_\alpha J_{t1}{}^{-1}\theta_{t1}(x_\Omega)\Big]\chi^2 + 1 - q_{t1}\chi^2 A_3^*(K)G_{t1,t1} + \hat{i}\Big(2\zeta_{t1}\chi - q_{t1}\chi^2 A_2^*(K)G_{t1,t1}\Big), \tag{10e}$$

$$\{\mathbf{E}_\Omega\}_{2,3} = \hat{i}\Omega_J^* J_\alpha{}^{-1}(K_\alpha/K_{t1})\theta_{t1}(x_\Omega)\chi, \tag{10f}$$

$$\{\mathbf{E}_\Omega\}_{3,1} = 0, \tag{10g}$$

$$\{\mathbf{E}_\Omega\}_{3,2} = -\{\mathbf{E}_\Omega\}_{2,3}, \tag{10h}$$

$$\{\mathbf{E}_\Omega\}_{3,3} = -\chi^2 + (K_\alpha/K_{t1})^2 + \hat{i}2\zeta_\alpha K_\alpha/K_{t1}\chi. \tag{10i}$$

Since U is initially unknown, so are K and ω, or χ i.e., the ratio between coupled-flutter and $t1$ torsional-mode frequencies. Complex matrix $\mathbf{E}_\Omega$ and vector $\bar{\boldsymbol{\xi}}_\Omega$ depend on the two unknowns, K and χ. Therefore, an iterative procedure is required to solve $\mathbf{E}_\Omega(K, \chi)\bar{\boldsymbol{\xi}}_\Omega = 0$. The non-trivial solution can be found iteratively by varying K over a suitable interval and enforcing the simultaneous vanishing of the real and imaginary parts of the determinant of $\mathbf{E}_\Omega$ as a function of χ.

3 Numerical Results

A numerical investigation is conducted on the full-scale Golden Gate Bridge.

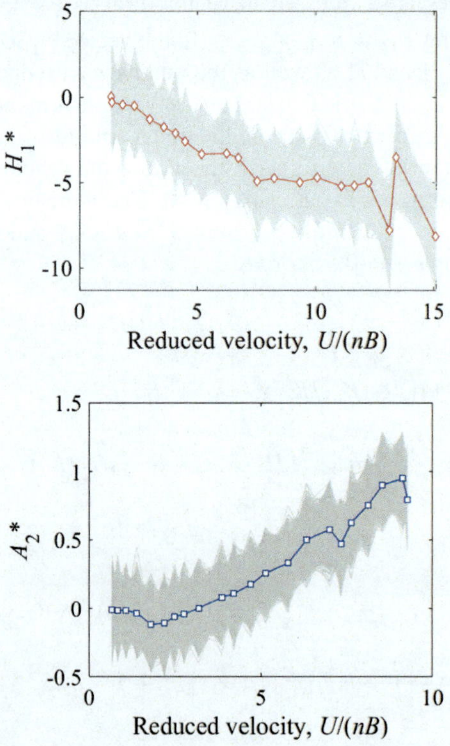

Fig. 2. Derivatives of the Golden Gate Bridge perturbed with independent Gaussian random noises: (a) $H_1{}^*$ derivative, (b) $A_2{}^*$ derivative.

The stabilizer is installed on the main span in the deck section at ¼ span length, $x_\Omega = 0.25\ell$, to maximize its effects since deck modes are skew-symmetric. Critical flutter wind speed has been calculated for two different constant gyricities, $\Omega = 37700$ kg m^2 rad s^{-1} and $\Omega = 75400$ kg m^2 rad s^{-1}. Structural properties and experimental data of the Scanlan derivatives are derived from Jain [8].

Derivatives are represented through piecewise linear segments between any two adjacent experimental data points. Each experimental data point is perturbed from its original (mean) value with independent, zero-mean Gaussian random noises with coefficient of variation of 10% referenced to the absolute maximum value of each derivative [11]. Figure 2 exemplifies two relevant derivatives, $H_1{}^*$ and $A_2{}^*$; reference experimental values are depicted with a continuous line with marker, whereas the "cloud" of the other lines represent a sample of 20000 realizations.

Preliminary two-mode flutter analysis results are shown in Fig. 3 that illustrates the empirical probability distribution of the critical flutter speed at both low and moderate gyricity values.

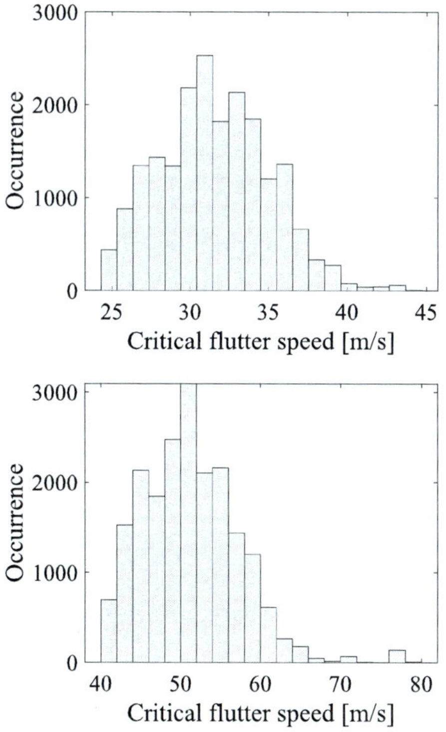

Fig. 3. Empirical probability distribution of the critical flutter speed (Golden Gate Bridge case study 20000 sampling): (a) low gyricity $\Omega = 37700$ kg m^2 rad s^{-1}, (b) moderate gyricity $\Omega = 75400$ kg m^2 rad s^{-1}. Note the difference of scale on the x-axis.

Results confirm the beneficial effects of the gyroscopic stabilizer in terms of increment of critical flutter speed, compared to the reference solution by Jain [8], where critical flutter velocity was determined as equal to 22.0 m/s. The median value of critical flutter speed is 31.4 m/s for $\Omega = 37700$ kg m^2 rad s^{-1} whereas it becomes 50.8 m/s for $\Omega = 37700$ kg m^2 rad s^{-1}. Nevertheless, the lower tail of the empirical flutter speed distribution is precariously close to uncontrolled bridge value.

Critical flutter speed increases even in the case of Fig. 3a with low gyricity but the lower tail of the empirical distribution is close to the critical flutter speed of the bridge without stabilizer. By contrast, in the second case with moderate gyricity (Fig. 3b) the lower tail of the empirical distribution reaches high values of critical speed close to 40 m/s.

Research is still ongoing. Future studies will further investigate the behavior of the gyroscopic device for structural reliability. Among future research scopes we note the study of the gyroscope as an active stabilizer against buffeting, forcing the bridge deck through active impulsive forces that limit displacement thresholds typical of serviceability limit states.

4 Concluding Remarks

Numerical results indicate that introduction of variability in the aeroelastic load may be detrimental even though it may still postpone flutter occurrence, since it requires a mean flow speed of magnitude usually 20% larger compared to the scenario without Scanlan derivative uncertainty.

However, higher gyricity values improve the performances of the bridge by drastically increasing the critical flutter speed (doubling it) and despite the introduction of variability in the aeroelastic load.

Additional studies should be conducted to confirm the influence of uncertainty in the aeroelastic loads, as a function of the standard deviation of the Scanlan derivatives. The final goal of this research is the identification and exclusion of flutter failure risk by more rigorous probabilistic analysis. Anticipated results may be employed in conjunction with future wind tunnel verification.

References

1. Scanlan, R.H., Tomko, J.J.: Airfoil and bridge deck flutter derivatives. J. Eng. Mech. ASCE **97**(EM6), 1717–1737 (1971)
2. Kan, Y., Psinh, Y., Lee, A.-C.: Investigation on the steady-state response of a symmetric rotors. J. Vib. Acoust. **114** (1992)
3. Giaccu, G.F.: Modeling a gyroscopic stabilizer for the improvement of the dynamic performances of slender monopole towers. Eng. Struct. **215**, 110607 (2020)
4. Garau, M., Carta, G., Nieves, M.J., Jones, I.S., Movchan, N.V., Movchan, A.B.: Interfacial waveforms in chiral lattices with gyroscopic spinners. Proc. R. Soc.A 474, 20180132 (2018)
5. Carta, G., Jones, I.S., Movchan, N.V., Movchan, A.B., Nieves, M.J.: Gyro-elastic beams for the vibration reduction of long flexural systems. Proc. Roy. Soc. A Math. Phys. Eng. Sci. **473**, 1–17 (2017)

6. Giaccu, G.F., Caracoglia, L.: A gyroscopic stabilizer to improve flutter performance of long-span cable-supported bridges. Eng. Struct. **240**, 112373 (2021)
7. D'Eleuterio, G.M.T.: Dynamics of gyroelastic vehicles. Institute for aerospace studies, Toronto (1986)
8. Jain, A.: Multi-mode aeroelastic and aerodynamic analysis of long-span bridges. Ph.D. Dissertation. Baltimore, Maryland, USA: Johns Hopkins University (1996)
9. Jones, N.P., Scanlan, R.H.: Theory and full-bridge modeling of wind response of cable-supported bridges. J. Bridge Eng. ASCE **6**(6), 365–375 (2001)
10. Caracoglia, L.: Wind-structure oscillations on long-span suspension bridges. Ph.D. Dissertation. Trieste, Italy: University of Trieste (2001)
11. Sarkar, P., Caracoglia, L., Haan, F.L., Sato, H., Murakoshi, J.: Comparative and sensitivity study of flutter derivatives of selected bridge deck sections. Part 1: Anal. inter-laboratory Exp. Data. Eng. Struct. (2008)

Bridge Flutter Stability in Turbulent Flow

Niccolò Barni[1](✉), Massimiliano Gioffrè[2], and Claudio Mannini[1]

[1] CRIACIV/Department of Civil and Environmental Engineering, University of Florence, Florence, Italy
{niccolo.barni,claudio.mannini}@unifi.it

[2] CRIACIV/Department of Civil and Environmental Engineering, University of Perugia, Perugia, Italy
massimiliano.gioffre@unipg.it

Abstract. Despite the many studies on suspension bridge flutter stability, the influence of turbulence has not been completely understood yet. Experimental and numerical results have shown that atmospheric turbulence may have either a stabilising or a destabilising influence. This work investigates the alteration of the flutter stability threshold due to the parametric modulation of self-excited forces due to large-scale turbulence. Such an effect is considered through the 2D rational function approximation model for self-excited forces. First, a Monte Carlo method addresses the stochastic stability of a three-degree-of-freedom two-dimensional model of the Hardanger Bridge, Norway, subjected to different turbulent wind conditions. Such an approach highlights the sensitivity of flutter stability to turbulence parameters, but does not allow any interpretation of the instability mechanism. Then, Floquet multipliers are used to study the stability of the system obtained by assuming a harmonic slowly-varying angle of attack. This simplified scenario is particularly suitable to underline parametric effects that are often hidden in case of random broad-band parametric excitation. For the considered case study, the paper emphasises a significant destabilising effect of large-scale turbulence and a stabilising role played by the lateral degree of freedom in the simplified time-periodic system.

Keywords: Suspension bridges · Flutter stability · Turbulent flow · Parametric self-excited forces · Time-variant system · Monte Carlo approach · Floquet multipliers

1 Introduction

Coupled flutter stability has thoroughly been investigated in the last decades, also under the pressure of unstable bridge vibration occurrences, such as the catastrophic event for the Original Tacoma Bridge Narrows or the large-amplitude oscillations recorded for the Golden Gate Bridge [1]. In spite of that, the question about turbulence effects on stability remains largely unanswered. Indeed, experimental evidence and numerical simulations have shown either stabilising or destabilising effects of wind turbulence, which have only partially been explained. So far, this has not caused critical problems in the design of suspension bridges; nevertheless, due to the recent increase in demand for long-span crossings, this lack of knowledge risks to represent a serious limit.

P. Schito and A. Zasso (Eds.): IN VENTO 2022, LNCE 461, pp. 248–259, 2024.
https://doi.org/10.1007/978-3-031-53059-3_22

Generally, bridge flutter is modelled based on self-excited forces, usually considered unaffected by atmospheric turbulence. Nevertheless, the latter may introduce different sources of time dependence in the parameters of self-excited forces. Lin and Ariaratnam [2] faced the problem of the random time change of the wind velocity at the beginning of the '80s by studying the almost sure stability of a bridge through statistical moment boundaries. This was one of the first attempts to account for the influence of longitudinal turbulence on bridge stability (considering only the torsional motion in this case) by assuming a white-noise random variation of the wind velocity, parametrically affecting the aerodynamic damping and stiffness through the wind kinetic pressure. Such a model for turbulence leads to a bridge response that can approximately be considered a Markov vector, thus simplifying the analytical complexity of the problem, which is then governed by a Itô-type stochastic differential equation. Lin and Ariaratnam concluded that turbulence might cause the instability of an otherwise stable motion, contrasting with some previous experimental experiences [3]. Along this line, Bucher and Lin [4] studied the stability of the second statistical moment of a two-degree-of-freedom (2-dof) system, considering the first torsional and vertical bending modes, as well as different bridge cross sections. They found that the random variation of the wind velocity due to longitudinal turbulence has a stabilising effect with regard to coupled flutter since random parametric excitation promotes an energy transfer from the critical modes to more stable modes. Then, using the stochastic average approach, Bucher and Lin [5] made a step forward in the efficient assessment of moment stability. The simplified assumption of white-noise turbulence was later removed in [6], introducing a new model for turbulence, which can reproduce different broad-band processes (including white noises) by suitably setting some model parameters. Later, Poirel and Price [7, 8] studied a similar stability problem for a 2-dof airfoil, obtaining the stochastic stability boundaries by numerical estimation of the largest Lyapunov exponent. The sensitivity of stochastic stability to turbulence parameters, such as intensity and integral length scale, as well as airfoil structure parameters, was also addressed. The results were consistent with those previously obtained by [2], confirming the destabilising role of longitudinal turbulence. In particular, more unstable flutter boundaries were obtained by increasing the turbulence length scale, highlighting the crucial role of low-frequency turbulent fluctuations. In [7, 8], parametric resonances due to narrow-band excitation were also thoroughly investigated using the Bogoliubov and Mitropolsky average method. This extreme simplification of turbulence effects allowed highlighting the primary resonance of parametric fluctuations at twice the flutter frequency, typically observed in Mathieu-Hill-type time-variant dynamical systems.

The works mentioned above focus on the random effects due to the fluctuation of the wind velocity magnitude appearing in the wind kinetic pressure. Howxever, this analysis is only partial since large-scale turbulence can also change the instantaneous angle of attack, which is known to affect the aerodynamic derivatives, and therefore implies a random variation of bridge aerodynamics (e.g., [9, 10]). In this respect, the problem of modelling the self-excited forces in turbulent flow has been widely addressed in the last thirty years through several nonlinear force models accounting for the low-frequency modulation due to large-scale turbulence [10–12]. In particular, Barni et al. [9, 13] set up a framework to assess the nonlinear buffeting response of a suspension bridge with reasonable computational effort. This approach includes self-excited force modulation produced by the variation of the angle of attack due to large-scale turbulence and considers the loss of spanwise correlation along the girder. It was also clearly shown that turbulence anticipates the flutter instability threshold.

The present paper draws upon previous work [13] to offer a first and simplified assessment of bridge flutter stability considering the parametric variation in time of the angle of attack due to large-scale turbulence. First, the stability of a simplified three-degree-of-freedom (3-dof) 2D model of the Hardanger Bridge, Norway, is determined through a Monte Carlo approach, aiming to underline the sensitivity of the flutter threshold to turbulence intensity and integral length scale. The Monte Carlo approach considers the whole large-scale turbulence parametric excitation (no white-noise or other approximations are needed), but it provides limited physical insight into the instability mechanism. Consequently, the stability of the system is then addressed by assuming a sinusoidal variation of the angle of attack. In this case, the stability boundaries can be obtained according to the theory of Floquet multipliers. This approach to the authors' best knowledge is used here for the first time to study bridge stability in turbulent flow and sheds some light on some stabilising and destabilising contributions of turbulent fluctuations and their interaction with bridge motion.

2 Time-Variant Dynamic System

According to the 2D rational function approximation (RFA) model [9], in presence of low-frequency wind velocity fluctuations due to large-scale turbulence, the transfer function between self-excited forces and bridge motion components maintains the same form valid for a linear system in smooth flow, but it becomes a function of the slowly-varying angle of attack (in addition to the reduced frequency of oscillation). Since the turbulent wind field varies not only in time but also in space, this angle of attack generally changes along the girder; clearly, this is not considered for the 2D bridge model assumed here (Fig. 1). Then, applying the inverse Fourier transform to the rational function approximation of the transfer function, self-excited forces (per unit length) can be expressed in the time domain, as follows:

$$\mathbf{q}_{se} = \left[q_y q_z q_M\right]^T = \frac{1}{2}\rho V_m{}^2\left(\mathbf{A}_1(\tilde{\alpha})\mathbf{r} + \frac{B}{V_m}\mathbf{A}_2(\tilde{\alpha})\dot{\mathbf{r}} + \sum_{l=1}^{N-2}\mathbf{A}_{l+2}\left(\tilde{\alpha}\right)\boldsymbol{\psi}_l\right) \tag{1}$$

$$\dot{\boldsymbol{\psi}}_l = -\mathbf{d}_l\left(\tilde{\alpha}\right)\frac{V_m}{B}\boldsymbol{\psi}_l + \dot{\mathbf{r}} \tag{2}$$

where:

$$\mathbf{A}_l = \begin{bmatrix} a^{\zeta}_{l,Dy}\left(\tilde{\alpha}_y\right) & a^{\zeta}_{l,Dz}\left(\tilde{\alpha}_z\right) & Ba^{\zeta}_{l,D\theta}\left(\tilde{\alpha}_\theta\right) \\ a^{\zeta}_{l,Ly}\left(\tilde{\alpha}_y\right) & a^{\zeta}_{l,Lz}\left(\tilde{\alpha}_z\right) & Ba^{\zeta}_{l,L\theta}\left(\tilde{\alpha}_\theta\right) \\ Ba^{\zeta}_{l,My}\left(\tilde{\alpha}_y\right) & Ba^{\zeta}_{l,Mz}\left(\tilde{\alpha}_z\right) & B^2a^{\zeta}_{l,M\theta}\left(\tilde{\alpha}_\theta\right) \end{bmatrix} \quad \mathbf{d}_l = \begin{bmatrix} d_l\left(\tilde{\alpha}_y\right) & & \\ & d_l\left(\tilde{\alpha}_z\right) & \\ & & d_l\left(\tilde{\alpha}_\theta\right) \end{bmatrix}$$

$$\alpha(t) = \operatorname{atan}\left(\frac{w(t)}{V_m + w(t)}\right) \tag{3}$$

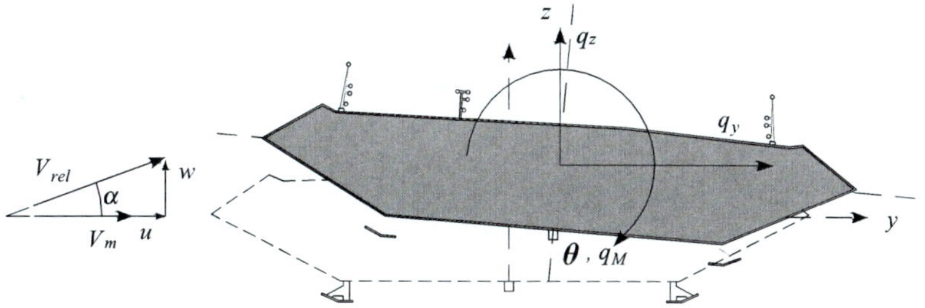

Fig. 1. Reference system for bridge displacements, forces and wind velocities.

The operator tilde indicates a low-pass filter applied to a certain quantity. V_m is the mean wind speed, ρ is the air density, B is the bridge cross-section width, and $\mathbf{r} = \left[y\, z\, \theta\right]^T$ represents the vector of the degrees of freedom of the bridge cross section. $\mathbf{A}_l$, $l \in \{1, 2, \ldots, N\}$, where $N - 2$ is the number of additional aeroelastic states, and $\mathbf{d}_l$, $l \in \{1, 2, \ldots, N - 2\}$, represent the rational function coefficients, which are polynomial functions of the slowly-varying angle of attack. q_y, q_z and q_M are the self-excited drag, lift and moment. $\boldsymbol{\psi}_l \in \mathbb{R}^3$, $l \in \{1, 2, \ldots, N - 2\}$, represents the additional aeroelastic states, which are defined as:

$$\boldsymbol{\psi}_l = \int_{-\infty}^{t} \exp\left(-\mathbf{d}_l\left(\widetilde{\alpha}\right)\frac{V_m}{B}(t-\tau)\right)\dot{\mathbf{r}}\, d\tau \tag{4}$$

The coefficients of the model can be identified through a constrained nonlinear least-squares algorithm to fit the aerodynamic derivatives measured for various angles of attack, as described in [9]. A crucial point in this self-excited load model is the choice of the cut-off frequency that defines the slowly-varying angle of attack due to turbulent fluctuations. The strategy of multiple cut-off frequencies for the self-excited force components associated with lateral, vertical and torsional motions is used here. In particular, three different slowly-varying angles of attack, denoted as $\widetilde{\alpha}_y$, $\widetilde{\alpha}_z$ and $\widetilde{\alpha}_\theta$, are determined fixing the cut-offs at the natural frequencies of the corresponding modes. Compared to the procedure proposed by Chen and Kareem [7], where the first bridge natural frequency is chosen as unique cut-off, the current approach allows considering the contribution to the modulating angle of attack of as many wind velocity fluctuations as possible, without violating the basic assumption of slow variation of the system parameters (for more details, see Barni et al. [13]). For a 3-dof dynamic system, the equations of motion can be written in a matrix form, as follows:

$$\ddot{\mathbf{r}} = -\mathbf{M}^{-1}\left(\mathbf{C} + \mathbf{C}_{ae}\left(\widetilde{\alpha}\right)\right)\dot{\mathbf{r}} - \mathbf{M}^{-1}\left(\mathbf{K} + \mathbf{K}_{ae}\left(\widetilde{\alpha}\right)\right)\mathbf{r} + \frac{1}{2}\rho V_m^2 \mathbf{M}^{-1}\sum_{l=1}^{N-2}\mathbf{A}_{l+2}\left(\widetilde{\alpha}\right)\boldsymbol{\psi}_l + \mathbf{q}_{ext} \tag{5}$$

where $\mathbf{C}_{ae} = -\frac{1}{2}\rho V_m^2 \mathbf{A}_2\left(\widetilde{\alpha}\right)$; $\mathbf{K}_{ae} = -\frac{1}{2}\rho V_m^2 \mathbf{A}_1\left(\widetilde{\alpha}\right)$

$\mathbf{M}$, $\mathbf{C}$ and $\mathbf{K}$ represent the structural mass, damping and stiffness matrices, respectively; while $\mathbf{C}_{ae}$, $\mathbf{K}_{ae}$ are the aerodynamic damping and stiffness matrice components directly deriving from the RFA approximation. The external buffeting load vector $\mathbf{q}_{ext}$ is defined as shown in [13]. Equation (5) is a second-order differential equation, the parameters of which depend on $\tilde{\alpha}$, which is a stochastic process. Therefore, applying a state-space transformation through $\boldsymbol{\gamma}_1 = \mathbf{r}$, $\boldsymbol{\gamma}_2 = \dot{\mathbf{r}}$ and $\boldsymbol{\gamma}_{l+2} = \boldsymbol{\psi}_l$, $l \in \{1, 2, \ldots, N-2\}$, the following system is obtained:

$$\begin{bmatrix} \dot{\boldsymbol{\gamma}}_1 \\ \dot{\boldsymbol{\gamma}}_2 \\ \dot{\boldsymbol{\gamma}}_{ad} \end{bmatrix} = \begin{bmatrix} 0 & \mathbf{I} & 0 \\ -\mathbf{M}^{-1}(\mathbf{K}+\mathbf{K}_{ae}) & -\mathbf{M}^{-1}(\mathbf{C}+\mathbf{C}_{ae}) & \mathbf{M}^{-1}\mathbf{Q} \\ 0 & \mathbf{I} & \mathbf{D} \end{bmatrix} \begin{bmatrix} \boldsymbol{\gamma}_1 \\ \boldsymbol{\gamma}_2 \\ \boldsymbol{\gamma}_{l+2} \end{bmatrix} + \begin{bmatrix} 0 \\ \mathbf{M}^{-1}\mathbf{q}_{ext} \\ 0 \end{bmatrix} \tag{6}$$

where: $$\mathbf{Q}\left(\tilde{\alpha}\right) = \frac{1}{2}\rho V_m^2 \left[\mathbf{A}_3\left(\tilde{\alpha}\right) \cdots \mathbf{A}_{l+2}\left(\tilde{\alpha}\right) \right] \in \mathbb{R}^{3\times[3\cdot(N-2)]}$$

$$\mathbf{D}\left(\tilde{\alpha}\right) = -\frac{V_m}{B}[\mathbf{I} \circ \mathbf{d}_l(\tilde{\alpha})] \in \mathbb{R}^{[3\cdot(N-2)]\times[3\cdot(N-2)]}$$

$\mathbf{I} \in \mathbb{R}^{3\times 3}$ is the identity matrix, and the symbol $\circ$ stands for the Hadamard product. In compact form, one can write:

$$\dot{\boldsymbol{\gamma}}(t) = \boldsymbol{\Omega}\left(\tilde{\alpha}\right)\boldsymbol{\gamma}(t) + \mathbf{B}\mathbf{q}_{ext}(t) \tag{7}$$

$\boldsymbol{\gamma} \in \mathbb{R}^{3(2+(N-2))}$ is the state vector, $\boldsymbol{\Omega}\left(\tilde{\alpha}\right) \in \mathbb{R}^{[3(2+(N-2))]\times[3(2+(N-2))]}$ is the time-variant state matrix of the system, and $\mathbf{B} = \left[\mathbf{0}\ \mathbf{M}^{-1}\ \mathbf{0}\right]^T \in \mathbb{R}^{[2N_{\text{mod}}+3\cdot(N-2)\cdot N_x]\times N_{\text{mod}}}$ is the input matrix. The system described by Eq. (7) is linear, albeit time-variant, so that the stability of the bridge is independent of the initial conditions. Therefore, only the parametric effect of turbulence associated with the slowly-varying angle of attack is able to affect the system stability. It is worth noting that this is traditionally considered a nonlinear feature, although the system remains linear in the state variables.

3 Stability Analyses

3.1 Broad-Band Variation of the Angle of Attack

This work does not have the ambition to formally study the stochastic stability of Eq. (7). Indeed, unlike the work of Lin and Ariaratnam [12] and subsequent developments (e.g., [5, 6]), the random variation of the system memory due to $\tilde{\alpha}$ does not allow approximating the bridge response as a Markovian process, which significantly simplifies the problem. Aiming at underlining the influence of turbulence on the stability threshold without any simplification in the parametric excitation, Lyapunov exponents (or even moment Lyapunov exponents) are a potential mathematical tool (e.g., [14]). Nevertheless, this is a task for future research, while here the problem is addressed in a heuristic way by calculating the nonlinear buffeting response of the 2D Hardanger Bridge dynamic system according to [13], up to the instability onset. The longitudinal turbulence intensity and integral length scale are varied from 10 to 30% and from 100 to 400 m, respectively. These ranges of parameters are chosen based on the wind field values assumed in the Hardanger Bridge design according to the N400 Norwegian code. The vertical turbulence intensity and the integral length scale are supposed to be equal to half and 10% of the longitudinal counterparts, respectively. The turbulent wind velocity components are generated as ten-hour long time series through the method reported in [15]. A sampling frequency of 6 Hz is chosen. The bridge response is then calculated considering two mean wind incidences, denoted as $\alpha_m = 0$ deg and $\alpha_m = 2.5$ deg. The latter is considered due to the upward inclination of the mean wind velocity between 2 and 2.5 deg, observed in the Hardanger Bridge site.

3.2 Narrow-Band Variation of the Angle of Attack

In the second part of this study, a sinusoidal parametric variation of the angle of attack is considered, defined by given frequency (f) and amplitude. Therefore, the system described by Eq. (7) becomes linear time-periodic, and its stability can be addressed by applying the Floquet theory. This can easily be done by numerically determining the Floquet state-transition matrix. The state-transition matrix is obtained by storing the system response due to a linearly independent complete set (base) of state variable initial conditions. The eigenvalues of the state-transition matrix evaluated at the period of the parametric excitation $(1/f)$ represent the Floquet multipliers λ_{FM}. The system is considered asymptotically stable if $|\lambda_{FM}| < 1$.

4 Case Study

4.1 Hardanger Bridge

The model previously introduced is applied to study the stability of a simplified 3-dof 2D model of the Hardanger Bridge in turbulent flow. This suspension bridge crosses the Hardanger Fjord, in Norway, and is characterised by a 1310 m-long main span and two 186 m-high reinforced concrete towers. It is the largest Norwegian bridge (the third one in Europe). The girder is a 3.2 m-high and 18.3 m-wide steel single-box deck (Fig. 1) composed of orthotropic plates. The three degrees of freedom (Fig. 1) are associated with the lateral, vertical and torsional modes that are most significant for the flutter stability, and the structural parameters are obtained from the modal analysis reported in [13]. In particular, the analysis considers modes 1, 4, and 15, which are calculated by linearising the structure around the dead load condition. The entries of the mass matrix **M** are the classical equivalent masses, calculated as the modal masses divided by the integral over the bridge deck of the square of the modal displacements; then, the three components of the vector **r** express the contributions to the deck response of the corresponding mode, read in the point of maximum modal displacement. A structural damping ratio of 0.5%

is assumed for all modes. The stiffness and damping matrices **K** and **C** in Eqs. (5)–(6) are also defined based on the modal properties of the bridge.

$$\mathbf{M} = \begin{bmatrix} m_{y,eq} & & \\ & m_{z,eq} & \\ & & I_{\theta,eq} \end{bmatrix}$$

$$m_{y,eq} = 10470\,\text{kg/m};\ m_{z,eq} = 12820\,\text{kg/m};\ I_{\theta,eq} = 426000\,\text{kgm}^2/\text{m}$$

$$f_y = 0.05\,\text{Hz};\ f_z = 0.14\,\text{Hz};\ f_\theta = 0.36\,\text{Hz}$$

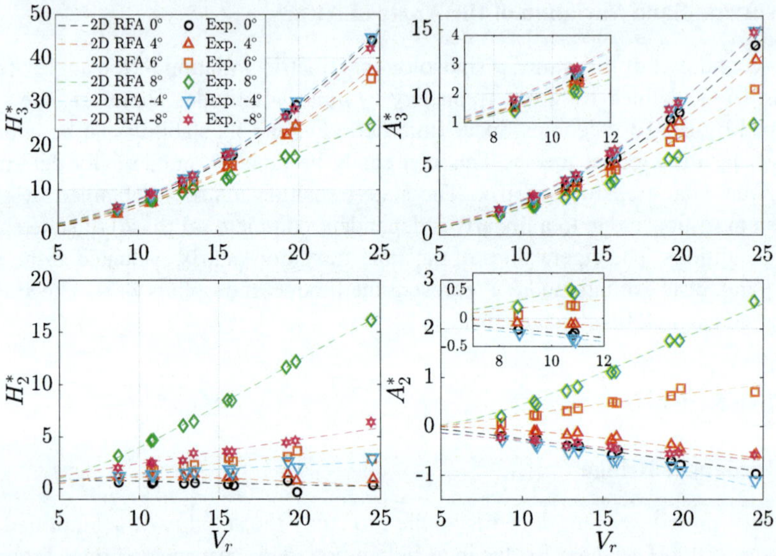

Fig. 2. Lift and moment aerodynamic derivatives associated with the torsional motion, reported as a function of the reduced wind velocity for various mean angles of attack.

4.2 Aerodynamic Coefficients

The aerodynamic derivatives, which were measured for different mean angles of attack with forced vibration tests and reported in [1, 2], are here approximated with 2D rational functions consisting of $N = 4$ terms, adopting 5th-order polynomials for all $\mathbf{A}_l$ and d_l parameters. Therefore, 228 coefficients are determined, ensuring high accuracy in the experimental data fitting (see Fig. 2). The 2D RFA approximation is carried out considering 11 angles of attack (namely, 8, 7, 6, 5, 4, 2, 1, 0, −2, −4, −6, and −8 deg), in a reduced velocity ($V_r = V_m/(fB)$) range up to 55. In the calculations, when $\widetilde{\alpha}$ falls outside this range, the 2D RFA coefficients are kept equal to the values corresponding to either +8 or −8 deg. The lift and moment aerodynamic derivatives associated with the torsional motion are reported in Fig. 2 for a subset of angles of attack, together with the corresponding 2D rational function approximation. Two essential features must be underlined in Fig. 2. First, the aerodynamic coefficients show a strong dependence on the mean angle of attack, which implies a possible marked time-variant behavior of the dynamic system. Secondly, the important coefficient A_2^*, which represents a direct contribution to the aerodynamic damping in torsion, takes positive values (resulting in a negative contribution to the aerodynamic damping) for mean angles of attack larger than about 5 deg. This feature clearly emphasises the importance of a model that accounts for such a nonlinearity.

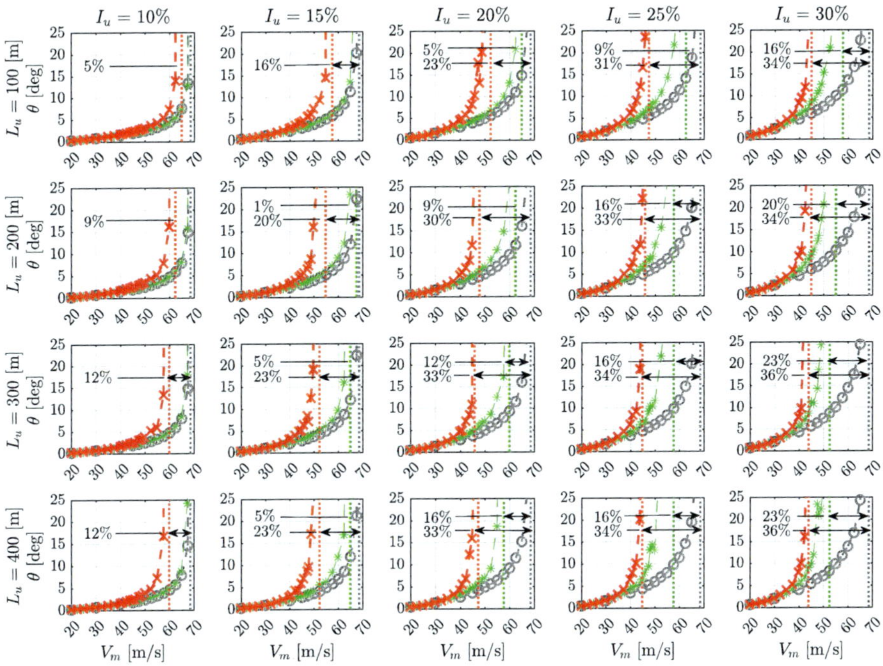

Fig. 3. RMS of the 3-dof nonlinear buffeting response obtained according to either a linear time-invariant (grey circles) or a linear time-variant (green asterisks for $\alpha_m = 0$ deg, and red crosses for $\alpha_m = 2.5$ deg) formulation of the self-excited forces.

Considering the aerodynamic derivatives for a null angle of attack, the classical deterministic flutter critical wind speed is $V_{cr} = 68.67$ m/s, and the critical frequency is $f_{cr} = 0.285$ Hz. This solution might also be obtained by imposing $\widetilde{\alpha}(t) \equiv 0$ in Eq. (7).

5 Discussion of Results

5.1 Stability Under Broad-Band Turbulence

In this section, the impact of turbulence on the flutter stability threshold is determined by comparing the results of Monte Carlo simulations with and without the parametric effect of turbulence. Figure 3 shows the linear time-invariant (grey circles) and time-variant

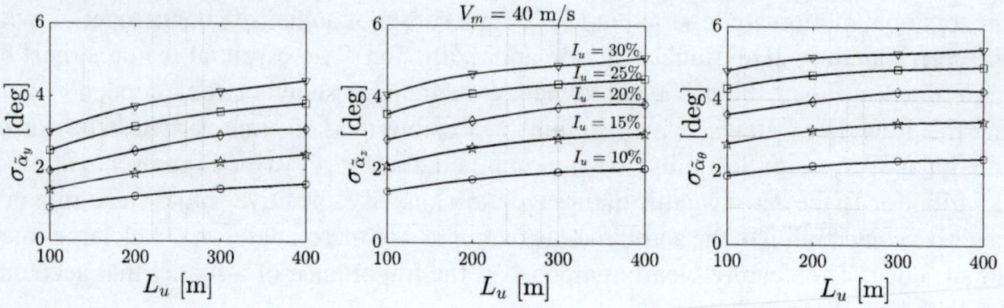

Fig. 4. RMS of the slowly-varying angles of attack associated with the three different motion components for different turbulence intensities and integral length scales ($\alpha_m = 2.5$ deg).

(green asterisks and red crosses for $\alpha_m = 0$ and 2.5 deg, respectively) torsional response of the bridge (in terms of root-mean-square, RMS) for different turbulent wind fields. In general, flutter stability is systematically lowered by the parametric random variation of the angle of attack $\tilde{\alpha}$, although this reduction is small for a null mean wind angle of attack and low to moderate turbulence intensity. This loss of stability can mainly be ascribed to the time-variation of the aerodynamic damping in torsion associated with the flutter derivative A_2^*. Indeed, an unstable behavior is promoted by a slowly-varying angle of attack when this becomes larger than about 5 deg (see Fig. 2). This argument explains the monotonic decrease of the stability onset with the turbulence intensity, since a higher turbulent intensity promotes higher fluctuations of the angle of attack. On the other hand, when a larger integral length scale of turbulence is considered, the power of low-frequency velocity fluctuations increases and so does the variance of lowpass filtered $\tilde{\alpha}(t)$, thus further reducing the system stability. However, the destabilising effect of the integral length scale of turbulence tends to saturate for large values of L_u, since the increments of variance for the slowly-varying angles of attack associated with the three different motion components become negligible (Fig. 4). In contrast, the RMS of the angles of attack almost linearly increases with the turbulence intensity.

5.2 Stability Under Sinusoidal Gusts

The analysis for a sinusoidal variation in the angle of attack is carried out here only for a null mean wind angle of attack ($\alpha_m = 0$ deg). Figure 5 shows the evolution of the flutter stability threshold through different curves (one for each mean wind velocity considered), representing the values of amplitude and frequency of the angle of attack oscillation for which a Floquet multiplier reaches and then crosses the value of one. Clearly, the dynamical system exhibits a more stable or unstable behavior depending on the sinusoidal gust frequency and amplitude. Typical features of nonlinear systems are apparent, such as parametric resonances for forcing frequencies that are either multiple of the torsional frequency or algebraic sums of multiples of the latter and of the lateral or vertical bending frequencies. In particular, some secondary parametric resonances of the torsional frequency can be recognised, such as $2/3f_\theta, f_\theta/2, f_\theta/3$ and others, clearly reducing the gust amplitude needed to destabilise the system, especially for low mean wind velocities. It is also worth pointing out the effect of the aerodynamic stiffness in torsion (mainly associated with the aerodynamic derivative A_3^*), which progressively reduces the torsional frequency as a function of the mean wind velocity. Figure 5 also highlights primary and secondary parametric resonances with stabilising effect. Among these, those associated with lateral-torsional frequency compositions ($f_\theta - f_y$ and $f_\theta + f_y$) are the most important, increasing the gust amplitude needed to destabilise the system: they even make appear stable islands for a wind velocity of 70 m/s, for which the time-invariant system is already unstable.

Previous observations are confirmed by Fig. 6, where the stability map is obtained disabling the parametric excitation in those components for which the sinusoidal gust frequency becomes higher than the associated natural frequency. This is done to comply with the main underlying assumption of the 2D RFA model, namely that the variation of the angle of attack is slow enough to produce a quasi-steady modulation of the self-excited force parameters. In this case, all the parametric resonances associated with lateral motion disappear (in this case, the parametric excitation is disabled beyond 0.05 Hz), while the main patterns of the stability boundaries remain almost unvaried. In general, we still expect that the variation of the stability threshold is strongly related to the parametric variation of the aerodynamic coefficient A_2^*. Other complicated features emerge from a careful inspection of the stability maps, such as sub-horizontal segments of the boundaries changing with the mean wind speed, which suggest the need for additional studies in this direction.

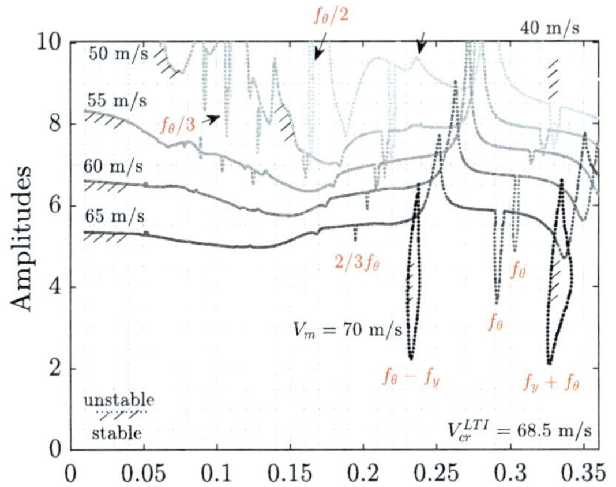

Fig. 5. Stability map of the 2D model of the Hardanger Bridge for different amplitudes and frequencies of the harmonic variation of the angle of attack modulating the self-excited forces and for various mean wind speeds ($\alpha_m = 0$). V_{cr}^{LTI} represents the time-invariant flutter threshold. f_y, f_z and f_θ are the cut-off frequencies, in this case equal to the still-air modal frequencies.

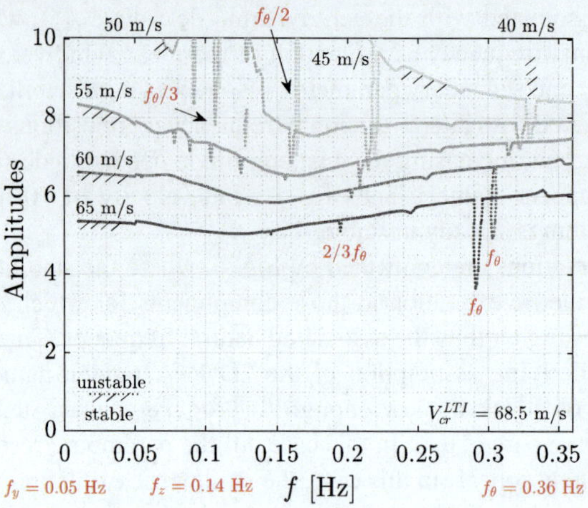

Fig. 6. Stability map obtained as in Fig. 5 but disabling the parametric excitation in those components for which the sinusoidal gust frequency becomes higher than the natural frequency.

6 Conclusions

The present work addresses the bridge flutter stability of a simplified 3-dof 2D bridge model considering self-excited force modulation due to the angle of attack produced by large-scale turbulence. This problem is modelled with a linear time-variant state-space model and solved with a Monte Carlo approach for different turbulent wind fields. Moreover, the Floquet theory is used to study the system stability parametrically forced by a sinusoidal variation of the angle of attack. Eventually, the following conclusions can be drawn:

- The parametric excitation due to large-scale turbulence can significantly destabilise the considered case study bridge, namely the Hardanger Bridge, especially for high turbulence intensity and integral length scale.
- Floquet analysis emphasises the potential destabilising effect of specific contributions to the parametric excitation, but it also points out the stabilising role of force modulations related to the lateral bending mode, which drain energy from the unstable torsional motion, in agreement with what observed by [5] for the parametric excitation associated with fluctuations of the wind kinetic pressure.

It is worth noting that, despite the variation of the angle of attack is assumed here to be perfectly correlated (simplified 2D bridge model), a clear impact on the flutter stability complying with the current results has also been highlighted in [13] for a partially-correlated random wind field.

Finally, the linear time-variant formulation of the problem presented herein seems suitable for a future more rigorous analysis of sample or moment stochastic stability of a bridge structure.

References

1. Vincent, G.S.: A summary of laboratory and field studies in the United States on wind effects on suspension bridges. In: International Conference on the Wind Effects on Buildings and Structures, vol. 20. H.M. Stationery, Teddington (1963)
2. Lin, Y.K., Ariaratnam, S.: Stability of bridge motion in turbulent winds. J. Struct. Mech. **8**(1), 1–15 (1980)
3. Irwin, H.P.A.H., Schuyler, G.D.: Wind effects on a full aeroelastic bridge model. In: ASCE Spring Convention and Exhibit, Penn., Pittsburgh (1978)
4. Bucher, C.G., Lin, Y.K.: Stochastic stability of bridges considering coupled modes. J. Eng. Mech. **114**(12), 2055–2071 (1988)
5. Bucher, C.G., Lin, Y.K.: Stochastic stability of bridges considering coupled modes: II. J. Eng. Mech. **115**(2), 384–400 (1989)
6. Lin, Y.K., Li, Q.C.: New stochastic theory for bridge stability in turbulent flow. J. Eng. Mech. **119**(1), 113–127 (1993)
7. Poirel, D.C., Price, S.J.: Structurally nonlinear fluttering airfoil in turbulent flow. AIAA J. **39**(10), 1960–1968 (2001)
8. Poirel, D., Price, S.J.: Random binary (coalescence) flutter of a two-dimensional linear airfoil. J. Fluids Struct. **18**(1), 23–42 (2003)
9. Barni, N., Øiseth, O., Mannini, C.: Time-variant self-excited force model based on 2D rational function approximation. J. Wind Eng. Ind. Aerodyn. **211**, 104523 (2021)
10. Diana, G., Rocchi, D., Argentini, T.: An experimental validation of a band superposition model of the aerodynamic forces acting on multi-box deck sections. J. Wind Eng. Ind. Aerodyn. **113**, 40–58 (2013)
11. Chen, X., Kareem, A.: Aeroelastic analysis of bridges: effects of turbulence and aerodynamic nonlinearities. J. Eng. Mech. **129**(8), 885–895 (2003)
12. Diana, G., et al.: Comparisons between wind tunnel tests on a full aeroelastic model of the proposed bridge over Stretto di Messina and numerical results. J. Wind Eng. Ind. Aerodyn. **54**, 101–113 (1995)
13. Barni, N., Øiseth, O., Mannini, C.: Buffeting response of a suspension bridge based on the 2D rational function approximation model for self-excited forces. Eng. Struct. **261**, 114267 (2022)
14. Caracoglia, L.: An Euler-Monte Carlo algorithm assessing Moment Lyapunov Exponents for stochastic bridge flutter predictions. Comput. Struct. **122**, 65–77 (2013)
15. Shinozuka, M.: Monte Carlo solution of structural dynamics. Comput. Struct. **2**(5–6), 855–874 (1997)

Aerodynamic Loads on Wind Turbine Towers Arranged in Groups with and Without Helical Strakes

Claudio Mannini[1(✉)], Tommaso Massai[1], Andrea Giachetti[1], and Alessandro Giusti[2]

[1] CRAICIV-University of Florence, Florence, Italy
claudio.mannini@unifi.it
[2] Siemens Gamesa Renewable Energy A/S, Brande, Denmark

Abstract. This work investigates the aerodynamic behavior of wind turbine towers when they are temporary placed in groups at small distance at the quayside of harbors. In this phase, many different arrangement geometries are possible, thus making wind tunnel tests a precious tool to pursue the economic design of the tower supporting structures. The effects of helical strakes, often employed to mitigate vortex-induced vibrations, is assessed through models at two different scales. The wind tunnel test results revealed complicated features of the problem and stressed the importance of the correct simulation in the experiments of the high Reynolds number regime. In particular, the strakes were found to reduce the mean base moment in the subcritical regime but to increase it in the developed supercritical regime.

Keywords: Tower Groups · Interference · High Reynolds Number · Aerodynamic Loads · Helical Strakes

1 Introduction

Offshore wind turbine towers are usually assembled at the quayside of pre-assembly harbors in groups including up to 10–12 towers at small distance, prior to being loaded onto special ships for the final offshore installation. Pre-assembly activities usually last 6 to 12 months. An accurate estimate of static and dynamic wind loads (in particular, the resultant tower base moment) in this temporary condition is crucial for the optimization of the design of the tower supporting structures (e.g., interfaces and foundation).

The literature does not adequately cover this important engineering problem, as most of the available results refer to groups of infinite circular cylinders in smooth flow (e.g., [1–4]) rather than finite towers in turbulent shear flow. Moreover, the wind loads are strongly dependent on the specific geometric configuration of the group, especially in terms of number of towers, grid arrangement and center-to-center spacing, thus making the support of wind tunnel tests extremely important.

However, an obstacle to the experimental study of the aerodynamics of groups of towers is the very high Reynolds number expected at full scale for the design wind speed,

P. Schito and A. Zasso (Eds.): IN VENTO 2022, LNCE 461, pp. 260–269, 2024.
https://doi.org/10.1007/978-3-031-53059-3_23

Fig. 1. View of a group of four towers at full scale.

which cannot be matched in the wind tunnel (e.g. [5]). This makes the laboratory results uncertain and probably overconservative.

Another peculiar feature of these structures is that they can be equipped with helical strakes to mitigate vortex-induced vibration [6]. The effect of these aerodynamic appendices is well known for isolated circular cylinders and towers, but it has not been sufficiently studied for groups of closely spaced towers.

The present work investigates the aerodynamic behavior of small groups of tapered towers (up to four towers), arranged according to a square mesh, with and without helical strakes in terms of base force and moment coefficients.

2 Case Study

The wind turbine towers considered herein are located in Taichung, Taiwan (Fig. 1). They have a cylindrical lower portion (0.586 H, where H is the height of the tower) and a linearly tapered upper part (0.414 H). The diameter at the top of the tower is 0.694 D_b, where D_b is the bottom diameter. Moreover, it results: $H = 14.5\ D_b$.

The three helical strakes cover the top 80% of the tapered part of the towers and are characterized by a pitch of 4.333 D_b. They have a triangular shape with a height of 0.025 D_b; therefore, they are rather small compared to the usual practice, for instance in chimneys [7, 8].

In addition to the free-standing configuration, a row of two towers (denoted as R2), a two-row pack of four towers (G4), and an L-shaped configuration of three towers (L3) are considered. In all cases, the center-to-center spacing of the towers is 1.333 D_b.

3 Wind Tunnel Tests

3.1 Experimental Facility, Set-Up and Models

The experimental campaign was carried out in the CRIACIV (Inter-University Research Centre on Building Aerodynamics and Wind Engineering) boundary layer wind tunnel in Prato, Italy. The overall length of the open-circuit facility is about 22 m, while 11 m is the available fetch to develop the boundary layer flow. The test section is 2.4 m wide and 1.6 m high. Air is drawn by a motor with a nominal power of 156 kW, and the flow speed can be varied continuously up to about 30 m/s.

The models of the towers were reproduced at two different scales, namely 1:130 (Fig. 2) and 1:75. The former was used for all the tests in atmospheric boundary layer flow, whereas the latter was employed for the isolated tower in smooth flow to further investigate Reynolds number and strake effects. Both models consist of two parts. The first one is made of steel and realized by a CNC lathe machine. It reproduces the cylindrical lower portion of the towers. Above this height, the diameter of the steel part abruptly reduces to constitute a spine for the tapered segment of the model. The latter is finely 3D printed in sintered Nylon, both with and without strakes, and wedged onto the steel spine.

The target turbulent wind profile was satisfactorily reproduced both in terms of mean wind speed and turbulence intensity (Fig. 3). The measuring model was connected to a six-component high-frequency force balance placed below the wind tunnel floor to record base forces and moments. The other (dummy) models in the tower groups were attached onto a base steel plate disconnected from the measuring model.

(a) (b)

Fig. 2. Model of an isolated tower (a) and of a group of four towers in the wind tunnel (b).

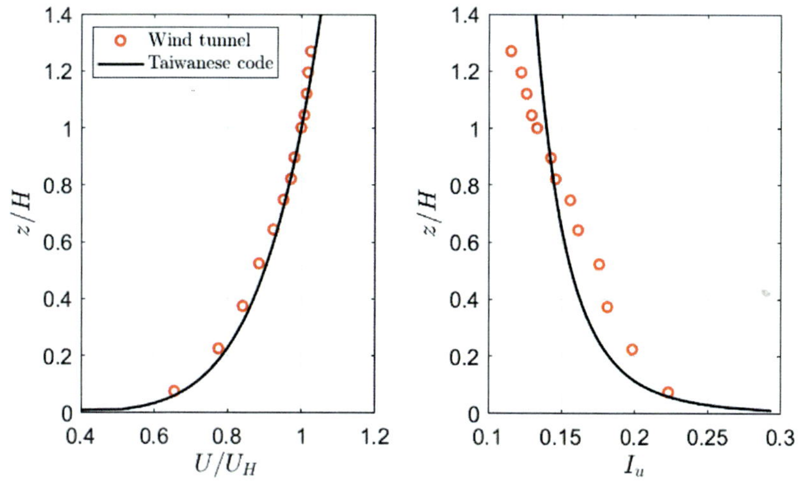

Fig. 3. Comparison of measured and target mean wind velocity and longitudinal turbulence intensity profiles. U_H is the undisturbed mean wind speed at the height of the top of the tower.

3.2 Simulation of the Supercritical Reynolds Number Regime

The crucial issue in this study was the simulation of the behavior of the towers, either isolated or arranged in groups, at a Reynolds number (Re) of the order of 10^7 (based on an average of the tower diameter and the mean wind speed at the height H), whereas the maximum value attainable in the wind tunnel was well below 10^5. A developed supercritical Reynolds number regime was then simulated by distributing discontinuous small strips of sandpaper over the surface of the models (see Fig. 2).

The effectiveness of this measure was verified by recording the pressure distribution on a circular cylinder in a confined turbulent flow [9], for which several data for high Reynolds numbers are available in the literature. In those tests, the circular cylinder had a diameter close to that of the 1:130 tower model. The results are reported in Fig. 4(a) along with the target pressure distribution provided by Eurocode 1 [10]. While a subcritical pressure coefficient pattern was found for the smooth cylinder, when surface roughness is employed, the pressure coefficient distribution is stable in the high wind speed range (say, beyond a wind tunnel Reynolds number of about $5 \cdot 10^4$) and very close to the target one. This is confirmed by the drag coefficient reported in Fig. 4(b).

The same strategy of simulation of the supercritical Reynolds number regime was employed for the three-dimensional towers that are the main object of the current work. The reliability of this approach was corroborated by the results obtained in smooth flow for the 1:75 scale model of the isolated tower (see Sect. 4.1).

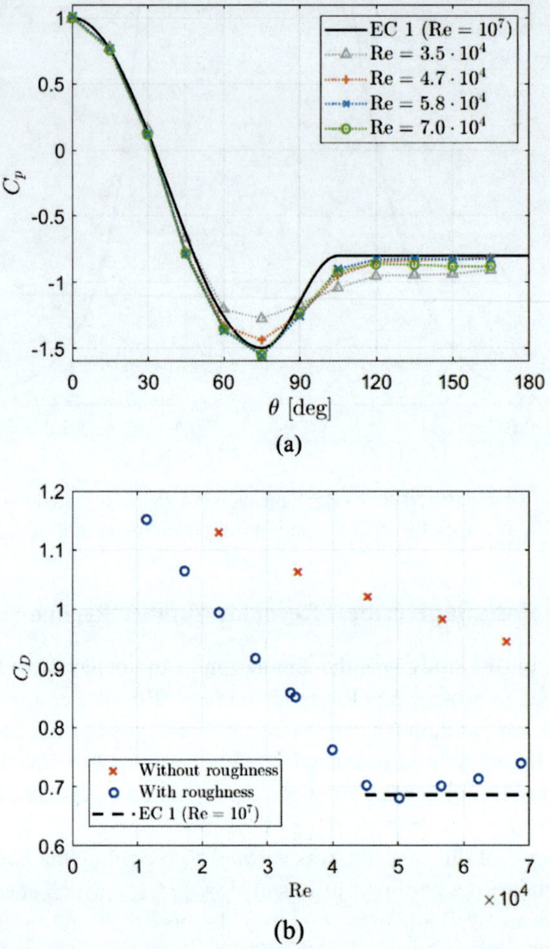

Fig. 4. Pressure coefficient distribution (a) and integrated drag coefficient (b) for the confined turbulent flow configuration of a circular cylinder. The Reynolds number in abscissa is simply calculated based on wind tunnel velocity and cylinder diameter and does not correspond to the simulated one when surface roughness is employed. EC 1 stands for Eurocode 1 [10].

4 Discussion of Results

In the following, the results are reported in terms of mean value of the base resultant moment coefficient, defined as follows:

$$C_M = \frac{M}{q_H \int_0^H D(z) z dz} \tag{1}$$

where M is the mean base resultant moment, q_H is the mean wind velocity pressure of the undisturbed flow at the height of the top of the tower, and $D(z)$ denotes the diameter of the tower along its height. Nevertheless, similar trends were found also for the mean value of the base resultant shear force coefficient. The measurements have always been repeated a few times to have an idea of the unavoidable uncertainty in the coefficients.

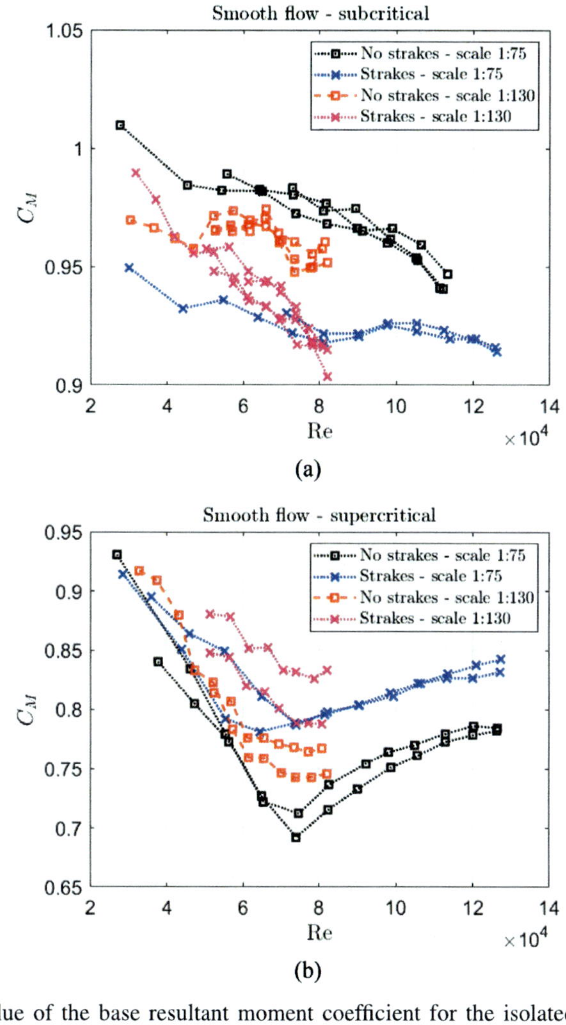

(a)

(b)

Fig. 5. Mean value of the base resultant moment coefficient for the isolated tower models in smooth flow without (a) and with technical roughness (b). The Reynolds number in abscissa is simply calculated based on wind tunnel velocity and an equivalent diameter of the tower and does not correspond to the simulated one when surface roughness is employed.

4.1 Isolated Tower

A first set of tests was performed in smooth flow on the 1:75 and 1:130 scale models of the tower without any surface roughness to simulate a higher Reynolds number (a configuration identified in the following as "subcritical"). Though not very useful from the practical design point of view, this configuration is interesting to understand the role of the helical strakes and the effectiveness of the roughness. The results are reported in Fig. 5a. At low wind velocity (that is, at low Reynolds number) the measurements are more uncertain because of the reduced accuracy of the force balance when the load is small. It seems that a transitional region is encountered around the highest wind speed reached in the wind tunnel. To interpret the results, it should also be pointed out that the small boundary layer that always develops on the smooth floor of the wind tunnel (having a height of about 15 cm) is nearly negligible for the 1:75 scale model, but it is not for the 1:130 one. Nevertheless, the most important feature to highlight is that the strakes reduce the mean value of the moment coefficient. This is a consequence of the lower amount of vorticity in the wake, caused by the suppression or attenuation of the vortex shedding, as revealed by the analysis of the lateral force spectra.

The developed supercritical Reynolds number regime is then forced through the addition of the previously mentioned sandpaper strips. The inspection of the results in Fig. 5b reveals that, as expected, the regime transition induced by the surface roughness on the model is slightly different for the model at scale 1:75 compared to the one at scale 1:130. Nevertheless, the value of the moment coefficient at the maximum wind tunnel velocity is rather close in the two cases. This suggests that the surface roughness adopted for the smaller tower model (which is the only one that can be employed for the measurements in atmospheric boundary layer flow) is effective in simulating the behavior at high Reynolds number. Moreover, though the helical strakes are small (and their manufacturing challenging) at the scale 1:130, their aerodynamic effect seems to be properly captured in the tests. In the supercritical Reynolds number regime, vortex shedding is less violent, and therefore the presence of the strakes implies a non-negligible increase of the base moment, instead of a reduction of it. In particular, the data reported in Fig. 5b show an increment of the load of about 6–6.5%, which complies with the increase of the effective tower diameter due to the presence of the strakes.

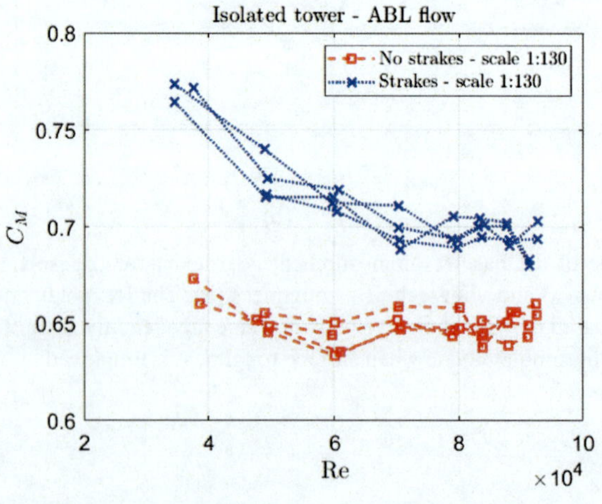

Fig. 6. Mean value of the base resultant moment coefficient for the isolated tower 1:130 scale model equipped with sandpaper strips in atmospheric boundary layer (ABL) flow.

Finally, the 1:130 scale model of the isolated tower was tested in the simulated atmospheric boundary layer flow. The resulting moment coefficient is reported in Fig. 6. One can note a general reduction of the load compared to the smooth flow condition. At the highest flow velocities reached in the wind tunnel, the mean value of the moment coefficient is quite constant. A load increase of about 6.5% due to the presence of the strakes is also apparent in this realistic wind condition.

4.2 Groups of Towers

The results for a group of two and four towers are reported in Fig. 7 for various wind directions β, whereas those for a group of three towers are shown in Fig. 8. For the considered tower spacing, an apparent increase of the loads compared to the isolated tower is observed. Moreover, the rise of the loads due to the presence of the helical strakes is non-negligible also in the group configurations. Indeed, for some wind directions the increase of the base moment coefficient attains about 10%, and therefore is even larger than for the isolated tower. The highest loads are obtained for the row of two towers (R2) for a wind direction between about $-10°$ and $-20°$ (considering the tower position P1 highlighted in Fig. 7), and for the external tower (either position P1 or position P3) of the L-shaped group of three towers (L3, see Fig. 8) and a similar wind direction. The comparison between the moment coefficients obtained for the configuration L3 and the configurations R2 and G4 is also reported in Fig. 8. It is apparent that the maximum mean loads for the L3 group of towers are very close to those for the physically similar configurations of groups R2 and G4.

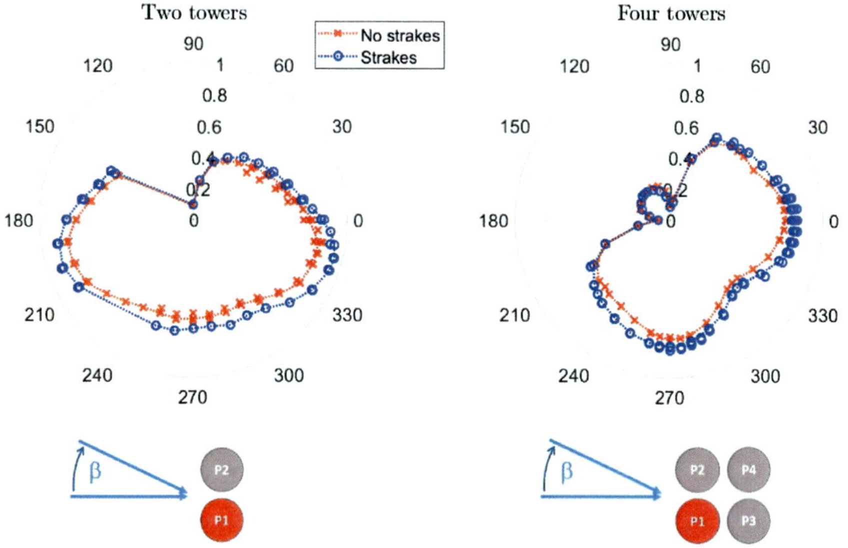

Fig. 7. Mean base resultant moment coefficient for the highlighted tower (position denoted as P1) in a group of two (R2) and four towers (G4).

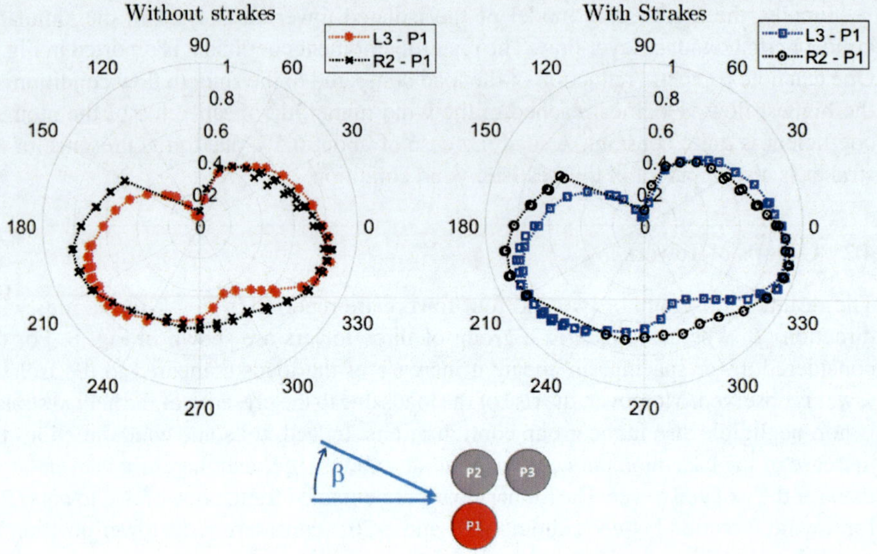

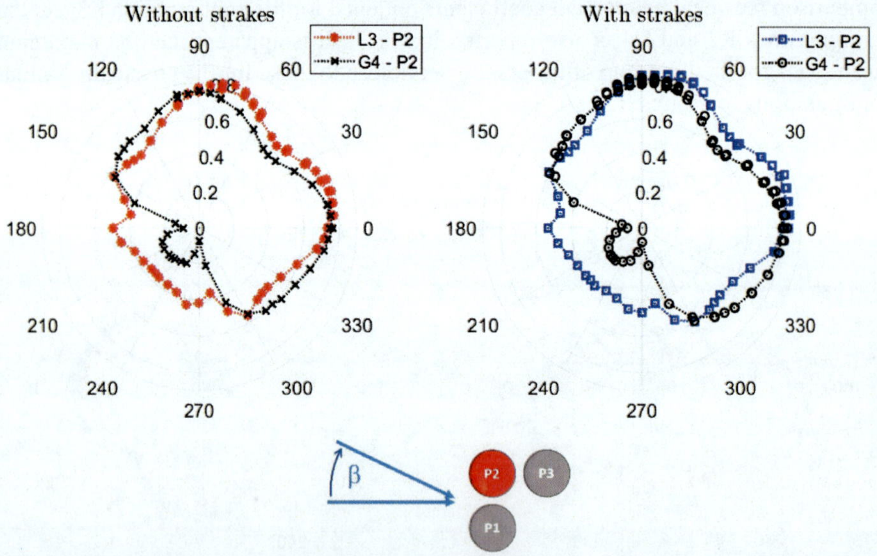

Fig. 8. Comparison of the mean values of the base resultant moment coefficient for the P1 tower in the L3 and R2 configurations (top row) and for the P2 tower in the L3 and G4 configurations (bottom row). Refer to the schematics in Fig. 7 for the positions P1 and P2 in the R2 and G4 configurations.

5 Conclusions

This study reveals the complicated aerodynamic behavior of groups of towers at small distance and emphasizes the importance of a correct simulation in the wind tunnel of the high Reynolds number regime expected at full scale.

Despite the limited height of the considered helical strakes, their effect on the aerodynamic loads is clearly detectable, as confirmed by the measurements carried out on a larger scale model of the isolated tower in smooth flow. In the low Reynolds number regime, the reduction of vorticity in the wake due to vortex-shedding mitigation is dominant, and the base shear and moment coefficients reduce when the strakes are present. In contrast, in the high Reynolds number regime, both in smooth and in turbulent boundary layer flow, a rise of the wind mean loads is observed due to the increased effective diameter of the tower.

The load increment due to the helical strakes is slightly more pronounced for the towers arranged in a group than for the isolated tower, with a growth of the design value of the base moment coefficient of the order of 10%.

References

1. Zdravkovich, M.M.: Interference between two circular cylinders forming a cross. J. Fluid Mech. **128**, 231–246 (1983)
2. Price, S.J., Paidoussis, M.P.: The aerodynamic forces acting on groups of two and three circular cylinders when subject to a cross-flow. J. Wind Eng. Ind. Aerodyn. **17**(3), 329–347 (1984)
3. Sayers, A.T.: Flow interference between four equispaced cylinders when subjected to a cross flow. J. Wind Eng. Ind. Aerodyn. **31**(1), 9–28 (1988)
4. Sumner, D.: Two circular cylinders in cross-flow: A review. J. Fluids Struct. **26**(6), 849–899 (2010)
5. Kareem, A., Kijewski, T., Lu, P.-C.: Investigation of interference effects for a group of finite cylinders. J. Wind Eng. Ind. Aerodyn. **77–78**, 503–520 (1998)
6. Zdravkovich, M.M.: Review and classification of various aerodynamic and hydrodynamic means for suppressing vortex shedding. J. Wind Eng. Ind. Aerodyn. **7**(2), 145–189 (1981)
7. Simiu, E., Yeo, D.: Wind Effects on Structures: Modern Structural Design for Wind, 4th edn. Wiley, Oxford (2019)
8. Dyrbye, C., Hansen, S.O.: Wind Loads on Structures, 1st edn. Wiley, Chichester (1997)
9. Mannini, C., Massai, T., Giachetti, A., Giusti, A.: Aerodynamic loads on groups of offshore wind turbine towers stored on quaysides during the pre-assembly phase. J. Wind Eng. Ind. Aerodyn. **242**, 105569 (2023)
10. EN 1991-1-4: Eurocode 1: Actions on structures - Part 1–4: General actions (2005)

Comparison of Wake Oscillator Models to Predict Vortex-Induced Vibration of Tall Chimneys

Saba Rahman[1](✉), Arvind K. Jain[1], S. D. Bharti[2], and T. K. Datta[2]

[1] Indian Institute of Technology, Delhi, India
sabarahman807@gmail.com, akjain@civil.iitd.ac.in
[2] Malaviya National Institute of Technology, Jaipur, India
sdbharti@mnit.in

Abstract. A comparative study has been carried out for tapered chimney considering various wake oscillator models available in the literature. These models are Facchinetti, Farshidian and Dolatabadi and Skop and Griffin. Tip displacement of 210m tall, tapered chimney is evaluated. The structure is analyzed using the stated models in first mode oscillation, and their results are compared with Eurocode method-I response. The nonlinear approach is adopted for analyzing wake oscillator models where the solution of the coupled equation is assumed to be non-harmonic, unlike the analytical models, and is solved using the ODE solver of MATLAB. The nonlinear method is then compared with the analytical solution proposed by the mentioned models, confirming that the nonlinear approach should be chosen for analyzing the lightly damped slender structures. It is observed that except for the models proposed by Farshidian and Dolatabadi, other models predict nearly the same responses. The parameters causing the variation among the models are also discussed.

Keywords: Deterministic model · Van der Pol · Vortex-induced vibration · Eurocode

1 Introduction

Chimneys are tall and slender structures with circular cross-sections, which are highly prone to wind forces. Wind exerts pressure on the chimneys' walls, producing unwanted forces. Vortex-induced vibration is one of such excitations which can lead to the failure of the chimneys. The vortex-induced force grows due to the shedding of vortices from alternating sides of the structures. It can be modeled as a deterministic or random process depending upon the flow regime. When the fundamental structural frequency comes close to the vortex shedding frequency, self-excited oscillation occurs. The self-excited vibration produces a large amplitude of vibration. Many empirical, semi-empirical and experimental studies have been conducted so far to understand the vortex-induced vibration of the structures [1]. Although conducting experiment studies is tedious, many researchers developed empirical and semi-empirical methods. These methods provide

P. Schito and A. Zasso (Eds.): IN VENTO 2022, LNCE 461, pp. 270–279, 2024.
https://doi.org/10.1007/978-3-031-53059-3_24

comparable responses as obtained from the experiment and can also be used to find the response of arbitrarily shaped chimneys. With technical and infrastructure advancement, chimneys with arbitrary cross-sections or complex modifications have been developed, which is harder to analyze experimentally.

The present study compares the semi-empirical wake-oscillator models used to predict structures' vortex-induced vibration. These models include those proposed by Facchinetti [2], Farshidian and Dolatabadi [3] and Skop and Griffin [4]. These models combine a wake oscillator model resembling the Van der Pol oscillator model and a single degree of freedom oscillation model. In order to use these models for estimating the top displacement of chimneys, the first mode vibration of the chimneys is only considered. The modal equation of the chimney constitutes the single degree of freedom model (SDOF). The equations of the wake oscillator model and the SDOF are simultaneously solved using an iterative procedure. The ODE solver of MATLAB is used to carry out the iterative solution. To carry out the comparative study, a tall concrete chimney height of 210 m has been chosen with a base diameter of 20m, top diameter of 12 m and thickness as 0.3 m. The responses of the chimney are also determined using the deterministic model given in Eurocode.

The dynamics of fluid-structure interaction were analytically examined to provide a closed-form solution for vortex-induced vibration and provide responses comparable with the experimental outcomes. Three coupled Van der Pol models, namely, displacement coupled model, velocity coupled model and acceleration coupled model, were used for the wake oscillator equation to predict the fluctuating lift coefficient. It was established by analyzing the coupled models for circular cylinders that the most suitable coupled forcing term was proportional to the acceleration of the structure. Therefore, for this study, only acceleration coupled models have been considered.

2 Wake Oscillator Models

2.1 Structural Oscillator

The tapered concrete circular chimney of varying diameters is shown in Fig. 1. The structure is subjected to wind flow and is free to oscillate in the transverse direction. The displacement equation of motion of the chimney considering the lumped mass approach is given below.

$$m\ddot{Y} + r\dot{Y} + hY = S \tag{1}$$

Where m, r and h are the effective mass, damping and stiffness of the chimney, respectively. The Eq. (1) of transverse flow (Y) is differentiated with respect to time. The effective mass and damping consist of both structure and fluid flow such that $m = m_s + m_f$ and $r = r_s + r_f$, where $m_f = \frac{C_M \pi \rho D^2}{4}$, $r_f = \gamma \overline{\Omega} \rho D^2$ and $r_s = 2m\xi\Omega_s$. The parameters ρ, C_M, γ and Ω_s is the density of the flow, added mass coefficient, stall parameter and structural angular frequency, respectively. The reference angular frequency ($\overline{\Omega}$) can be identified as the Strouhal frequency as it depends on the Strouhal number, such as,

$\overline{\Omega} = \Omega_f = 2\pi S_t U/D$. The term S is the forcing term controlled by vortex lift force acting on the structure. Equation (1) can be further simplified as below

$$\ddot{Y} + \left(2\xi\Omega_S + \frac{\gamma}{\mu}\Omega_f\right)\dot{Y} + \Omega_S^2 Y = S/m \tag{2}$$

$$S = \frac{1}{2}\rho U^2 D C_L \tag{3}$$

2.2 Wake Oscillator

The classical wake oscillator model is based on the Van der Pol model, which was modified and included various wind tunnel-based parameters. The classical oscillator can be expressed as

$$\ddot{q} + \varepsilon\Omega_f\left(q^2 - 1\right)\dot{q} + \Omega_f^2 q = F \tag{4}$$

Here, F is the forcing term coupled with the transverse equation of the motion of structure through acceleration, velocity, and displacement term. It is coupled with the motion as it interacts with the wake region of the structure during fluid-structure interaction. Since the dimensionless wake variable (q) is the function of the vortex lift coefficient (C_L) and can be divulged as the following

$$\frac{q}{q_0} = \frac{C_L}{C_{L0}} \tag{5}$$

Where, C_{L0} and q_0 are defined as reference lift coefficient and amplitude of quasi-harmonic vibration at the structural angular frequency (Ω_s). Krenk and Nielsen [5] developed the relationship between the q_0 and the transverse velocity of fluid flow, while some authors like Facchinetti and de Langre [6] assumed the initial value of q_0 as two based on observation. On the other hand, Fardasshian used optimization tools to find its value.

In order to couple both the structural and wake oscillator equations, Eqs. (2), (4) are reduced to dimensionless form. The dimensionless oscillator equations are expressed as

$$\ddot{y} + \left(2\xi\delta + \frac{\gamma}{\mu}\right)\dot{y} + \delta^2 y = s \tag{6}$$

$$\ddot{q} + \varepsilon\left(q^2 - 1\right)\dot{q} + q = f \tag{7}$$

The parameter δ is known as reduced angular frequency and can be defined as.

$$\delta = \frac{\Omega_s}{\Omega_f} = \frac{1}{S_t U_r};\ U_r = \frac{2\pi}{\Omega_s}\frac{U}{D} \tag{8}$$

The forcing term s in Eq. (6) entails the interaction of flow near the wake region of the bluff structure and can be expressed as a

$$s = \frac{S}{mD\Omega_f^2} = S\frac{D}{4\pi^2 S_t^2 U^2 m} \tag{9}$$

In unitless form, Eq. (9) can be written as

$$s = Mq, M = \frac{C_{L0}}{2}\frac{1}{8\pi^2 S_t^2 \mu} \tag{10}$$

Where M is identified as a dimensionless mass number. The coupling term f given in Eq. (7), which is generated due to the vorticity of fluid-structure interaction, is coupled with the motion of the structure . Earlier, Hartlen and Currie established a velocity-based coupled term later modified by Facchinetti and de Langre to understand the behavior of acceleration coupled and displacement coupled interactions. Here only the acceleration coupled term is evaluated as it successfully predicts all the vortex-induced oscillation features. The parameters A and ε (in Eq. (7)) are acknowledged as Van der Pol parameters. Peak response is observed near lock-in state when the Strouhal frequency approaches near structural frequency, i.e., $\omega = \delta = \Omega \cong 1$. Peak displacement and closed form solution of analytical models given by authors Facchinetti, Farshidian and Dolatabadi and Skop and Griffin are given in Table 2.

Table 1. Wake structure oscillator for all the models along with their closed-form solution

Wake-structure Oscillator Equations	Closed-form Solution
Facchinetti	
$\ddot{y} + \left(2\xi\delta + \frac{\gamma}{\mu}\right)\dot{y} + \delta^2 y = s$ $\ddot{q} + \varepsilon\left(q^2 - 1\right)\dot{q} + q = f$	$\frac{y_0}{q_0} = \frac{M}{\left[(\delta^2-\omega^2)^2+\left(2\xi\delta+\frac{\gamma}{\mu}\right)^2\omega^2\right]^{0.5}}$ $q_0 =$ $2\left[1 + \frac{AM}{\varepsilon}\frac{C}{(\delta^2-\omega^2)^2+\left(2\xi\delta+\frac{\gamma}{\mu}\right)^2\omega^2}\right]^{0.5}$
Farshidian and Dolatabadi	
$\ddot{y} + \left(2\xi\delta + \frac{\gamma}{\mu}\right)\dot{y} + \delta^2 y = s$ $\ddot{q} + \varepsilon\left(\beta q^2 - 1\right)\dot{q} + q = f$	$\frac{y_0}{q_0} = \frac{M}{\left[(\delta^2-\omega^2)^2+\left(2\xi\delta+\frac{\gamma}{\mu}\right)^2\omega^2\right]^{0.5}}$ $q_0 =$ $\frac{2}{\sqrt{\beta}}\left[1 + \frac{AM}{\varepsilon}\frac{C}{(\delta^2-\omega^2)^2+\left(2\xi\delta+\frac{\gamma}{\mu}\right)^2\omega^2}\right]^{0.5}$

(continued)

Table 1. (*continued*)

Wake-structure Oscillator Equations	Closed-form Solution
Skop and Griffin	
$\frac{\ddot{y}}{d} + 2\xi\omega_n\frac{\dot{y}}{d} + \omega_n^2\frac{y}{d} = \frac{\rho U^2 L}{2M}C_L$ $\ddot{C}_L + \omega_s^2 C_L -$ $\left[C_{L0}^2 - C_L^2 - \left(\frac{\dot{C}_L}{\omega_s}\right)^2\right]\left(\omega_s G\dot{C}_L - \omega_s^2 HC_L\right) =$ $\omega_s F\left(\frac{\dot{y}}{d}\right)$	$A = \frac{\frac{BC_{L0}}{S_G}}{(\delta^2+4)^{\frac{1}{2}}}$

Table 2. Maximum amplitude of vibration of analytical models

Facchinetti	$y_M = \frac{C_{L0}/2}{S_G+4\pi^2 St^2\gamma}\sqrt{1 + \frac{A}{\varepsilon}\frac{C_{L0}/4}{S_G+4\pi^2 St^2\gamma}}$
Farshidian and Dolatabadi	$y_M = \frac{C_{L0}/2}{S_G+4\pi^2 St^2\gamma}\sqrt{\frac{1}{\beta}\left(1 + \frac{A}{\varepsilon}\frac{C_{L0}/4}{S_G+4\pi^2 St^2\gamma}\right)}$
Skop and Griffin	$A = \frac{\frac{BC_{L0}}{S_G}}{(\delta^2+4)^{\frac{1}{2}}}$

3 Methodology

3.1 Analytical Method

The authors have assumed structural and wake coefficients to be harmonic in nature for simplification. Although it reduces the computational efforts, the approach fails to include aeroelastic effects. Authors Facchinetti, Farshidian and Dolatabadi and Skop and Griffin proposed the following solution to obtain the result of the closed form equation given in Table 1.

$$y(t) = y_0\cos(\omega t);\ q(t) = q_0\cos(\omega t - \phi) \tag{11}$$

Both structural and wake oscillators are time-dependent parameters that depend on the common angular frequency (ω). On substituting Eq. (11) into Eqs. (6), (7) and normalizing the oscillator equation, the closed-form solution was achieved. Farshidian and Dolatabadi modified the classical oscillator equation given by Facchinetti by introducing a new parameter (β) in the oscillator equation. This attempt was made to obtain a more realistic analytical response to the structure.

3.2 Numerical Integration Method

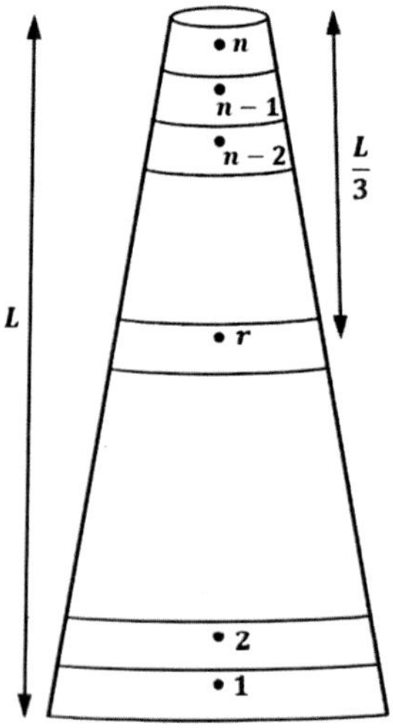

Fig. 1. Discretized chimney representation

The closed form solution given by mentioned authors was developed for circular cylinders; implementation of those into the actual chimney is complicated and unreliable as the analytical solution overlooks the aeroelastic action. Flexible structures such as tall chimneys show strong nonlinearity, especially near a locked-in state and low damping ratio. Consequently, a numerical integration technique was developed based on normal mode theory to investigate the tall chimney. This mathematical model incorporates the aeroelastic effects efficiently and can also be used for chimneys of various heights and shapes.

The mathematical model was developed for numerical integration using the MATLAB ODE solver. The chimney is discretized into n number of nodes, and the mass at each node is lumped and expressed as the diagonal mass matrix (M) of size $n \times n$. Taking the first mode and assuming the vortex lift force acting at the top one-third height of the chimney, the following expressions are obtained

$$\overline{m}_1 = \phi_1^T M \phi_1 \tag{12}$$

$$s_1 = \phi_1^T S \tag{13}$$

$$S_i = \frac{1}{2}\rho U_i^2 D_i C_L(t);\ i = 1, 2, \ldots r \ldots n \tag{14}$$

where, ϕ_1, $\overline{m}_1$ and s_1 are the first mode shape of size $n \times 1$, generalized mass and modal vortex lift force. And i, r and n represent the node number as shown in Fig. 1. The equation of motion for the first mode lock-in oscillation is given by

$$\ddot{y}_1 + \left(2\xi\delta + \frac{\gamma}{\mu}\right)\dot{y}_1 + \delta^2 y_1 = \frac{S_1}{\overline{m}_1} \tag{15}$$

where $y = Y/D_{av}$; $D_{av} = \frac{1}{n}\sum D_i$; D_i is the outer diameter of the ith segment Equation (15) is a modified structural oscillator that, along with Eq. (7), is used to obtain the solution of structure.

4 Results and Discussion

This paper concerns the comparative study of circular chimneys based on analytical and numerical integration methods. A 210 m tall, tapered chimney has been considered to carry out the study. The Strouhal number is assumed to be 0.2, and the first fundamental mode shape of the chimney is used. The mass number and effective diameter of the chimney are 6.10×10^{-4} and 21.33 m, respectively. The fundamental structural frequency of the chimney is 0.35 Hz. Due to the unavailability of Van der Pol parameter for 210 m chimney, it is assumed to be that given by Facchinetti for analytical and numerical methods. These values are 12 and 0.3 for A and ε. To capture nonlinearity, 0.5%, 1% and 2% structural damping have been used. The chimney is analyzed as a 2D structure as the analytical models given by the authors were developed for the 2D cylinder to avoid various complexities.

The maximum amplitude of oscillation obtained by the analytical and numerical methods is given in Table 3. Figures 2 and 3 show the variation of the chimney's peak displacement with the reduced velocity considering numerical and analytical approaches, respectively. Figure 2 is plotted for the three damping cases. It can be concluded from Figs. 2 and 3 that among the three models, the Farshidian and Dolatabadi model shows comparable results between numerical and analytical methods. The reason is the incorporation of a coefficient (β) and assumption of the value of q_0 as eight instead of two, which is taken in the classical Fachhiniti model. Although, if the value of the coefficient is established as $\beta = 1$, then the predictions of Farshidian and Dolatabadi's models would be the same as the classical Facchinetti's model. During the numerical integration method, the initial value of q_0 is taken the same as specified by models, but with repetitive iteration, its value changes, and the final converged value decides the oscillation amplitude. Also, the wake variable (q), which is assumed to be harmonic by the mentioned models, does not remain to be purely harmonic. It becomes a time-dependent variable. However, the Farshidian and Dolatabadi model over conserve as it magnifies the response of the chimneys.

As the damping increases, the peak amplitude decreases for the three models. The difference in response obtained from numerical and analytical methods is nonlinear. The higher damping case shows less percentage difference between the two approaches. This can be interpreted as the numerical method efficiently capturing the nonlinearity. Therefore for 2% damping, the difference between the numerical and analytical methods is less. The numerical method should opt for the low damping case.

The response obtained from Eurocode method-I is mentioned in Table 4. Eurocode method-I [7] developed its provision based on Ruscheweyh's model. Also known as the deterministic model and it considers correlation length. The final expression of the peak amplitude of vibration given by Eurocode method-I is expressed as

$$\frac{y_{max}}{D} = \frac{1}{S_c}\frac{1}{S_t^2}KK_wC_{Lat} \tag{16}$$

Where S_c, S_t and C_{Lat} are the Scruton number, Strouhal number, and RMS value of lift force per unit length. D is the diameter of the chimney. The coefficient K is the natural mode of vibration parameter, and for the cantilever chimney, it is specified as 0.13. The coefficient K_w correlation length parameter. The Eurocode method-I incorporates the aeroelastic effect in terms of the correlation length parameter (K_w). The upper threshold value of K_w is given as 0.6 and is obtained from experimental studies considering the fluid-structure interaction. From Table 4, it can be inferred that the Eurocode method-I overestimates the response for low damping case and should be used for structure with high damping (Table 3).

Table 3. Peak amplitudes (m/m) of the chimney obtained by the analytical and numerical method

	Damping-0.5%		Damping-1%		Damping-2%	
	Analytical	Numerical	Analytical	Numerical	Analytical	Numerical
Facchinetti	0.07	0.09	0.04	0.06	0.02	0.030
Farshidian and Dolatabadi method	0.37	0.42	0.22	0.24	0.12	0.130
Skop and Griffin	0.05	0.06	0.05	0.06	0.03	0.034

Table 4. Peak displacements (m/m) of the chimney obtained from Eurocode method-I, classical Fachinetti model

Damping	210 m chimney		
	0.5%	1%	2%
Scruton Number	4.89	9.79	19.57
Eurocode method-I	0.15	0.07	0.03
Facchinetti (Analytical)	0.07	0.04	0.02
Facchinetti (Numerical)	0.09	0.06	0.03

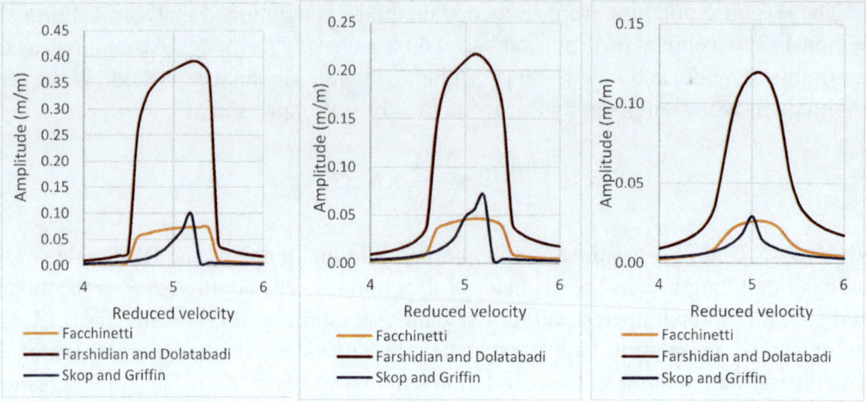

Fig. 2. Plot of amplitude of vibration obtained from numerical analysis considering (a) 0.5% damping; (b) 1% damping; (c) 2% damping

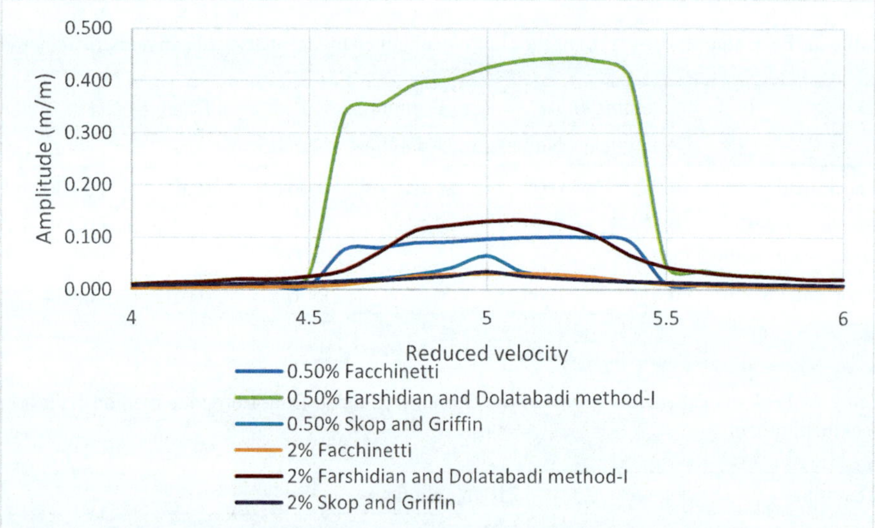

Fig. 3. Plot of amplitude of oscillation obtained from analytical model for 0.5% and 2% damping.

5 Conclusions

A comparative study where analytical model and numerical integration model are investigated for 210 m tapered chimney. It can be concluded that the responses predicted by the Facchinetti model, and the model proposed by Skop and Griffin are nearly the same, while the model proposed by Fashidian and Dolatabadi predicts a higher response. Skop and Griffin are the only model which showed the peak response at a reduced velocity greater than one; further, the peak region of the oscillation for all the models except Skop and Griffin displayed a plateau-like feature. It means Strouhal frequency gets locked to structural frequency for a broader range, which is more realistic. Therefore, Skop and Griffin's model is less reliable.

The effect of damping is also examined, and it can be concluded that vibration amplitude decreases for higher damping. Also, high nonlinearity appears at low damping, which can be proved through high differences between the numerical and analytical methods. The numerical method predicts a higher response than the analytical method as it incorporates the nonlinearity very efficiently. Although, for high damping, the difference between the methods diminishes as nonlinearity reduces at higher damping.

Eurocode method-I should be opted for the chimney with high damping as it reasonably predicts response.

References

1. Rahman, S., Jain, A.K., Bharti, S.D., Datta, T.K.: Comparison of international wind codes for across wind response of concrete chimneys. J. Wind Eng. Ind. Aerodyn. **207**(3), 104401 (2020)
2. Facchinetti, M.L., Langre, E., Biolley, F.: Coupling of structure and wake oscillators in vortex-induced vibrations. J. Fluids Struct. **19**(4), 123–140 (2004)
3. Farshidianfar, A., Dolatabadi, N.: Modified higher-order wake oscillator model for vortex-induced vibration of circular cylinders. Acta Mech. **224**(7), 1441–1456 (2013)
4. Skop, R.A., Griffin, O.M.: A model for the vortex-excited resonant response of bluff cylinders. J. Sound Vibr. **27**(2), 225–233 (1973)
5. Krenk, S., Nielsen, S.R.K.: Energy balanced double oscillator model for vortex-induced vibrations. J. Eng. Mech. **125**(3), 263–271 (1999)
6. de Langre, E.: Frequency lock-in is caused by coupled-mode flutter. J. Fluid Struct. **22**(6), 783–791 (2006)
7. Eurocode 1: Actions on structures - Part 1–4: General actions - Wind actions, 2nd edn. (2010)

The Reliability Analysis of Chimneys Due to Random Vortex Shedding

Saba Rahman[1](✉), Arvind K. Jain[2], S. D. Bharti[3], and T. K. Datta[4]

[1] Research Scholar, Indian Institute of Technology, Delhi, India
sabarahman807@gmail.com
[2] Indian Institute of Technology, Delhi, India
akjain@civil.iitd.ac.in
[3] Malaviya National Institute of Technology, Jaipur, India
sdbharti@mnit.in
[4] Malaviya National Institute of Technology, Jaipur, India

Abstract. The literature presents the applicability of reliability principles to wind-sensitive structures like a tall chimney. Many uncertainties involved in the unreliability of any structure subjected to wind-induced vibration are illustrated in the literature. These uncertainties include the randomness of incoming inflow and the precision of models used to predict wind loadings. A 210 m tall, tapered chimney is considered an illustrative example. The reliability analysis is carried out considering the Vickery and Basu spectral analysis for vortex shedding force. Spectral analysis is carried out in the frequency domain considering only the fundamental mode. It is concluded from the study that threshold crossing reliability of the concrete chimney is greatly affected by parameters like randomness or turbulence of the incoming low, structural damping and other critical parameters involved in the spectral analysis. Strouhal number appears to be not imparting any influence on the reliability of the chimney. The study aims to understand the chimney's behavior from a reliability point of view.

Keywords: Fragility curve · Reliability analysis · Vortex-induced vibration

1 Introduction

Chimneys are generally tall and slender structures with circular cross-sections, which are highly prone to wind forces. Wind exerts pressure on the chimneys' walls, producing unwanted forces. Vortex-induced oscillation is one such excitation that can lead to the failure of the chimneys. Therefore, vortex-induced oscillation of chimneys is of great concern to researchers and practitioners [1, 2] since many failures of chimneys due to vortex shedding have occurred in the past. Consequently, extensive research has taken place on the subject over decades. Many laboratory experiments have been performed to verify the theoretical models proposed to predict vortex-induced forces, including aeroelastic effects. Comparatively, very few proto-type measurements have been recorded to verify the proposed theoretical models. Because of this reason, the theoretical models

P. Schito and A. Zasso (Eds.): IN VENTO 2022, LNCE 461, pp. 280–287, 2024.
https://doi.org/10.1007/978-3-031-53059-3_25

developed with the help of experimental laboratory data are utilized for analyzing the chimneys for vortex-induced forces [3, 4]. This calls for reliability analysis of the predictions of the responses of the chimneys produced due to vortex shedding phenomena. Although several works [5, 6] of literature exist on the vortex-induced oscillation of chimneys, including code provisions, the reliability analysis of chimneys against failure caused due to vortex shedding is scanty.

In the present study, the reliability analysis of chimneys against vortex shedding failure is presented, assuming the uncertainty in vortex shedding phenomena to be more significant than other uncertainties, and hence, the latter is ignored. The vortex shedding is modeled as a stationary random process and is represented by a power spectral density function (PSDF). The PSDF of the tip displacement and vortex force of the chimney is obtained by performing a frequency domain spectral analysis using a matrix approach. For this purpose, both chimney and random wind forces are discretized over several points along with the chimney's height. The method of analysis duly accounts for the aeroelastic effects. The double barrier threshold crossing level, as proposed by Vanmarcke [7], is used for determining the probability of crossing different threshold levels of the tip displacement of the chimney. The reliability estimate is derived from the fragility curve. A 210m tall concrete chimney with a base diameter of 20m, top diameter of 12m and thickness of 0.3m has been taken as an illustrative example. From the study, it is clear that turbulence intensity highly governs the reliability analysis. Other parameters like structural damping and correlation length can also lead to structural failure. Strouhal number, on the other hand, does not affect the structure's response.

2 Theory

Understanding across wind oscillation is complex because the aeroelastic lift force that causes the chimney's transverse motion depends on the structure's motion. Although aeroelastic response can be obtained from a wind tunnel experiment, the full-scale experiment is difficult for the chimney of high Reynolds number ($10^{\wedge 7}$). Therefore, among various available models, Vickery & Basu model [8] is most preferred to obtain the response of the structure. Many parameters cause uncertainties in the response. These parameters will be discussed later in Sect. 4.

2.1 Vickery and Basu Model

Among various available models to predict the aeroelastic response of the structure, the Vickery & Basu model is the most reliable and majorly used by many international wind codes. The formula of power spectral density function (PSDF) of lift force specified by the Vickery & Basu model is represented as

$$S_f(\omega) = \frac{\left(\frac{1}{2}\rho V^2\right)^2}{\sqrt{\pi}\,\omega_s}\left[\frac{\left(D\sigma_f\right)^2}{b} exp\left\{-\left(\frac{1-\frac{\omega}{\omega_s}}{b}\right)^2\right\}\right] \tag{1}$$

where ρ is the fluid density, V is the critical velocity, D is the effective diameter of the tapered concrete chimney, σ_f is RMS lift coefficient, b is the width of the lift spectrum of the chimney, which depends on the turbulence intensity (I) by the following relation,

$$b = I + 0.1 \tag{2}$$

The above relation of the b and turbulence intensity is also used by the CICIND code [9]. The simplified form given by Vickery and Basu in Eq. (1) is for chimney with no or little taper $\left(\frac{d_t}{d_b} > 0.5\right)$. Since the chimney problems taken in this study have little taper, therefore the following formulation is valid, and chimney is considered as no-tapered chimney with effective diameter as average of top and bottom diameter at any section. On the other hand, the Euro code [10] provides the formulation for the concrete chimney's response corresponding to zero turbulence intensity.

2.2 Spectral Analysis

The concrete chimney is discretized, and lumped mass approach is used to analyze the tapered chimney and mass matrix (M) at node is obtained. The stiffness matrix (K) of the concrete chimney is generated from the flexibility matrix approach. The Rayleigh damping method is used to find out the damping matrix (C_s) and modal mass is obtained from lumped mass matrix (M) of the chimney, given by the following equation.

$$M_n = \phi_i.M.\phi_i^T \tag{3}$$

where, ϕ_i is the ith mode shape of the chimney. Mechanical admittance function (H_L) which is given in Eq. (5), is employed to generate the response variance of the chimney, given as

$$\sigma_y^2 = H_L.S_f(f_n).H_L^{*T}, y = k_p\sigma_y \tag{4}$$

$$H_L = \left[K - M\omega^2 + iC_s\omega\right]^{-1} \tag{5}$$

where ω_s is the Strouhal frequency and ξ is the total damping of the chimney. RMS of Eq. (4) responds to the structure, which, when multiplied by the peak factor (k_p), , gives the peak amplitude of oscillation (y). The peak factor is assumed to be four, as given by Vickery & Basu.

2.3 Reliability Analysis

The probability of finding excursion above a particular level can be understood in terms of up-crossing or threshold probabilities. The probability of failure can be framed as

$$P_f = \int_0^\infty 1 - L(T)dt \tag{6}$$

T is the duration of lift force acting on the concrete chimney and $L(T)$ is the probability of the response given in Eq. (10). The duration of lift force is kept upto 3600s. The result of the first excursion analysis is tough but incorporating a few simplified assumptions can be obtained. These assumptions are:

1) Structure recovers immediately after undergoing failure.
2) Structural failure constitutes the Poisson process, i.e., failure occurs independently.
3) The probability of failure within a time interval is the random time T distribution function.

Incorporating these assumptions, the first excursion study can be effortlessly obtained. The zeroth, first and second moment of the response is essential. These moments are specified as

$$\lambda_n = \int_0^\infty \omega^n S_f(\omega) d\omega \tag{7}$$

where, $S_f(\omega)$ is the PSDF of the lift force acting on the structure. With the help of Eq. (7), the following parameters can be obtained.

$$\gamma_0 = \frac{1}{2\pi}\sqrt{\frac{\lambda_2}{\lambda_2}} \tag{8}$$

$$q = \sqrt{1 - \frac{\lambda_1^2}{\lambda_0 \lambda_2}} \tag{9}$$

The mean rate of exceeding the threshold level can be given as.

$\gamma_{M0} = \gamma_0 e^{-\frac{\psi^2}{2}}; \psi = \frac{M_0}{\sigma_{M0}}$

The probability of a response under a specified threshold decays exponentially with time, which can be expressed as

$$L(T) = L_0 e^{-\alpha_D T} \tag{10}$$

where, L_0 is the probability of the starting of the threshold, which is given as

$$L_0 = 1 - e^{-\frac{\psi^2}{2}} \tag{11}$$

α_D is the double barrier decay rate, which is assumed as $\alpha_D = 2\gamma_{M0}$.

3 Parametric Study

To understand the behavior of tapered concrete chimney, 210 m tall chimney has been considered. The top and bottom diameters of the chimney are 10m and 20m, respectively. The spectral analysis is carried out for first mode vibration, and a comparative study is presented with various parameters. These parameters are turbulence intensity, Strouhal number, correlation length and structural damping. The Strouhal number is taken as 0.2 for all the calculations except for the comparative study where its value varies from 0.1 to 0.3.

4 Result and Discussion

As mentioned above, many parameters lead to unreliability in the structure. The most uncertain parameters are turbulence, structural damping, correlation length and lift coefficient. Turbulence in the incoming flow and lift coefficient add the most complexities to the analysis.

4.1 Turbulence Intensity

The section deals with the effect of turbulence intensity on the aeroelasticity of the concrete chimney. The effect of turbulence intensity comes indirectly in the form of width of lift spectrum, the relation between the two is given in Eq. (2). The influence of turbulence intensity in the structure's wake is highly unpredictable. As the turbulence intensity increases, the wake region of the chimney gets destroyed, reducing the structure's transverse motion. Turbulence intensity also decreases the correlation length of the structure. Reliability analysis of chimneys considering various turbulence intensity has been carried out and shown in Fig. 1. It is clear that the higher the turbulence intensity, lower the response of the structure; thus, the probability of failure of the structure is meager. Turbulence intensity taken for the study is according to the formulation of the Euro code. Euro code gave the following equation for finding out the turbulence, which depends on the terrain condition (z_0).

$$I_v = \frac{k_1}{c_0.\ln(\frac{z}{z_0})} \tag{12}$$

where z is the height of the structure since the turbulence intensity here is obtained at the tip of the chimney, k_1 is the turbulence factor whose value is taken as 1, c_0 is the orographic value, which is assumed to equal to 1.

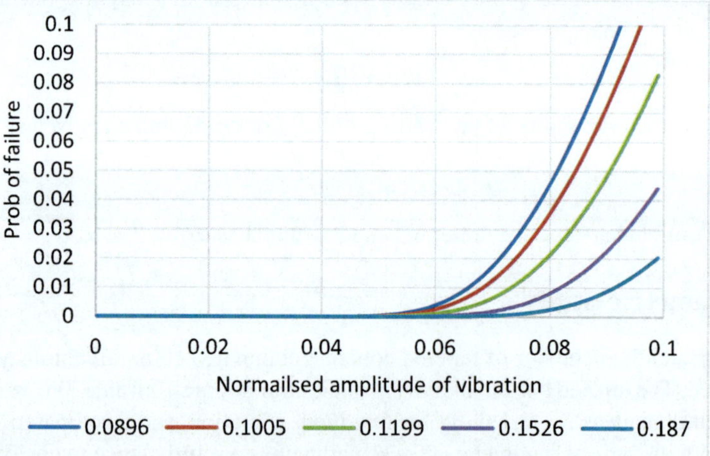

Fig. 1. Probability of failure vs. amplitude of oscillation considering different turbulence intensity

4.2 Structural Damping

As the structural damping increases, the energy dissipation capacity of the structure also increases, leading to more reliability in the structure. Figure 2 shows that with an increment in damping, the probability of failure decreases, and the structure becomes more reliable.

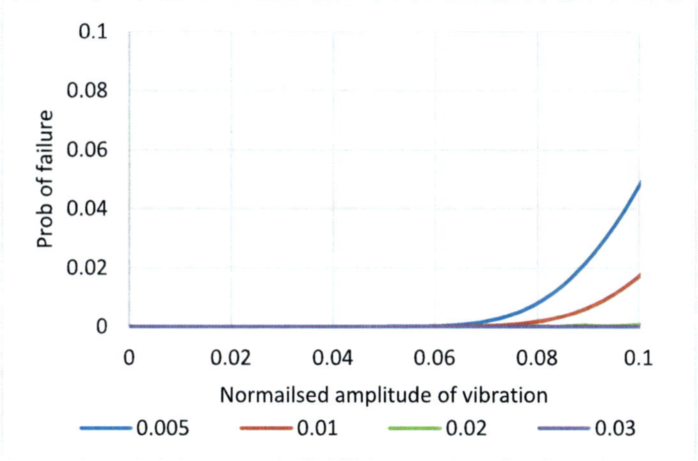

Fig. 2. Probability of failure vs. amplitude of oscillation considering different structural damping

4.3 Correlation Length

The lift force acting on a section of the structure is not perfectly correlated with the lift force at another section, but for simplification, it is assumed to be perfectly correlated for a limited section along the height of the structure. The length is known as correlation length, generally expressed in terms of the diameter of the structure. Vickery & Basu has assumed the correlation of one diameter. High correlation length implies that the eddies formation in the wake region of the chimney is correlated, leading to a narrow band vortex lift spectrum, thus higher response of the structure. Figure 3 shows that the higher the correlation length, the higher the probability of structural failure.

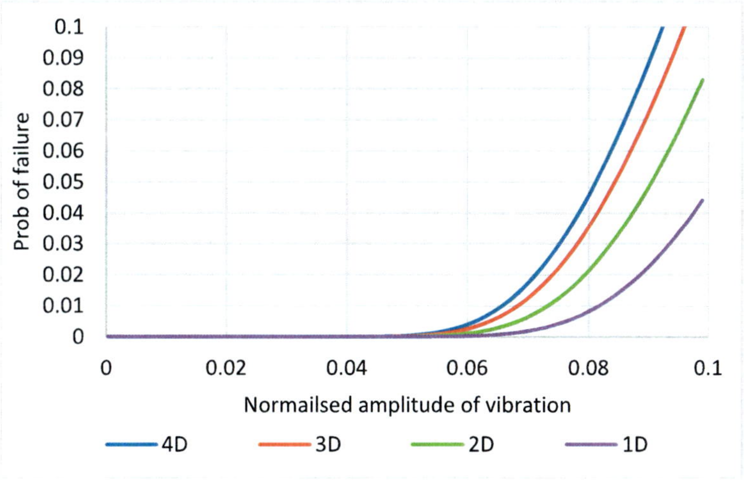

Fig. 3. Probability of failure vs. amplitude of oscillation considering different correlation length

5 Strouhal Number

High vibration amplitude is observed when vortex frequency gets locked into structural frequency. The vortex frequency depends on the Strouhal number, which ranges from 0.1 to 0.3 depending on Reynold's number. However, Fig. 4 shows that the Strouhal number does not much contribute to the oscillation of the chimney. Therefore, the fragility curve does not show much variation. This so because, for a uniform wind velocity distribution, Strouhal number can change the critical velocity at which Strouhal law becomes valid but the response of structure remains same.

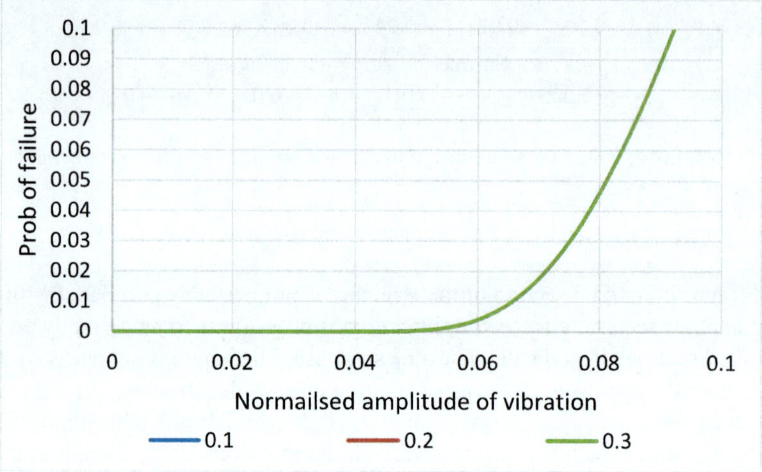

Fig. 4. Probability of failure vs. amplitude of oscillation considering different Strouhal number

6 Conclusions

The study aims to understand the influence of parameters affecting the vortex shedding phenomenon with the fragility curve. The results of the study show that the assumed structural damping significantly influences the threshold crossing reliability of the peak displacement of the concrete chimney, turbulence intensity, correlation length, and lift coefficient. It can also be observed from the paper that the influence of the Strouhal number is negligible on the fragility curve. The influence of turbulence intensity is also examined for the high-rise concrete chimney. The turbulence intensity, correlation length and structural damping play vital parameters in the development of the spectral wind force. Further, the aeroelastic effect influences the reliability estimate to a great extent for small structural damping.

References

1. Anusha, S., Sowjanya, G.V.: Madhusudhan: Reliability Analysis of Circular RC Chimney. Inter. J. Innovative Res. Sci. Eng. Technol. **7**(5), 12274–12280 (2016)

2. Menon, D., Rao, P.S.: Uncertainties in codal recommendations for across-wind load analysis of R/C chimneys. J. Wind Eng. Ind. Aerodyn. **72**, 455–468 (1997)
3. Madsen, O.H.: First Order vs. Second Order Reliability Analysis of Series Structures. Structural Safety **2**(8), 207–214 (1985)
4. Kareem, A., Hseih, J.: Reliability analysis of concrete chimneys under wind loading. J. Wind Eng. Ind. Aerodyn. **25**(8), 93–112 (1986)
5. Kareem, A.: Wind effects on structures: a probabilistic viewpoint. Probabilistic Eng. Mech. **2**(4), 166–200 (1988)
6. Milford, R.V.: Structural reliability and crosswind response of tall chimneys. Eng. Struct. **4**(8), 263–270 (1982)
7. Solari, G.: Wind-excited response of structures with uncertain parameters. Probabilistic Eng. Mech. **12**(2), 75–87 (1997)
8. Vickery, B.J., Basu, R.: Simplified approaches to the evaluation of the across-wind response of Chimneys. J. Wind Eng. Ind. Aerodyn. **14**(3), 153–166 (1983)
9. Modal Code for Concrete Chimney (CICIND). 3rd edn. Germany (2011)
10. Eurocode 1: Actions on structures - Part 1–4: General actions - Wind actions. 2nd edn. (2010)

Check for updates

A Bivariate Statistical Model of Large Pressure Peaks Measured in Wind Tunnel Tests

Vincent Denoël[1], Thibaud Bastin[1], Giulia Pomaranzi[2](✉), and Alberto Zasso[2]

[1] Structural and Stochastic Dynamics, University of Liège, Liege, Belgium
v.denoel@uliege.be

[2] Department of Mechanical Engineering, Politecnico Di Milano, Milan, Italy
giulia.pomaranzi@polimi.it

Abstract. A statistical approach for the treatment of large negative peak pressures is discussed. It is based on spatial correlation so that these large peak values, which are difficult to handle in cladding design, could be smoothed out by invoking the characteristic size associated with these events. The proposed method hinges on a mixture model of bicubic translation processes. Development of the method is summarized and applied to the computation of the area-average pressure on the building façade of a model tested in wind tunnel facilities at Politecnico di Milano.

Keywords: building façades · bicubic translation model · Bayesian regression · mixture model · size effect

1 Introduction

Several experimental evidence have shown that very high negative pressures could be measured on the facades of a building in the wind tunnel [1, 2]. They are being more frequently observed because of the widespread tendency to use higher frequency sampling and number of pressure taps in wind tunnel testing. The management of these major suction peaks remains an important issue today. Indeed, if these were to be treated in complete transparency, experimentalists would not be able to provide reasonable pressure peak values for the dimensioning of the structures. The main reason is that the time-space equivalence used today to smooth out the peak pressures does not hold at these small scales.

The same experimental evidence also show that the spatial extent of the peak pressure is relatively limited. This effectively justifies that building façades do not need to be designed accounting these extreme values since integration of pressure over a finite (non-zero) area would result in a smaller average pressure. A very high density of pressure taps is required to establish the size effect related to the spatial extent of this phenomenon. The data processing techniques traditionally used to limit the impact of these large negative pressure peaks consists of averaging in time rather than in space. The standard approach is to use the well-known TVL method [3]. This approach does not offer a uniformly valid solution in areas with large negative peaks, which is expected

P. Schito and A. Zasso (Eds.): IN VENTO 2022, LNCE 461, pp. 288–297, 2024.
https://doi.org/10.1007/978-3-031-53059-3_26

given that its origins are based on a stationary (in time) and homogeneous (in space) pressure field.

The origin of large negative peak pressures on the facades of the building corresponds to physical phenomena different from those generally observed in the shear layer. They presumably correspond to small hairpin vortices while a typical shear layer is composed of larger vortices [4]. Therefore, it is possible to model the measured pressure as a mixture of two components, one related to the small-scale phenomenon, and one related to the large scale [5].

The separation of the measured pressure into two components then becomes an important objective. While it is virtually impossible to attribute the pressure measured at a given time to any one of the two components, the de-mixing of the two processes can be achieved in a statistical sense. There exist several methods to de-mix mixtures of random variables (e.g. [6]) and also random processes [7], possibly by taking advantage of the different timescales in the two phenomena [8]. With these methods, it is possible to set up a methodology for decomposing a measured pressure signal into two components.

But the question of the spatial extent of these two components has not been addressed yet. This question is addressed in this paper by considering the mixture of different physics in joint distributions of pressure at several neighboring taps. Ultimately, this will answer the question of the spatial extent of each component and help derive appropriate techniques to determine peak value distribution from wind pressure data, after averaging on finite size (non-zero) areas.

The illustration is based on very high-fidelity experimental data, measuring pressure on the facades of a building with a very high density but the proposed methodology is applicable to scarcer tap density. The article is organized as follows: Sect. 2 presents the mathematical model which is fitted to the given data. Section 3 briefly describes the experimental setup. Section 4 presents the results of the study. Section 5 provides a discussion and conclusions of this preliminary research.

2 Methods: The Statistical Model

The proposed statistical model hinges on the cubic translation [9, 10] which expresses the pressure p_i each pressure tap as a cubic transformation of a Gaussian variable u

$$p_i(u) = h(u, \pi_i) = \frac{\alpha}{b}\left[\frac{u^3}{3} + au^2 + (b-1)u - \mathrm{a}\right] + \mu \tag{1}$$

and where the coefficients $\pi_i = (\mu, \sigma, a, b)$ need to be adjusted to the data. It is possible to express the Probability Density Function (PDF) of the transformed variable $f_i(p_i; \pi_i$ as a function of these coefficients [11]. For brevity, its expression is omitted here. So, a suitable way to adjust them is to find the optimal values that minimize the (e.g. least-square) error between the PDF of the experimental data, obtained from a scaled histogram, and the PDF predicted by the model [8].

As introduced earlier, the cubic translation model is not suitable to model the pressure at taps with large negative peaks, where several physics are captured. Instead, a mixture model is used. It is chosen to mix two such cubic translations, such that the PDF at critical pressure taps is modeled as:

$$f_{mix,i}(pi;\,\theta_i) = w_i f_i(pi;\,\pi_i^A) + (1-w_i) f_i(pi;\,\pi_i^B) \tag{2}$$

Where π_i^A and π_i^B represent two sets of 4 parameters, one for each component of the model, and where w_i represents the mixing ratio. This formulation is sufficiently general and also covers the case where either component is Gaussian, since the cubic translation model is adjustable to Gaussian data, when the transformation is linear. In total, the unilateral PDF at each pressure tap is represented by a set of 9 parameters collected in θ_i, corresponding to twice four parameters plus the mixing ratio.

In this paper, we exploit an extension of the cubic translation model, that had been coined *bicubic translation model* in [12] as it applies to a joint PDF between two different processes which are represented by cubic translation models each. Specifically, the considered joint PDF model $f_{ij}\left(p_i,p_j \prod_{ij}\right)$ of the pressures measured at taps i and j is expressed as a function of the 4 parameters of each pressure, plus their correlation coefficient. The nine parameters, again, in total are gathered in vector $\prod_{ij}$ to simplify notations. For brevity, the complete expression of f_{ij} is also omitted.

Finally, the model is further extended in order to include two components, in the sense of a mixture model. As such, the joint PDF of the pressures recorded at two neighboring pressure taps is represented by:

$$f_{mix,ij}(p_i,p_j;\,\Theta_{ij}) = w_{ij} f_{ij}\left(p_i,p_j \prod_{ij}^{A}\right) + (1-w_i) f_{ij}\left(p_i,p_j \prod_{ij}^{B}\right) \tag{3}$$

Where $\Theta_{ij} = \left(\prod_{ij}^A, \prod_{ij}^B, w_{ij}\right)$ gathers a set of twice nine, plus one, i.e. the 19 parameters of the bicubic model. In summary, a bicubic translation model is composed of 19 parameters; it generalizes the concept of cubic translation model, while allowing to account for correlation of two non-Gaussian processes, at least to some extent. It falls in the family of parametric models to estimate the statistical properties of experimental data.

3 Experimental Setup

Figure 1 depicts some details of the test setup that has been used to illustrate the findings of this research. It consists of a regular building with size 2000 × 1000 × 200 mm (H × W × b) at wind tunnel scale, equipped with a set of 254 pressure sensors which are distributed with variable density over the specimen. Specifically, the area located in an upper corner of the building (see Fig. 1-a, in blue) is equipped with a very high density of pressure taps. The smallest distance between two neighboring taps is as small as 3.4 mm. The closeup view in Fig. 1-c allow appreciating the high density of the mesh.

The wind pressures have been measured on this building for various wind exposures at a wind tunnel sampling frequency of 500 Hz, and for a total duration of 5 min of measurement (N = 150,000 sample points per channel). The reference pressure used to determine pressure coefficients is the dynamic pressure at the top of the building. In total, a large number of configurations have been tested, with various wind exposures,

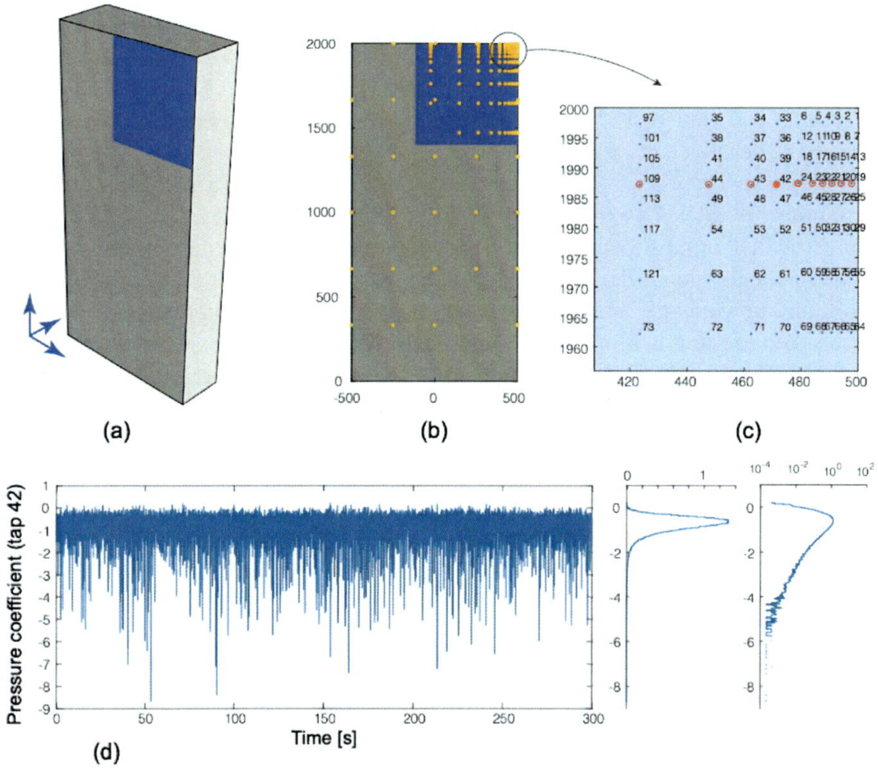

Fig. 1. (a) Sketch of the experimental setup, (b) Positions of the pressure taps (units are in mm of the model), (c) Numbering of pressure taps (d) example of pressure coefficient (measured at tap 42) and its histograms.

two different terrain categories, higher tap density on three different tiles (A, B, C), and with or without additional buildings in the near vicinity of the tested building. The full data set is available from an open repository [13]. From this very large data set, only one configuration is considered in the following illustration. It concerns configuration A02B00, which stands for (i) tile A (higher density of pressure taps in the upper corner of the specimen, as shown in Fig. 1-a), (ii) terrain category II, (iii) test without secondary building (B00). Also, the considered exposure is equal to 10°, which corresponds to a wind attacking the specimen from the back with an incidence of 10° with respect to the large face of the building.

4 Statistical Analysis of Large Negative Pressures

Figure 1-d shows the time series of the pressure coefficient measured at pressure tap 42. The corresponding PDF is also shown on the right in both linear and logarithmic scales. The average pressure coefficient is −0.79 and its standard deviation is 0.45. However, values as large as −8.68 have been registered by the scanner. These large negative values are not isolated events since the experimental PDF regularly drops at high negative pressure coefficients. The non-Gaussian nature of the pressure is visible in the log plot of the PDF, which does not decrease as a quadratic (as it would be the case if the process was Gaussian), but less fast, which is responsible for the small occurrences of the large negative pressures.

The pressure coefficient at pressure tap 42 can be modelled as a mixture of two cubic translation processes. Its experimental PDF (i.e. the scaled histogram) is adjusted to the parametric PDF of the mixture model. This operation is advantageously led by means of a Bayesian regression technique implementing the famous Monte Carlo Markov Chain algorithm known as the Metropolis-Hastings algorithm [14]. Alternative approaches, such as the Expectation Maximization algorithm [6] have also been used, but they have led to poorer estimates [15]. The results of the de-mixing are represented in Fig. 2 for pressure tap 42. This figure shows the posterior distribution of the model parameters π_i^A and π_i^B as well as the PDF resulting from this 9-parameter univariate model. The agreement with the experimental data is almost perfect in both the linear and log scales.

The posterior distribution is obtained as the likelihood of Bayesian regression since uninformative prior was considered. A set of 7,500 jumps of the Metropolis-Hastings algorithm has been used with a burn-in size of 2,500 samples. The parameters of Component A (with the smallest average pressure in absolute value) are globally identified with a better confidence.

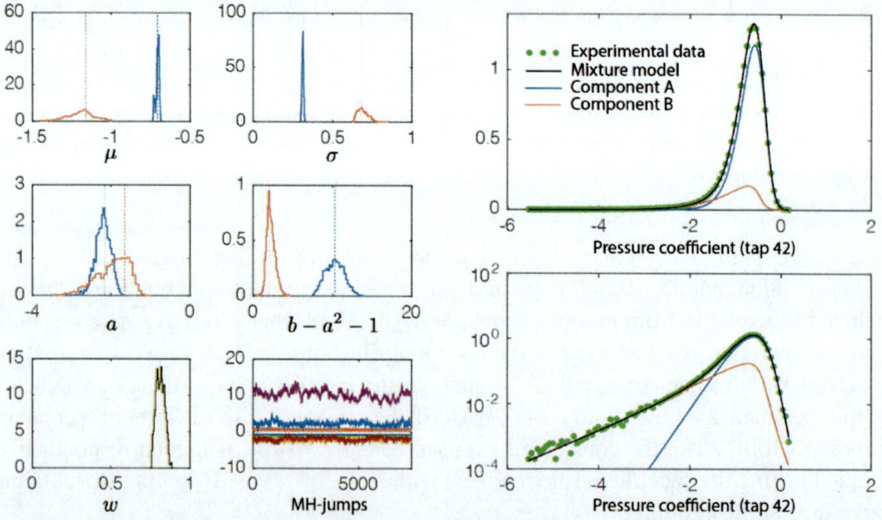

Fig. 2. Result of the de-mixing of the pressure coefficient measured at pressure tap 42. (Left) Posterior distributions of the $(4 \times 2 + 1 =)$ 9 model parameters and samples of the Metropolis-Hastings algorithm, (Right) Experimental and identified PDF of the pressure coefficient at tap 42.

The numerical values (median) of the 9 identified model parameters are summarized in Table 1 for pressure tap 42 as well as for some other pressure taps located along the same horizontal line, see Fig. 1 for tap numbering. These values show that the identified

Table 1. Median estimates of model parameters for the pressure coefficients measured at several taps.

Tap	μ_A	σ_A	a_A	Δb_A	μ_B	σ_B	a_B	Δb_B	w
Tap 109	−0.707	0.300	−3.80	18.4	−1.33	0.535	−7.34	410	0.97
Tap 44	−0.661	0.307	−2.64	13.2	−1.38	0.525	−2.71	4.50	0.95
Tap 43	−0.736	0.295	−3.40	17.7	−1.59	0.643	−0.54	2.18	0.92
Tap 42	**−0.703**	**0.311**	**−2.18**	**10.3**	**−1.18**	**0.688**	**−1.87**	**2.22**	**0.81**
Tap 24	−0.643	0.310	−2.03	9.15	−1.19	0.774	−3.15	2.60	0.84
Tap 23	−0.648	0.276	−2.09	11.6	−1.13	0.641	−4.72	3.87	0.75
Tap 22	−0.627	0.254	−2.38	14.6	−1.10	0.515	−6.26	8.90	0.72
Tap 21	−0.617	0.257	−2.11	11.6	−1.14	0.407	−8.41	32.2	0.78
Tap 20	−0.614	0.256	−2.03	11.1	−1.16	0.390	−13.2	132	0.78
Tap 19	−0.610	0.229	−2.18	14.2	−1.17	0.391	−14.0	97.1	0.71

values of the mixture model are more or less smooth in space, although they have been independently identified. In particular, the average pressures μ_A and μ_B in both components and the mixing ratio w all evolve very regularly along the considered line of pressure taps. The mixing ratio increases as one moves from right to left, i.e. going away from the edge of the building. This indicates that the isolated events corresponding to the tail of the distribution (the large negative peaks) vanish in a regular manner.

Figure 3 illustrates the joint PDF between the pressure coefficients at tap 42 and 24, which are two neighboring taps. From the univariate analysis it is known that each of them can be decomposed into two components, A and B. The same development is also possible with the joint PDF: the mixture model is a weighted sum of two bicubic translation processes, and the 19 parameters of this model are adjusted in order to match the PDF of experimental data. Distributions on the right of Fig. 3 are related to these parameters.

Interestingly, the mixing ratio w is again around 0.8, which corresponds to the value obtained with the unilateral analysis. Moreover, the correlation coefficients associated with the two components, A and B are respectively equal to 0.95 and 0.82, which indicates that the former phenomenon is more correlated over space that the latter, i.e. the typical turbulence of the shear layer is more correlated than the large negative peaks.

The same analysis is reported in Fig. 4, for the correlation between tap 42 and tap 19, which are still on the same line but a little more away from each other. Both correlations drop and, in particular, the correlation associated with the large negative peaks is now around 0.4.

Ultimately, the correlation coefficients between pressure tap 42 (used as a reference in this study), and the sequence of neighbors, i.e. taps 24 to 19 is recapitulated in Fig. 5. On the left, this figure represents the distributions of the correlation coefficients between the pressure at tap 42 and various other taps (24 to 19). The thin lines refer to component A (bulk) while the thicker lines refer to component B (tail). The maximum likelihood

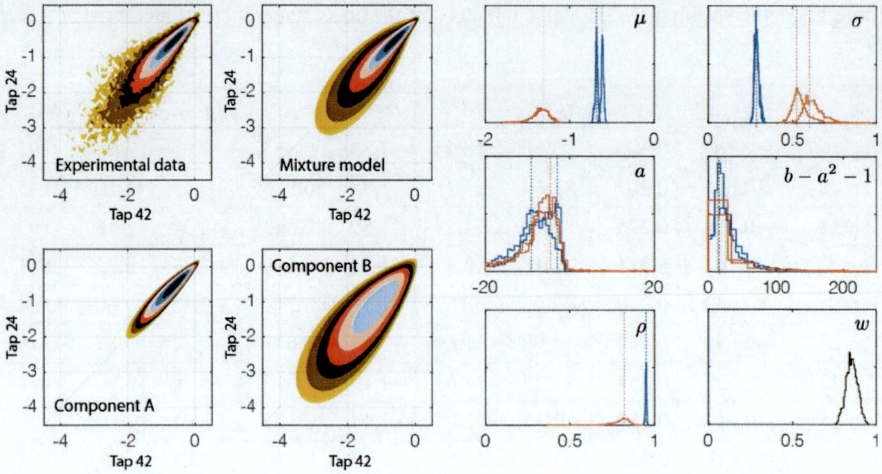

Fig. 3. Results of the joint de-mixing of the pressure coefficients measured at pressure tap 42 and 24. (Right) Posterior distributions of the 19 model parameters (Left) Experimental and identified joint PDF.

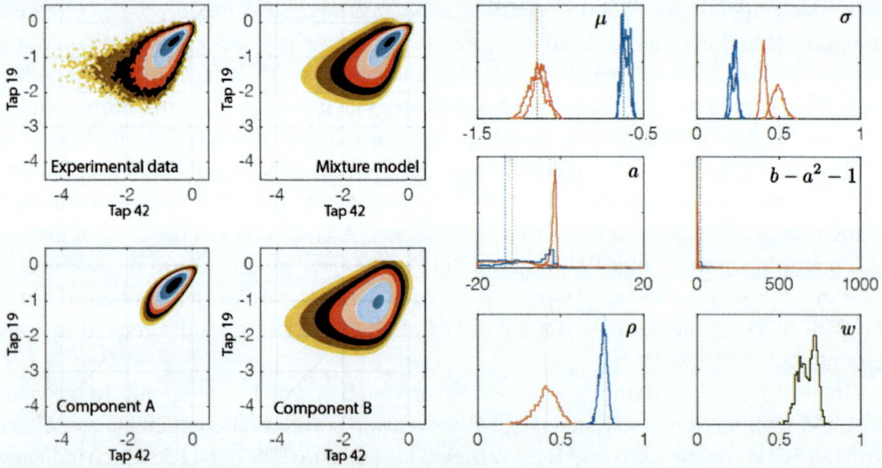

Fig. 4. Results of the joint de-mixing of the pressure coefficients measured at pressure tap 42 and 19. (Right) Posterior distributions of the 19 model parameters (Left) Experimental and identified joint PDF.

estimation (mode of each these distributions) is represented on the right, for both components A and B, and as a function of the relative distance between sensor 42 and its neighbors. The regular drop of the correlation is again visible especially the correlation coefficient associated with component B which corresponds to the large negative peaks. This indicates the small size of these phenomena in space. The intermediate dashed line represents the correlation coefficient of the full data set, considered as a whole, i.e. before de-mixing is performed.

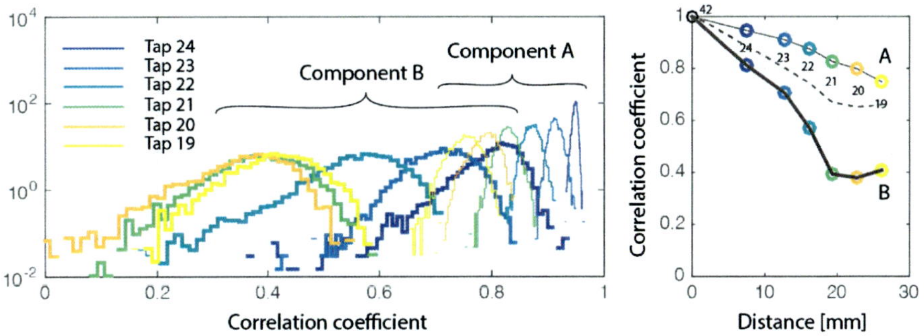

Fig. 5. Correlation coefficients of components A and B between pressures at tap 42 and various other taps (24 to 19). The left plot shows the posterior distributions of the identified correlation coefficients; the right plot shows the maximum likelihood estimators as a function of the distance between the two sensors.

5 Application of the Statistical Model to Area Averaging of Pressures

The statistical framework presented in the previous section is a good tool to analyze, understand and predict the design pressure on façade elements. In particular, it is possible to provide a proxy of the area-averaged pressure on a surface with finite size. To illustrate this point, we have computed the extreme value of the average pressure along lines of continuously growing size. Specifically, the time series of the average pressure between taps 42 and 24, then 42, 24 and 23, ..., until 42, 24, 23, ...19 are established. The average of a type-I extreme value distribution of the minimum pressure coefficients on 10 windows of equal sizes is used as an experimental estimator of the extreme pressure along a finite length. This reference solution, available here because of the large tap density, is compared to a proxy which is based on the mixture model, see Fig. 6.

More precisely, the extreme pressure averaged on the surface is obtained as the product of (i) the experimental estimator of the extreme pressure at a single tap (tap 42), (ii) a decreasing function of the total length of averaging, starting from 1 and monotonically decreasing. We notice that (i) is always available since it corresponds to usual statistical treatment of pressure taps, while (ii) is here constructed as the average of the squared correlation coefficient associated with component B, see Fig. 5. This choice of function is justified by the fact that the correlation in space of component B (the one related to large negative peaks) drives the size effect related to area averaging.

It is also emphasized that the use of a smooth estimate of the correlation vs. distance relationship (like a decreasing exponential) instead of a high-density information would give a simple means to predict extreme values of area-averaged pressures. Additional work is required on this question, in order to assess the strictly required mesh density, in order to be able to properly capture the size effect. One possibility would be to use a scarcer tap mesh than in this study (hopefully similar to standard distances), and to adjust of model for the correlation coefficient of the tail component (B) with the possibly low values.

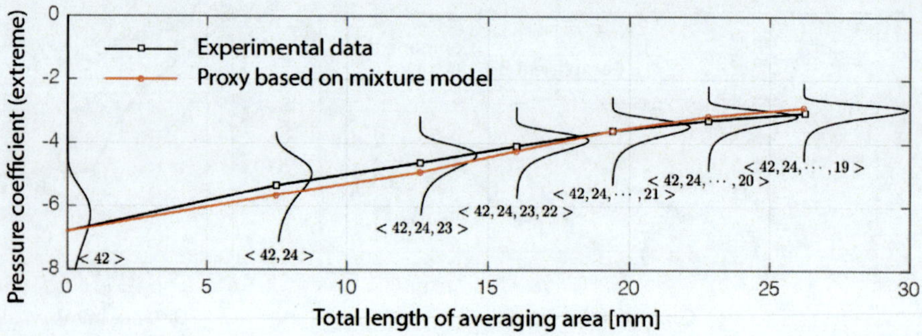

Fig. 6. Extreme pressure coefficient: comparison of experimental data and prediction of the model, based on the correlation coefficient of component B.

6 Discussion and Conclusions

We have presented a statistical analysis of large negative peak pressures. For the first time, wind tunnel pressures are analyzed by means of joint bicubic distributions with the aims to focus on large extremes, decompose the measured processes in tail and bulk components and quantify the spatial correlation of each of them. While a mixture of the usual cubic translation process is used for the univariate case, the bicubic extension is used for the 2-by-2 joint statistics. There is a slight inconsistency in the model since the marginal distributions obtained from the joint PDFs do not exactly correspond to the univariate cases. The simplifying choice to allow this slight inconsistency was made to lighten the considered model and avoid mixing more than 2 components.

The 2-by-2 correlation is the necessary and sufficient information to quantify the statistics of the average of the pressure at 2 taps. As soon as more taps are averaged, higher correlation would, in principle, be necessary. However, it was found that the length scale extracted from the correlation in the tail distribution does a decent job in estimating the size effect. Therefore, it appears that the use of more complex models, e.g. non-Gaussian copulas and mixtures thereof, is not fully justified.

Despite providing decisive information about the characteristic scales in the problem, the model can be used to predict area-averaged pressure coefficients that do not require such a dense tap meshing. The current results are very encouraging, and the quality is explained by the fact that nowhere in the proposed formulation we need to formulate homogeneity in space, which is unfortunately required in classical methods to estimate the reduction of pressure on finite size non-zero areas.

Future works will aim at developing the same analysis to surface-average, possibly apply the same technique to structures with corner vortices [12].

References

1. Lamberti, G., Amerio, L., Pomaranzi, G., Zasso, A., Gorlé, C.: Comparison of high resolution pressure measurements on a high-rise building in a closed and open-section wind tunnel. J. Wind Eng. Ind. Aerodyn.Aerodyn. **204**, 104247 (2020)

2. Pomaranzi, G., Amerio, L., Schito, P., Lamberti, G., Gorlé, C., Zasso, A.: Wind tunnel pressure data analysis for peak cladding load estimation on a high-rise building. J. Wind Eng. Ind. Aerodyn.Aerodyn. **220**, 104855 (2022)
3. Lawson, T.: The design of cladding. Build. Environ. **11**(1), 37–38 (1976)
4. Simiu, E., Scanlan, R.H.: Wind effects on structures. John Wiley and Sons, New York (1996)
5. Rigo, F., Andrianne, T., Denoël, V.: A de-mixing approach for the management of large negative peaks in wind tunnel data. J. Wind Eng. Ind. Aerodyn. **206**, 104279 (2020)
6. Moon, T.K.: The expectation-maximization algorithm. IEEE Signal Process. Mag. **13**(6), 47–60 (1996)
7. Comon, P., Jutten, C. (ed.). Handbook of Blind Source Separation: Independent component analysis and applications. Academic press (2010)
8. Rigo, F., Andrianne, T., Denoël, V.: Mixture model in high-order statistics for peak factor estimation on low-rise building. In: Conference of the Italian Association for Wind Engineering, pp. 613–629. Springer (2018). https://doi.org/10.1007/978-3-030-12815-9_47
9. Grigoriu, M.: Crossings of non-Gaussian translation processes. J. Eng. Mech. ASCE **110**(4), 610–620 (1984)
10. Winterstein, S.R.: Moment-based Hermite models of random vibration, Lyngby, Report No. 219. Department of Structural Engineering, Technical University of Denmark, Lyngby, Denmark (1987)
11. Rigo, F., Andrianne, T., Denoël, V.: On the use of the cubic translation to model bimodal wind pressures. Mathematical Model. Civil Eng. **15**(2), 20–32 (2019). https://doi.org/10.2478/mmce-2019-0006
12. Blaise, N., Andrianne, T., Denoël, V.: Assessment of extreme value overestimations with equivalent static wind loads. J. Wind Eng. Ind. Aerodyn.Aerodyn. **168**, 123–133 (2017)
13. Pomaranz,i G., Amerio, L., Zasso, A.: Politecnico di Milano - Wind tunnel test data on high-rise building, High-resolution pressure data recorded in wind tunnel tests performed at the Politecnico di Milano wind tunnel on a generic prismatic high-rise building (June 2020). https://zenodo.org/record/3948348#.YO0_nC0ivfY
14. Hastings, W.K.: Monte carlo sampling methods using markov chains and their applications. Biometrika **57**(1), 97–109 (1970)
15. Bastin T.: Development and application of a separation algorithm for wind pressures on building façade panels, Master thesis, Univertisy of Liège (2021)

Drag Coefficient Limitations as Reference Parameter to Estimate Tangential Wind Load Affecting VGS

G. Barbotti[1](✉) and C. A. Castiglioni[2]

[1] DABC, Architecture, Built Environment and Construction Engineering, Politecnico di Milano, Via Giuseppe Ponzio 31, 20133 Milano, Lombardy, Italy
giovanni.barbotti@polimi.it

[2] Fincon Consulting Italia Srl, Via Volturno 46, 20124 Milano, Italy
c.castiglioni@finconitalia.it

Abstract. Vertical green systems (VGS) are emerging as an alternative to curtain wall for the design of building envelopes for retrofitting or new projects, since they are implemented to reduce and mitigate the Urban Heat Islands phenomenon and enhance biodiversity. The research opens with a reflection that raises questions and looks forward: if vegetation will assume a fundamental role as response of urban space threats, it will be necessary to manage all related aspects, including its design and consequently structural verification to comply with in the building field. However, currently no design guideline supporting designers addressing the structural response under wind actions is available. A more accurate estimation of tangential wind action affecting VGS might be considered at least as a baseline guide for designers, but a severe lack of studies about drag coefficient is highlighted. The author demonstrates after a literature review analysis, the limits if this aerodynamic coefficient would be implemented as design parameter and the gap of knowledge in the research existing.

Keywords: drag coefficient · façade engineering · tangential action · VGS · wind motion

1 Introduction

Designing envelope of buildings is currently accounted by façade engineering field. The requirements to comply with are linked also to the safety and durability of the system, although façade and curtain wall more generally are not defined as a structural component according with UNI EN 13830:2015 "Curtain Wall. Guidelines product" since they are expected to transfer loads to the main structure (considering the EN1991-1-1:2002. Eurocode 1: Actions on structures – General actions) even if extreme loading configurations, like natural hazard events, are not properly accounted. Recently (2014) in Italy great attention has been grabbed by the challenges posed by the trees and shrubs on the hanged balconies in Milano (Bosco Verticale) verified for the mechanical resistance and safety in wind tunnel (respectively first analysis in Milano Politecnico and then

P. Schito and A. Zasso (Eds.): IN VENTO 2022, LNCE 461, pp. 298–309, 2024.
https://doi.org/10.1007/978-3-031-53059-3_27

in Miami). The vulnerability of these projects is linked with the unknown question of how tall buildings can tame strong winds, additionally if they are exposing innovative solutions such as trees or plants. Without deepening the typologies of vertical green systems (VGS) for instance green façade (GF) or living wall (Ogut et al., 2022), the core of the discussion is: could drag coefficient be a safe parameter being implemented for tangential action calculation? Furthermore, is the drag enough to fulfill calculations for both isolated plant and canopy? (Kerzenmacher and Gardiner, 1998; Shaw et al., 1982; Finnigan and Belcher, 2004)

Fig. 1. Picture taken by the author, in Leonardo Campus, Politecnico di Milano. The vertical green wall is the "Zero Gravity Eden" designed by ItalMesh company (BS). This typology of VGS is particularly at elevated level of complexity since the vegetation is combined with an expanded metal mesh.

1.1 Wind Loading on Plants

The whole contribution is aimed to introduce and highlight the limit of application if drag coefficient is implemented as parameter to quantify the tangential action. Currently, there is a lack of reliability to simulate and investigate the response and in general the effects which can occur if plants are installed on façade or roof since there are not worldwide official rules for designers, therefore the author proposal is to introduce the drag effects operating a simplification of the system affected by a series of equivalent static forces, and checking the effect compared to a case where the greenery is not present at all such as the curtain wall, characterized by glazed panel as surface mainly exposed to lateral wind action.

Firstly, let assume the previously mentioned wind-structure interaction as the consequence because of combining the response of a homogenous porous layer (wired mesh) and a second porous layer constituted by branches and leaves (the whole crown) and these last one subjected to a wide spectrum of natural species. Since it is not aim of the work to classify and identify each sample of botanical species, there will be some general considerations on the potentially common aspects of them accordingly to the purpose of providing a general base method to introduce in the structural calculations the effect induced due to the vegetated layer interacting with the sub-system of VGS. Another strong assumption here made by the authors accordingly with the studies carried out (Tadrist et al., 2018): foliage motion is observed to be prevalently influenced by individual leaf flutter at low wind velocities, on the other side the branch turbulence buffeting takes place at higher wind speed. Lastly, there is the simplification linked to the flexibility attitude observed by the leaf namely reconfiguration (Vogel S., 2000): if the single branch or trunk can still be assumed such a rigid stem subjected to a bending due to the load exerted by the wind action, this is not extendable or reasonable for the whole crown or in the greenery systems currently installed worldwide, often characterized by flowers, which are the ones more preferable also to satisfy aesthetical requirements.

2 Materials and Methods

The section here is intended to highlight the difficulty in identifying determined values for drag, given the vast number of plants that are used in the construction industry. The section is organized according to the two main quantities: the drag coefficient, which is briefly presented with one of the most comprehensive studies and accompanied by values from the literature. On the other hand, the tangential load becomes the basic calculation to graphically represent the particularly important variability that does not allow an immediate association of drag for each type of plant sample. Then, based on a previous work that had the clustering of a dataset, the division into ranks is considered, which determines threshold values, for the purposes of calculation to compare the plant case with a curtain wall.

2.1 Literature Review of Wind-Vegetation Interaction Studies

Even if the authors highlight a gap of research in the subject here below introduced, a mention goes to the work performed by several authors (Fischenich, and Dudley, 1999; Schindler, 2008; Västilä et al., 2013) and based on the contribution coming from previous research (Cionco, 1965; H. Lettau, 1969; Mayer, 1987; Rahmeyer et al., 1995): an interesting aspect highlighted by the papers is the role played by geometry of the plants in terms of branches, leaf ad trunk, together with the Reynold's number (Re) accounting different drag (density, shape) even comparing different plants species (Cao et al., 2012; Gillies et al., 2002).

It appears essential to isolate the terms of drag coefficient (C_D) which has not to be confused with the "dynamic coefficient" as referred in the CNR-DT207-2018. It is mathematically a dimensionless number which ranks the resistance of an object into a fluid environment, for instance H_2O or atmospheric air. A frequent findable form is:

$$C_D = \frac{2F_d}{\rho u^2 A} \tag{1}$$

$2F_d$ = drag force

ρ = mass density of the fluid
u = flow [1] speed of the object relative to the fluid
A= is the reference area

2.2 Tangential Action in the Standard

The previous findings and values have been implemented for the next calculation, which regards the lateral load referred to our exposed surface. Accordingly with the Italian Guideline, which does not include any reference coefficient in presence of VGS (Table 1) the tangential action (w_f) can be calculated as prescribed by the CNR standard so:

$$w_f(z) = q_p(z) \cdot c_f \left[\frac{N}{m^2} \right] \tag{2}$$

While the same standard proposes for the drag (c_D, here expressed as c_f) the following definition:

$$c_f = \frac{w_f}{\frac{1}{2} \cdot \rho \cdot V^2} \left[\frac{N}{m^2} \right] \tag{3}$$

- q_p (z) is the wind kinetic peak pressure[2] and defined positive if aligned with the wind flow direction.
- c_f is the drag coefficient, function of the surface roughness (Eq. 3).
- z is the reference height associated with the definition of c_f.
- V is the wind velocity, in this case the mean value ($V = v_m$).

Table 1. Taken by CNR-DT207–2008 and modified by the author, page. 137, paragraph G.12

Surface	*cf*
Steel, smooth concrete	0.01
Rough concrete, tar surfaces	0.02
Waved surfaces	0.04
VGS	Not available for designers

[1]Consider that the mean wind velocity might be for high urban density area, not a significant value
[2]See paragraph 3.2.7, pag.52, CNR-DT 207 R1/2018 ROMA – CNR 06 febbraio 2019 "Istruzioni per la valutazione delle azioni e degli effetti del vento sulle costruzioni".

2.3 Drag Coefficient Classification

As displayed in the previous section (Eq. 1), the reference area plays a role in the definition of the drag coefficient which has not to be underestimated, and it has grabbed attention into research field as discussed (Molina-Aiz F.D. et al., 2006).

Accordingly (Fischenich et al.,1999) an important contribution given to the whole analysis conducted by the author is the research based on the contribution and the mean drag values coming from previous research (Rahmeyer et al., 1995). An aspect highlighted by the paper is the geometrical factor: in fact, determining in an effective way the vegetation surface that will be lifted by wind motion can help to better estimate the configuration factor (Vogel exponent) and the initial porosity. However, as seen in the previous mathematical expressions for computing CD, the surface which is exposed upward can be estimated through different methods. Here the authors highlighted the concept of leaf area index (LAI) which is the sum of the total leaf area per unit area of ground. Further, it is also observed an influence on the results of drag itself. Lastly, a fundamental statement is that taking Re numbers and geometrical properties identical, the list of values presented below make able an adaption both for water and air flow. For instance:

- C_{DD}: is the drag considering density of the vegetation element.
- C_{Df}: is the shape.
- C_{Dw}: is the drag accounting the wetted area.

Table 2. The list is referring to the tangential action calculated by the author for each of the previous category identified by Rahmeyer et al., 1995.

Vegetation species		Tangential action (Eq. 2) [N/m²]					
		With leaves			Without leaves		
1	Staghorn Sumac	49.1	36.5	23.2	0.7	0.5	0.3
2	Arctic Blue Willow	78.2	47.5	45.1	0.7	0.5	0.4
3	Norway Maple	452.2	22.7	59.8	4.4	0.2	0.6
4	Western Sand	8.8	25.3	13.2	0.1	0.4	0.2
5	Common Privet	134.9	43.7	58.1	2.2	0.7	0.9
6	Blue Elderberry	218.1	68.9	48.5	2.1	0.7	0.5
7	French Pink	259.3	126.4	77.4	5.6	2.7	1.7
8	Sycamore	12.6	30.1	9.1	0.1	0.3	0.1
9	Dogwood 1–1	34.7	30.2	37.3	1.1	1.0	1.2
10	Dogwood 2–1	35.0	52.4	52.0	0.7	1.1	1.1
11	Euonymus	7.4	54.7	15.5	0.3	1.7	0.5
12	Dogwood 3–1	21.4	39.6	17.3	0.3	0.7	0.3

In the following part some briefly observations are accompanied by proper charts edited and plotted accordingly with two main categories proposed. Specifically, the drag with and without leaves and the tangential action they are expected to induce (Eq. 2). The purpose is to investigate their distribution, estimating a value for the drag which could be implemented for the shear stress assessment. It is predicted that (Koizumi et al., 2010) defoliated samples will display less drag coefficient than fully crowned (actually, less porous).

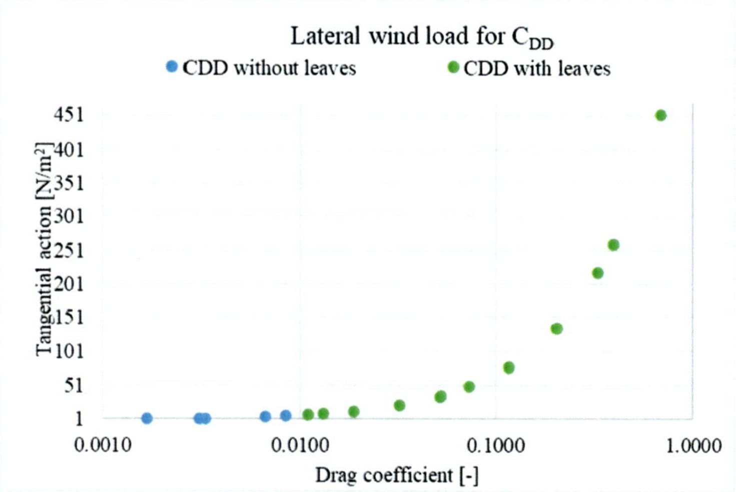

Fig. 2. Edited by the author, C_{DD}: is the drag considering density of the vegetation element.

For the "C_{Dd} with leaves" it is observed a not so satisfactory trend: the disagreement with the values is due to a significant wide range: almost three orders of magnitude.

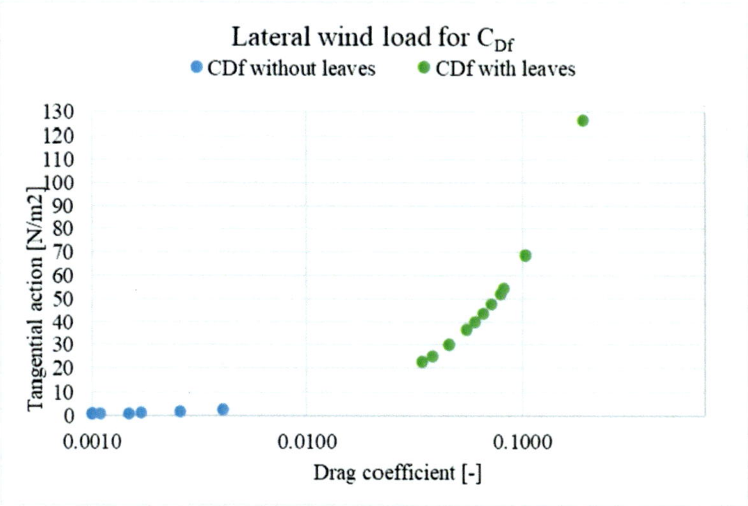

Fig. 3. Edited by the author. C_{Df}: is the drag considering form of the element.

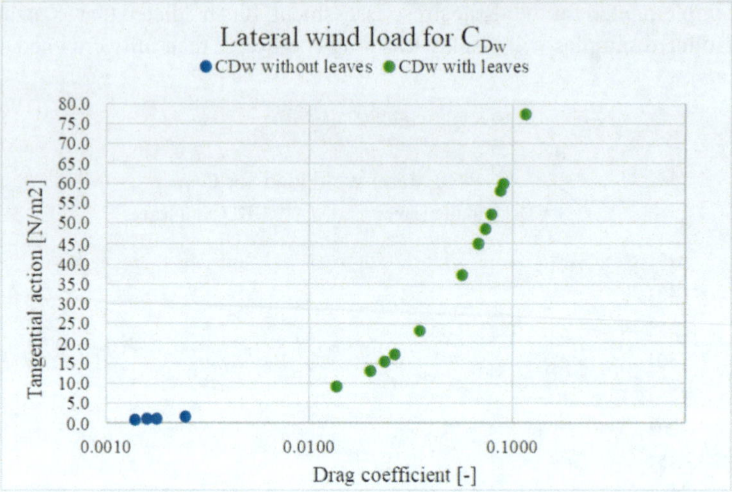

Fig. 4. Edited by the author, C_{Dw}: is the drag accounting the wetted area.

As expected, the case in which there is no presence of foliage shows values which are lower than with crown, and this is verified for all the three categories of C_D. There are then a series of values for different vegetative species which are classified accordingly with their order of magnitude since the properties of each species seems complex or not understood at all to be used to group them. For example, is significative the wide range (Rahmeyer et al., 1995) of values for the Norway Maple (*Acer Platainodes*) respectively: C_{Dd} 0.68; C_{Df} 0.034; C_{Dw} 0.10. More than one order of magnitude for the same sample is misleading for the purpose of the work.

2.4 Clustering of Drag Dataset

The second section of the paper is aimed at introducing the second approach to investigate the drag coefficient variability and having an insight on the criteria. Unfortunately, the aspects not allowing a clear understating that raised up during the literature review are the wide spectrum of porosity, linked with the frontal area and the negative relation in between wind speed and drag coefficient. Accordingly with a dataset of plants properties (J.N. Tzortzi et al., 2022), it is performed a k-means clustering using MATLAB, and for the sake of clarity are showed the two main significant cluster outcomes, chosen thanks to a reliability test namely "silhouette index". Clustering method can be defined as an iterative process, diffused as procedure to test and observe the outcoming distributions of a preassigned database of the subject of interest. Each chart displays its series of centroids.

It is provided in Table 1 the ranks obtained from the k-means clustering analysis here updated with the dataset proposed and then implemented to the Eq. (2):

It is noticed a wide range distribution: almost three orders of magnitude. Furthermore, it is evident that in between a shrubby species and a tall tree the drag studied

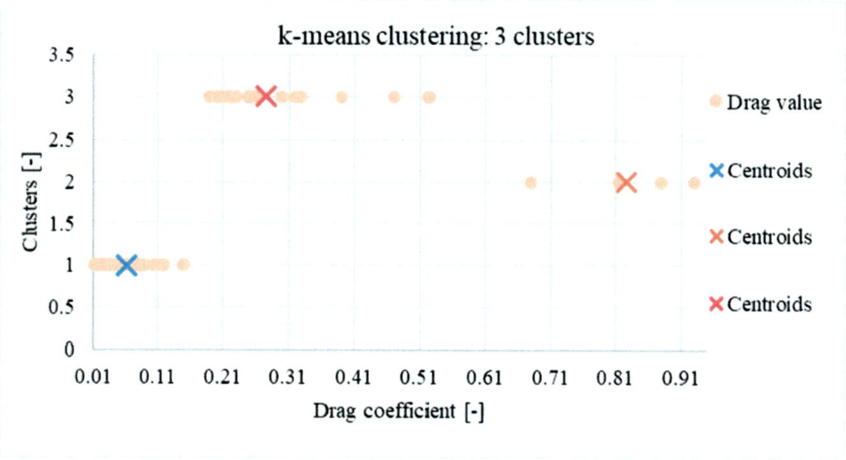

Fig. 5. The cluster process here well operates a distinction, highlighting four thresholds.

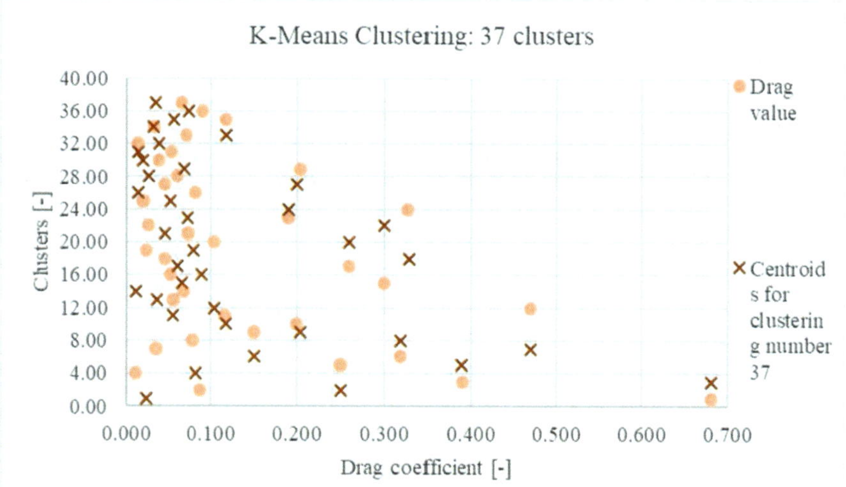

Fig. 6. Selecting 37 clusters out of a dataset of less than one hundred values lead to a dense cloud, where the centroids are for the majority under the threshold of 0.15, excluding higher drag coefficient values.

varies consistently but accordingly with few determined properties: the porosity (LAI estimated) the wind speed and Re number.

Table 3. Ranks determined by the author with the k-means clustering (table source: J.N. Tzortzi et al., 2022). The lowest values are the ones recorded in absence of foliage. The peak value is given by the Malayan Banyan (*Ficus macrocarpa*) as studied thanks to Senlin Zheng, et al., 2020.

1st	$0 \leq CD \leq 0.10$	[-]
2nd	$0.10 \leq CD \leq 0.35$	
3rd	$0.35 \leq CD \leq 0.70$	
4th	$0.7 \leq CD \leq 0.93$	

Table 4. Edited by the author, the ratio in between the greenery accounting the drag values and a standard glazed curtain wall.

		Unit of measure- ments
wf (z) vegetated wall (categories plotted in Table 1)	66	N/m2
	232	
	465	
	631	
w_f (z)	1.0	
q_p	664	
CD Greenery (Categories plotted in Table 1)	0.1	[-]
	0.35	
	0.7	
	0.93	
C_D Glazed curtain wall	0.0015	
c_e (z) Exposure coeff	1.7	[-]
ρ density of the fluid/gas	1.25	[kg/m^3]
c_r return coeff	1.00	[-]
v_b base wind velocity	25	[m/s]
k_r soil coefficient	0.20	[-]
k_a ref. Parameter to compute	0.40	[-]
c_a		
a_0 ref. Altitude to calculate c_a	1000	[m]
z_0 roughness length	0.10	[m]
z_{min} min. Height	5.00	[m]

Table 5. Ratio of the *wf* of greenery and curtain wall compared to the four different values of C_D.

Ratio *wf*	($C_D = 0.10$)	66	[-]
	($C_D = 0.35$)	232	
	($C_D = 0.70$)	465	
	($C_D = 0.93$)	631	

3 Results and Discussion

3.1 Comparison in Between Glazed and Vegetated Vertical Surfaces

The input values in Table 3 are the ones provided and extendedly explained in the guidelines to estimate and calculate the wind actions: here for the sake of simplicity it has been calculated a standard situation, implementing the outcoming ranks of paragraph 2.4.

The case which leads to sixty-six times more in terms of w_f is equal to 0.1, so two orders of magnitude more than the value of C_D assumed for the glazed façade as 0.0015. If this result is already mathematically significant, more severe becomes the C_D equal to 0.35 and in fact it displays a result of 232 times more. Increasing with 0.7 so closer to the threshold which has been assumed as reliable from the clustering classification, it is observed respectively a w_f very severe if it verified in reality: 465 times more means that our subsystem is facing a stressful case, which is increased considerably if the drag is 0.95, leading to a ratio which is far distant from the theoretical of the curtain wall.

4 Conclusions

The findings show the effect induced by the wind motion in different samples from the literature. The high variation of drag coefficient is confirmed furthermore it is showed the wide range of values and not associable with vegetation classes even when k-means clustering is applied to the dataset proposed (Table 2).

Moreover, it has been understood that the plant variety led to wide outcome in terms of values. As noticed the plants presence can generate different cases of C_D with more than one order of magnitude of difference. So, a reasonable research path to be paved seems to be the one to assume safe threshold values, such as "ultimate" value. Pragmatically it could be to assume a higher value of drag corresponding to a configuration mode which presents higher surface of the plant exposed to the wind.

A concern of the author is that this procedure requires to assume a strongly variation of the drag coefficient which might led to oversizing or underestimation during the calculation step, and it could affect the designer choice for fasteners, in case of building application.

At the end of the facts, the proposal is therefore, to evaluate the shear stress with real green panel conducting wind tunnel tests. Finally, even a CFD model could be considered to quantify the actual stress imposed on the sub-system. The full process might be an interesting development according with the Technical Standard CNR-DT 207 R1/2018 ROMA resulting in a computational wind engineering (CWE) model, to validate the numerical considerations so far achieved and giving insight even into the wind-structure interaction.

Acknowledgments. The full paper here presented is the result of the research carried out by the author. A mention goes to the Italian company ItalMesh (BS) that is active partner into the research on the VGS technology (Fig. 1 does display the patented Zero Gravity Eden). This activity received no additional external funding, furthermore the author declares no conflict of interest.

References

Cao, J., Tamura, Y., Yoshida, A.: Wind tunnel study on aerodynamic characteristics of shrubby specimens of three tree species. Urban Forestry Urban Greening **11**, 465–476 (2012). https://doi.org/10.1016/j.ufug.2012.05.003

Cionco, Ronald M. 1965. A Mathematical Model for Air Flow in a Vegetative Canopy. Journal of Applied Meteorology (1962–1982): 517–22. Accessed September 8, 2021. http://www.jstor.org/stable/26172935

Dirk Schindler, 2008. Responses of Scots pine trees to dynamic wind loading, Meteorological institute, University of Freiburg, D-79085 Freiburg, Germany

Etnier, S., Vogel, S.: Reorientation of daffodil (Narcissus: Amaryllidaceae) flowers in wind: drag reduction and torsional flexibility. Am. J. Bot. **87**, 29–32 (2000)

Finnigan, J.J., Belcher, S.E.: Flow over a hill covered with a plant canopy. Q. J. R. Meteorol. Soc. **130**, 1–29 (2004). https://doi.org/10.1256/qj.02.177

Fischenich, C., and Dudley, S.,1999. "Determining drag coefficients and area for vegetation," EMRRP Technical Notes Collection (ERDC TN-EMRRP-SR-08), U.S. Army Engineer Research and Development Center, Vicksburg, MS. www.wes.army.mil/el/emrrp

Gardiner, B., Berry, P., Moulia, B.: Wind impacts on plant growth, mechanics, and damage. Plant Sci. **245**, 94–118 (2016). https://doi.org/10.1016/j.plantsci.2016.01.006

Gillies, J.A., Nickling, W.G., King, J.: Drag coefficient and plant form response to wind speed in three plant species: Burning Bush (*Euonymus Alatus*), Colorado Blue Spruce (*Picea Pungens Glauca*), and Fountain Grass (*Pennisetum Setaceum*). J. Geophys. Res. **107**(D24), 4760 (2002). https://doi.org/10.1029/2001JD001259

Kerzenmacher, T., Gardiner, B.: A mathematical model to describe the dynamic response of a spruce tree to the wind. Trees **12**, 385–394 (1998). https://doi.org/10.1007/s004680050165

Koizumi A, Motoyama J, Sawata K., Sasaki Y., Hirai T., 2010. Evaluation of drag coefficients of poplar-tree crowns by a field test method, Journal of Wood Science, 189–193, 56 (3). doi:https://doi.org/10.1007/s10086-009-1091-8

Lettau H., 1969. Journal of applied meteorology, Volume 8- Note on an Aerodinamic Roughness Parameter Estimation on the basis of roughness element description Dept. of Meteorology, University of Wisconsin, Madison 2 April 1969 and 16 June 1969

Mayer, H.: Wind-induced tree sways. Trees **1**, 195–206 (1987). https://doi.org/10.1007/BF01816816

F.D. Molina-Aiz, D.L. Valera, A.J. Álvarez, A. Madueño, 2006. A Wind Tunnel Study of Airflow through Horticultural Crops: Determination of the Drag Coefficient, Biosystems Engineering. Volume 93, Issue 4, Pages 447–457, ISSN 1537–5110

Ogut, O., Tzortzi, N.J., Bertolin, C.: Vertical Green Structures to Establish Sustainable Built Environment: A Systematic Market Review. Sustainability **14**, 12349 (2022). https://doi.org/10.3390/su141912349

Rahmeyer, W., Werth, D., and Cleere, R., 1995. The Study of Resistance and Stability of Vegetation in Flood Channels. Utah State University Lab Report USU-376, Prepared for the USAEWES

Roger H Shaw, A.R Pereira, 1982. Aerodynamic roughness of a plant canopy: A numerical experiment, Agricultural Meteorology, Volume 26, Issue 1,1982, Pages 51–65, ISSN 0002-1571, https://doi.org/10.1016/0002-1571(82)90057-7

Tadrist, L., Saudreau, M., He´mon P, Amandolese X., Marquier A., Leclercq T., de Langre E.,: Foliage motion under wind, from leaf flutter to branch buffeting. J. R. Soc. Interface. **15**, 20180010 (2018). https://doi.org/10.1098/rsif.2018.0010

Tzortzi, N., Barbotti, G., Castiglioni C.A., Voza, A. 2022. Changing Cities for Resilience against Climate Changes: Architectural, Engineering and Human Health Implications the H2020 Project HARMONIA. In Book of Abstracts of the International Conference on CHANGING CITIES V Spatial, Design, Landscape, Heritage & Socio-economic Dimensions - ISBN: 978–618–84403–7–1

Västilä, K., Järvelä, J., Aberle, J.: Characteristic reference areas for estimating flow resistance of natural foliated vegetation. J. Hydrol.Hydrol. **492**, 49–60 (2013). https://doi.org/10.1016/j.jhydrol.2013.04.015

Observation of Flow Downstream of a Bridge Deck Model Using Cobra Probe and Lidars

Mohammad Nafisifard[1](✉), Shahbaz Pathan[2], Jasna B. Jakobsen[1], Mikael Sjöholm[2], Alberto Zasso[3], Stefano Giappino[3], Jonas Snæbjörnsson[1,4], and Jakob Mann[2]

[1] University of Stavanger, Stavanger, Norway
{mohammad.nafisifard,jasna.b.jakobsen}@uis.no
[2] Technical University of Denmark, Roskilde, Denmark
{shpa,misj,jmsq}@dtu.dk
[3] Politecnico di Milano, Milano, Italy
{alberto.zasso,stefano.giappino}@polimi.it
[4] Reykjavik University, Reykjavik, Iceland
jonasthor@ru.is

Abstract. Application of lidars in civil engineering is evolving in many areas, for instance in planning and estimating design loads for long-span bridges. The use of the lidar technology in the wind tunnel provides a new option for turbulence measurements around model scale structures. The paper presents a wind tunnel test campaign using a combination of two continuous wave lidars and Cobra probes to study the flow conditions downstream of a bridge deck model, within and outside of the wake region. The measurements took place in a boundary layer flow at the wind tunnel facility of Politecnico di Milano where the performance of the two lidars for studying bridge aerodynamics was tested and is herein validated by reference Cobra probe measurements. The wake characteristics around the bridge deck are explored with the aim of improving the understanding of the nature of the wind forces acting on a bridge girder and demonstrating the potential of the remote sensing technology for wind tunnel studies relevant to bridge design.

Keywords: Wind Tunnel Tests · Bridge Deck Section · Continuous-Wave Lidar · Cobra Probes

1 Introduction

Wind tunnel studies of long-span bridges are essential for a comprehensive insight into the aerodynamic loading of bridges in the atmospheric flow. Time-averaged force coefficients are commonly measured with static bridge deck models, using various types of force transducers. Occasionally, surface pressure measurements are performed to increase the understanding of the interaction of the airflow with a bridge deck, both with a fixed and a moving model (Rocchi et al., 2015; Argentini et al., 2012).

Wake flow characteristics of streamlined closed-box bridge girders have been investigated in wind tunnels by the means of surface pressure measurements (Ricciardelli and

P. Schito and A. Zasso (Eds.): IN VENTO 2022, LNCE 461, pp. 310–321, 2024.
https://doi.org/10.1007/978-3-031-53059-3_28

Hangan, 2001), high-resolution particle image velocimetry (PIV) (Chen et al., 2014), and numerical flow simulations (Kuroda, 1997; Fransos and Bruno, 2010; Kusano et al., 2021). In full-scale, pressure measurements on a bridge deck surface have been performed in a limited number of studies (Daniotti, 2022). The work of Daniotti (2022) also includes turbulence measurements by sonic anemometers in a point upstream and a point downstream of the bridge, at deck level. The use of remote sensing technology (Cheynet et al., 2017a) makes it possible to survey a much larger flow domain around a bridge structure than practically feasible using anemometers.

The present work introduces the potential use of lidars in wind tunnel investigations of bridge decks. Remote sensing of a wind field by continuous wave lidars inside a wind tunnel is a relatively unexplored subject. Previous lidar investigations in wind tunnels are limited to the validation of mean wind speed measurements by Pedersen et al. (2012), a study of the effect of droplets in an icy wind tunnel on the high-frequency velocity data by Sjöholm et al. (2017), and measurements downstream of wind turbine models by Dooren et al. (2017) and Sjöholm et al. (2017).

Herein, the lidars, in parallel with Cobra probes, are used to investigate the flow around a bridge deck model in the wind tunnel. The case studied concerns a suspension bridge across the Lysefjord in Norway, which is instrumented by an array of sonic anemometers for long-term wind monitoring (Snæbjörnsson et al., 2017). Several lidar measurement campaigns have also been performed at the bridge site (Cheynet et al., 2017a; Cheynet et al., 2016; Cheynet et al., 2017b; Nafisifard et al., 2021; Nafisifard et al., 2023a; Nafisifard et al., 2023b) motivating the present study.

An investigation of the flow conditions around the bridge deck model, as performed in the present study, aims to improve the understanding of the nature of the wind forces acting on a bridge girder.

2 Wind Tunnel Test Setup and Measurement Equipment

The tests were performed in the boundary layer test section of the closed-circuit wind tunnel at the Politecnico di Milano. The test section is 13.84 m wide, 3.84 m high, and 35 m long and the facility can generate mean wind speeds up to 16 m/s. The boundary layer flow was created by spires and floor roughness elements (Fig. 1a). The horizontal turbulence intensity was about 10%, which is relevant for a prototype bridge. A 4 m long section model of the bridge deck, which is 12.3 m wide and 2.76 m tall in full-scale, was fabricated in a 1:15 scale (Fig. 1b) and tested at a zero-pitch angle. A Pitot tube was used to monitor the wind velocity in the tunnel at deck level, 4.5 deck widths upwind from the leading edge of the bridge deck.

The lidar measurement equipment was provided by the Technical University of Denmark (DTU) in the form of two 2-inch-wide optical telescopes that were connected to a single continuous-wave laser (Fig. 2a) (Sjöholm et al., 2017). Special clamps were designed and 3D-printed at DTU, to fix the telescopes to a supporting aluminum beam. The telescopes could slide along the beam as well as rotate about a fixed axis. The beam was attached to a specially designed frame, providing support for the lidars in different measurement configurations (Fig. 2b). An aerosol generator was used to generate droplets of sufficient concentration for the lidar measurements (Fig. 2c).

The bridge deck wake characteristics were monitored by an array of four 4-hole Cobra probes manufactured by Turbulent Flow Instrumentation distributed at lateral distances up to 1B, where B = 820 mm is the model deck width. A traversing mechanism was used to obtain wake profiles at distances 0.16B, 0.5B, 1B, and 3.25B (profiles W1, W2, W3, and W4) downstream from the trailing edge of the deck model, see Fig. 3a and Fig. 5. The Cobra probe data was recorded with a sampling frequency of 2 kHz.

(a) Spires and floor roughness elements upstream of the model.

(b) The 1/15 scale bridge deck section model.

Fig. 1. The bridge deck model setup in the wind tunnel.

(a) Lidar telescopes.

(b) Aluminium supporting frame.

(c) Aerosol seeding system.

Fig. 2. Lidar measurement setup.

The wake profiles (mean wind speeds and turbulence) were also measured by dual lidar, in front of strip B (Point C, in Fig. 5) at several heights including five points above the trailing edge level (667 mm, 500 mm, 313 mm, 133 mm, and 53 mm), a point at the trailing edge level, and four points below the trailing edge level (−60 mm, −123 mm, −303 mm, and −480 mm). Lidar data 1B downstream from the trailing edge of the deck model (section W3, in Fig. 5) is presented herein.

(a) Array of four Cobra probes fixed to a traverse.

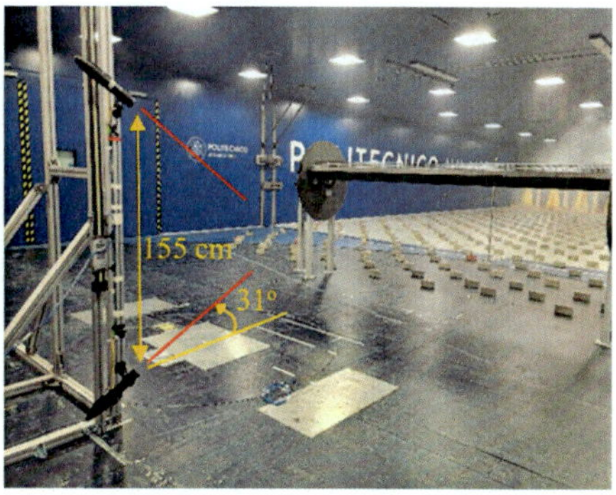

(b) Two lidar telescopes fixed to a traverse.

Fig. 3. Wake measurement setup.

As illustrated in Fig. 3b, the two telescopes were orientated at an angle of ± 31° from the horizontal, to provide information on the turbulent flow in the vertical plane. The measurement volume in this configuration was 3 cm (full width at half maximum, FWHM).

Measurements using Cobra probes and lidars were performed in separate runs, ensuring that the lidar measurements were free of any interference from the Cobra probe setup in the proximity of the target volume. However, a co-location of Cobra probes and lidars was also included for a single case, where simultaneous measurements by lidars and a Cobra probe were carried out, to enable a direct comparison of the turbulence time-series acquired by the two types of sensors. This setup is shown in Fig. 4, with the lidars focusing on a point 303 mm below the deck trailing edge, 800 mm downstream of the 820 mm wide bridge deck model (see Fig. 4a and 4b). Simultaneously, a Cobra probe was deployed at a lateral distance of 100 mm from the focus point of the lidars (see Fig. 4c).

Measurements of the lateral coherence, in both the disturbed and the undisturbed flow regions, were also undertaken by Cobra probes. Lateral coherence was also studied using lidars, with the two telescopes in a side-by-side arrangement, as shown in Fig. 2a. Simultaneous measurements by two lidars were carried out at three different inclination angles (−30°, 0°, and 30°), to enable the estimation of the lateral coherence of the u- and w-turbulence components. The two line-of-sight velocities were sampled at a frequency of approximately 600 Hz. Figure 5 outlines the experimental plan showing the location of the measurement points in sections across (Fig. 5a) and along (Fig. 5b) the deck.

(a) Side view of the lidar telescopes.

(b) Schematic of tailored scanning.

(c) Closeup front view of the instruments.

Fig. 4. Co-location measurement setup. For clarity, the laser beams are illustrated by red solid lines. Target points of lidars and Cobra probe are marked by red and green circles, respectively.

3 Data Analysis and Measurement Results

3.1 Cobra Probe Data Quality and Orientation

Two main steps were adopted in the initial processing of the Cobra probes recordings. The first one concerns the quality control of data to eliminate erroneous observations of the occasional flow directions outside the interval $\pm 45°$ from the Cobra probe x-axis, which Cobra probes cannot resolve properly. This kind of flow is relevant to vortices due to the flow-model interaction. This has been investigated by introducing a data quality factor Q:

$$Q = \frac{N_o}{N_{total}} \tag{1}$$

where N_o is the number of observations with zero value recorded by a Cobra probe and N_{total} is the total number of observations. The results in Fig. 6 show that, as expected, the poorest data quality is seen within $\pm$ 60 mm ($\pm$0.3H where H = 184 mm is the model height) from the trailing edge of the bridge deck, at sections W1 and W2. For the other measurement locations, further downstream of the bridge, the data quality from the Cobra probes was good, with Q factor values higher than 95%.

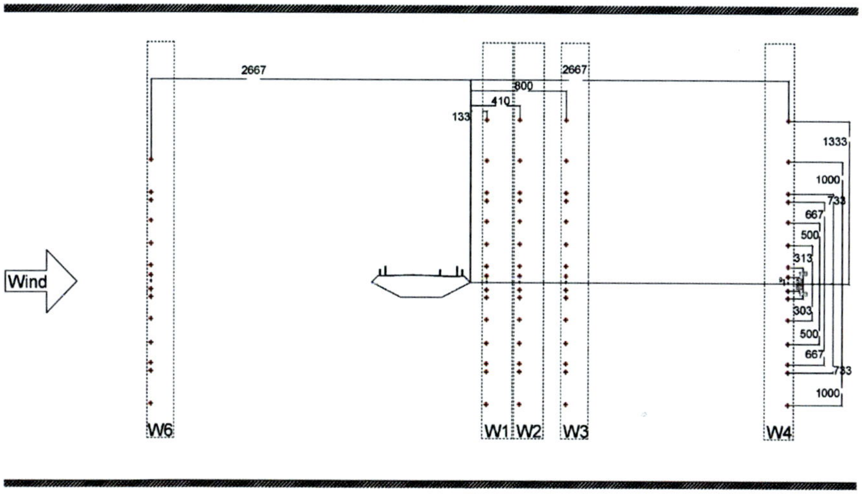

(a) Side view of target points.

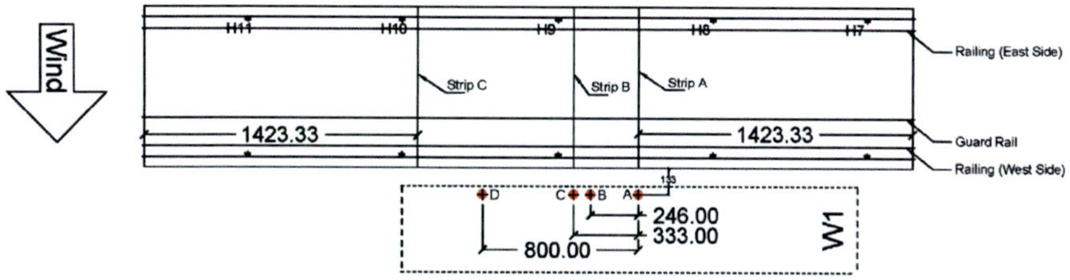

(b) Top view of target points. For sake of simplicity, only the W1 case is illustrated in the drawing.

Fig. 5. Target points for flow measurements.

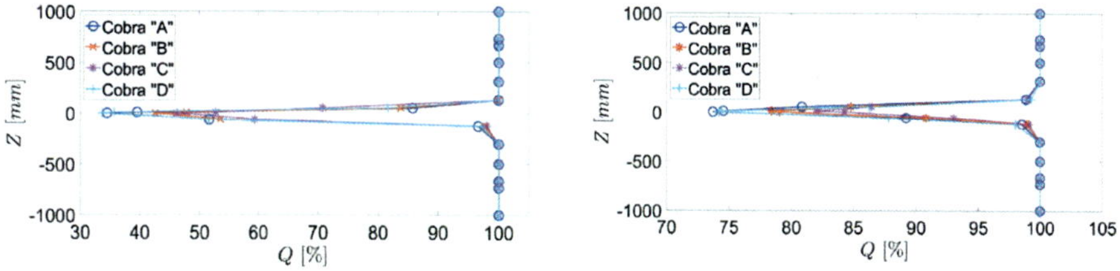

(a) the W1 case. (b) the W2 case.

Fig. 6. The data quality factor, Q, at different heights, Z, from the leading/trailing edge, for data recorded 0.15B (case W1) and 0.5B (case W2) downstream of the trailing edge of a B = 820 mm wide deck model.

The second consideration is the adjustment of slight misalignments of the Cobra probes, in such a way that the rotation of the turbulence data into a re-aligned coordinate system yields zero-mean v- and w- values at the top of the boundary layer in the wind tunnel. The recordings at the highest profile measurement point (1333 mm above the deck leading/trailing edge) were used to establish the pitch and yaw angles needed for the correction.

3.2 Wake Study

An illustration of the results from the wake measurements is given in Fig. 7, in terms of the mean wind speed profiles observed by a Cobra probe, in comparison to the profile upstream of the model (section W6 in Fig. 5). Around the trailing edge level, data values equal to 0 are disregarded in the calculations as discussed in Sect. 3.1.

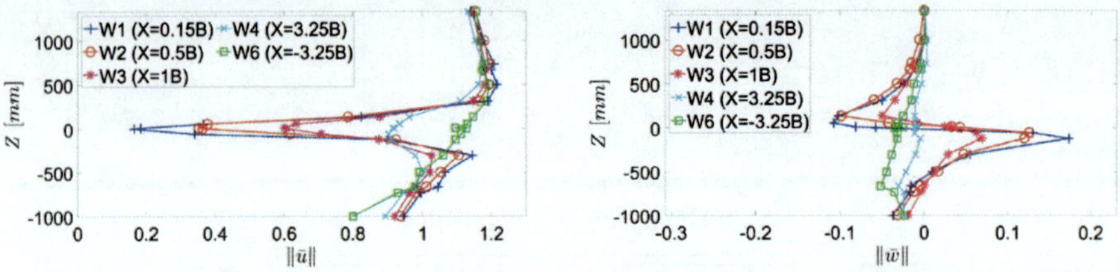

(a) Mean along wind speed normalized by pitot velocity. **(b) Mean vertical wind speed normalized by pitot velocity.**

Fig. 7. Normalized mean wind velocities recorded by Cobra probe A, at several distances X from the deck trailing edge.

The values are normalized by the horizontal mean wind speed recorded by the Pitot tube upstream of the model. The horizontal wind profiles clearly demonstrate the velocity deficit in the wake embedded in the boundary layer flow.

In Fig. 8, the associated turbulence intensities are displayed. At the three shortest distances from the model (sections W1, W2, and W3), both the horizontal and the vertical turbulence intensities exceed 30% at the trailing edge level. This is a triple and a fivefold increase compared to the turbulence level of the undisturbed flow (sections W4 and W6) for the two velocity components, respectively.

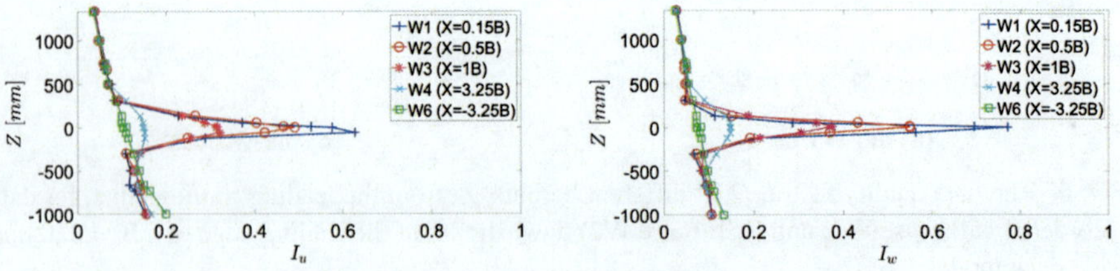

Fig. 8. Turbulence intensity of u- and w- components recorded by Cobra probe A.

In Fig. 9 a comparison between the wake results obtained by the Cobra probe and the lidars, one deck width downstream of the model trailing edge is given. It can be seen that Cobra and lidar profiles are in good agreement with each other except for few locations. In particular, for the vertical wind velocity profile some discrepancies can be seen close to the wake center, below the deck. A slight discrepancy in the along wind velocity is also noticeable above the deck. This can potentially be attributed to the finite sampling volume combined with a steep spatial gradient. Difference may also be related to irregularities in the wake pattern in crosswise direction due to the objects on the deck surface (posts, railings etc.). The profiles were not recorded simultaneously and using a different traversing positioning system, which may also affect the comparison in terms of the accuracy in the probe/lidar position and interference with the probe stands. Furthermore, the velocity in the tunnel was reduced during the lidar measurements, to improve performance, as higher wind velocities caused higher dispersion of the smoke required. A difference in the dynamic pressure by a factor of 2.4 between the lidar and the Cobra probe measurements may have also contributed to a slightly different flow-model interaction, especially just above the deck where railings and hanger posts disturb the flow pattern.

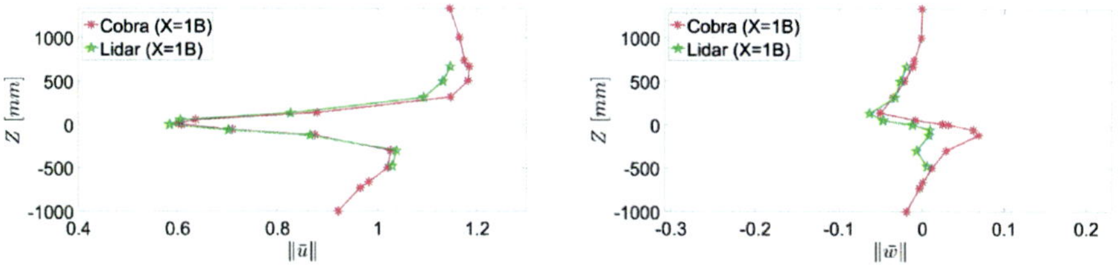

(a) Mean along wind speed normalized by pitot velocity.

(b) Mean vertical wind speed normalized by pitot velocity.

Fig. 9. Normalized mean wind velocities recorded by lidars and Cobra probe at one deck width downstream of the model trailing edge.

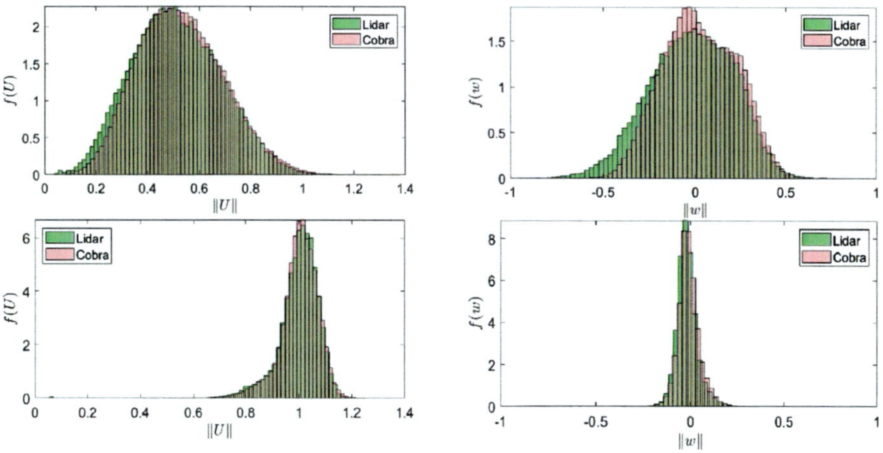

Fig. 10. Histogram of normalized along-wind (U) and vertical (w) velocity components recorded by Cobra probe D and lidar, for heights Z = 0 mm (top panels) and Z = 667 mm (bottom panels) at 800 mm (1B) downstream from the trailing edge.

The histograms of the along-wind (U) and vertical (w) velocity components measured by Cobra probe D and the lidars, inside (Z = 0 mm) and outside of the wake (Z = 667 mm), are displayed in Fig. 10. The velocities, monitored one deck width downstream of the model, are normalized by the along-wind mean speed at Z = 667 mm. In the nearly undisturbed flow at Z = 667 mm, turbulence variations are confined to a relatively small range of velocity values compared to the distributions at the trailing edge height (Z = 0 mm). A broader range of values at the trailing edge level is consistent with the elevated turbulence levels in the wake presented in Fig. 8. They also indicate an increased variability in the local flow directions within the wake, e.g., in relation to the vortex shedding process. In line with the results shown in Fig. 8, the average along-wind velocity is considerably reduced at the trailing edge height compared to the undisturbed mean wind velocity.

The frequency content of turbulence within and outside of the wake, one deck width from the trailing edge (line W3) is presented in Fig. 11. At the trailing edge level (Z = 0 mm), the spectral peak has a higher frequency (smaller scale turbulence), compared to that at Z = 733 mm, which indicates a flow-deck interaction. In particular, the vertical turbulence spectrum is more narrow-banded in the wake than outside the wake, indicating a vortex shedding process. The spectrum is centered at a Strouhal number, $St = \frac{fH}{U}$, close to 0.2. In full scale, St values from 0.15 to 0.25 have been observed, for turbulence intensities, I_u, ranging from 0.06 to 0.27 (Daniotti, 2022).

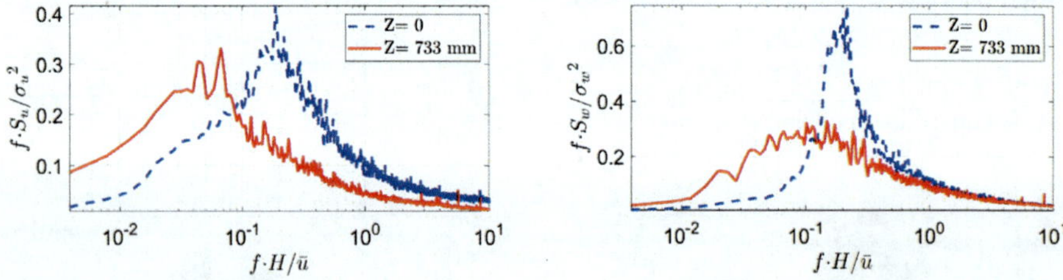

(a) u- component. **(b)** w- component.

Fig. 11. The turbulence spectra outside and within the deck model wake recorded by Cobra probes 0.98B downstream of trailing edge. Dashed lines belong to Z = 0 mm and solid lines belong to Z = 733 mm.

3.3 Comparison of Measurements by Collocated Cobra Probe and Lidars

The results from the simultaneous measurements by the collocated Cobra probe and the lidars obtained with the setup shown in Fig. 4, are provided in Table 1. The tests were carried out at wind tunnel power of 40%, at which the quality of the lidar measurements was found to be better than at higher wind speeds. At higher power, a lower number of aerosols was encountered in the measurement volume during the recording interval. The results show an overall good agreement but the results for the mean vertical velocity (w_m), are sensitive to the alignment of the instruments relative to the mean along wind velocity (U_m). Since the along wind component is in the order of 30 times larger than the vertical component, the difference of 0.12 m/s for the mean vertical velocity between the Cobra probe and the lidar results could correspond to an un-accounted for misalignment between the instruments of 1°. In the wake analysis, the Cobra probe misalignment was corrected by assuming that the mean vertical and transversal velocities were zero at the top measurement height of the profile, which rendered tilt corrections between 0.2° and 1.3° for the four Cobra probes at the four different downstream locations. Since the co-location setup was only based on one measurement point within the influence area of the surface roughness, such alignment correction was not possible. As the lidar data is based on two laser beams in a vertical plane, vertical misalignment should not be an issue, however, a minor horizontal misalignment is possible for the lidar setup. It is noteworthy that the standard deviation of the vertical velocity component is the same from both lidars and Cobra probe.

Table 1. Flow characterization by Cobra and Lidar at the co-location point for 40% power in the wind tunnel.

Sensor	Mean velocity (m/s)		Standard deviation (m/s)		Turbulence intensity		Reynolds shear stress (N/m2)
	U_m	w_m	σ_u	σ_w	I_u	I_w	r_{uw}
Cobra	6.27	0.12	0.89	0.54	0.141	0.086	−0.0376
Lidar	6.16	0.24	0.76	0.52	0.124	0.084	−0.0625

3.4 Correlation Study

Simultaneous measurements by four Cobra probes enabled the estimation of the flow correlation at six different lateral separations (see Fig. 5). In Fig. 12, the lateral correlation of the along-wind velocity at the leading/trailing edge height is displayed both for the undisturbed flow upstream of the model (W6) and disturbed flow downstream of the model (W1-W4). The correlation is seen to be strongly reduced in the wake, in which the smaller scale turbulence generated by the bridge deck dominates, which is in line with the reduction of the low frequency content shown in Fig. 11. In the undisturbed flow, the correlation of the along-wind turbulence at lateral separations drops to a value of $\exp^{(-1)}$ at separations corresponding to about one half of the deck width (0.5B). In the wake, one deck width downstream from the trailing edge (W3), the corresponding length scale reduces to about 0.1B.

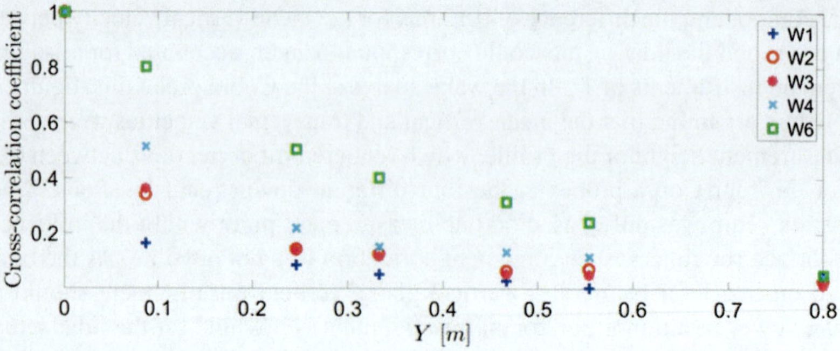

Fig. 12. Cross-correlation coefficient of the along-wind turbulence at the trailing edge height, as a function of lateral separation.

4 Discussion and Conclusion

The comparison of the collocated measurements from the lidars and the Cobra probe shows a good agreement despite a difference observed for the mean vertical velocity component, which is likely linked to the sensitivity of the vertical component to the alignment of the instruments relative to the along wind component.

Some discrepancies between the results from the different instruments can also be seen in the wake profiles which potentially could be attributed to the extended sampling volumes of the lidars in combination with steep gradients in the flow field. However, differences are also seen between the results from the four different Cobra probes indicating that the flow field is varying along the bridge model. This fact, in combination with slightly different measurement locations and separate measurement runs using different tunnel velocity, is also likely to influence the results.

A redistribution of the turbulence spectra due to the flow interaction with the bridge model is illustrated using the Cobra measurement data. In the wake, the deck "signature" turbulence is characterized by higher frequencies and smaller scales compared to those in the inflow. Correspondingly, the observed overall lateral correlation of turbulence is significantly reduced downstream of the model.

The paper describes the measurement campaign and documents an overall good agreement between the flow statistics observed by the Cobra probes and the lidar around a bridge deck model. While more detailed results can be retrieved from the measurement data in future studies, this work demonstrates the potential of lidar remote sensing technology for wind tunnel studies relevant to bridge design.

Acknowledgements. This work has received funding from the European Union Horizon 2020 through the Innovation Training Network Marie Skłodowska-Curie Actions: Lidar Knowledge Europe (LIKE (grant no. 858358)). We would like to thank Luca Ronchi and Umberto Spinelli for their assistance in the wind tunnel measurements and Lorenzo Rosa for taking photos of the experiments.

References

Argentini, T., Rocchi, D., Muggiasca, S., Zasso, A.: Cross-sectional distributions versus integrated coefficients of flutter derivatives and aerodynamic admittances identified with surface pressure measurement. J. Wind Eng. Ind. Aerodyn. **104–106**, 152–158 (2012)

Chen, W., Li, H., Hu, H.: An experimental study on the unsteady vortices and turbulent flow structures around twin-box-girder bridge deck models with different gap ratios. J. Wind Eng. Ind. Aerodyn. **132**, 27–36 (2014)

Cheynet, E., et al.: Application of short-range dual-Doppler lidars to evaluate the coherence of turbulence. Exp. Fluids **57**, 1–17 (2016). https://doi.org/10.1007/s00348-016-2275-9

Cheynet, E., et al.: Full-scale observation of the flow downstream of a suspension bridge deck. J. Wind Eng. Ind. Aerodyn. **171**, 261–272 (2017a)

Cheynet, E., Jakobsen, J.B., Snæbjörnsson, J.T., Reuder, J., Kumer, V., Svardal, B.: Assessing the potential of a commercial pulsed lidar for wind characterisation at a bridge site. J. Wind Eng. Ind. Aerodyn. **161**, 17–26 (2017b)

Daniotti, N.: Bridge deck aerodynamics: A case study in full-scale (PhD thesis). Stavanger: University of Stavanger (2022)

Fransos, D., Bruno, L.: Edge degree-of-sharpness and free-stream turbulence scale effects on the aerodynamics of a bridge deck. J. Wind Eng. Ind. Aerodyn. **98**, 661–671 (2010)

Kusano, I., Cheynet, E., Jakobsen, J. B., Snæbjornsson, J.: Aerodynamic study of a suspension bridge deck by CFD simulations, wind tunnel tests and full-scale observations. In: IOP Conference Series: Materials Science Engineering, p. 012007 (2021)

Kuroda, S.: Numerical simulation of flow around a box girder of a long span suspension bridge. J. Wind Eng. Ind. Aerodyn. **67**, 239–250 (1997)

Nafisifard, M., et al.: Dual lidar wind measurements along an upstream horizontal line perpendicular to a suspension bridge. In: IOP Conference Series: Materials Science Engineering, vol. 1201, p. 012008 (2021)

Nafisifard, M., Jakobsen, J. B., Snæbjörnsson, J. T., Sjöholm, M., Mann, J.: Lidar measurements of wake around a bridge deck. J. Wind Eng. Ind. Aerodyn. **240**, 105491 (2023a)

Nafisifard, M., Jakobsen, J.B., Snæbjörnsson, J.T., Sjöholm, M., Mann, J.: Triple-lidar measurements of wind across a virtual rotor plane over a sea surface. In: Journal of Physics: Conference Series, vol. 2626, p. 012022 (2023b)

Pedersen, A.T., Montes, B.F., Pedersen, J.E., Harris, M. Mikkelsen, T.: Demonstration of short-range wind lidar in a high-performance wind tunnel. In: Proceedings of European Wind Energy Conference and Exhibition, Copenhagen, Denmark (2012)

Ricciardelli, F., Hangan, H.: Pressure distribution and aerodynamic forces on stationary box bridge sections. Wind Struct. **4**(5), 399–412 (2001)

Rocchi, D., Argentini, T., Sbrosi, M.: Pressure distribution and global forces on a bridge deck section: experimental and CFD analysis of static aerodynamic forces. J. Bridge Eng. **20**(9) (2015)

Sjöholm, M., et al.: Lidars for wind tunnels - an IRPWind joint experiment project. Energy Procedia **137**, 339–345 (2017)

Snæbjörnsson, J.T., Jakobsen, J.B., Cheynet, E. Wang, J.: Full-scale monitoring of wind and suspension bridge response. In: IOP Conference Series: Materials Science Engineering, vol. 276, p. 012007 (2017)

van Dooren, M.F., Campagnolo, F., Sjöholm, M., Angelou, N., Mikkelsen, T. Kühn, M.: Demonstration and uncertainty analysis of synchronised scanning lidar measurements of 2-D velocity fields in a boundary-layer wind tunnel. Wind Energy Sci. **2**(1) 329–341 (2017)

Author Index

P. Schito and A. Zasso (Eds.): IN VENTO 2022, LNCE 461, pp. 323–324, 2024.
https://doi.org/10.1007/978-3-031-53059-3